Mechanical Behavior of Metallic Materials under Different Loading Conditions

Mechanical Behavior of Metallic Materials under Different Loading Conditions

Editors

Martin Heilmaier
Martina Zimmermann

Basel • Beijing • Wuhan • Barcelona • Belgrade • Novi Sad • Cluj • Manchester

Editors

Martin Heilmaier
Karlsruhe Institute of
Technology (KIT)
Karlsruhe, Germany

Martina Zimmermann
Technical University Dresden
Dresden, Germany

Editorial Office
MDPI
St. Alban-Anlage 66
4052 Basel, Switzerland

This is a reprint of articles from the Special Issue published online in the open access journal *Metals* (ISSN 2075-4701) (available at: https://www.mdpi.com/journal/metals/special_issues/mechanical_loading).

For citation purposes, cite each article independently as indicated on the article page online and as indicated below:

Lastname, A.A.; Lastname, B.B. Article Title. *Journal Name* **Year**, *Volume Number*, Page Range.

ISBN 978-3-0365-9168-1 (Hbk)
ISBN 978-3-0365-9169-8 (PDF)
doi.org/10.3390/books978-3-0365-9169-8

Cover image courtesy of Sebastian Schöne and Martina Zimmermann

© 2024 by the authors. Articles in this book are Open Access and distributed under the Creative Commons Attribution (CC BY) license. The book as a whole is distributed by MDPI under the terms and conditions of the Creative Commons Attribution-NonCommercial-NoDerivs (CC BY-NC-ND) license.

Contents

About the Editors vii

Preface ix

André T. Zeuner, Thomas Wanski, Sebastian Schettler, Jonas Fell, Andreas Wetzig,
Robert Kühne, et al.
Influence of a Pronounced Pre-Deformation on the Attachment of Melt Droplets and the Fatigue
Behavior of Laser-Cut AISI 304
Reprinted from: *Metals* **2023**, *13*, 201, doi:10.3390/met13020201 1

Thomas Wegener, Tao Wu, Fei Sun, Chong Wang, Jian Lu and Thomas Niendorf
Influence of Surface Mechanical Attrition Treatment (SMAT) on Microstructure, Tensile and
Low-Cycle Fatigue Behavior of Additively Manufactured Stainless Steel 316L
Reprinted from: *Metals* **2022**, *12*, 1425, doi:10.3390/met12091425 13

Matthias Droste, Sebastian Henkel, Horst Biermann and Anja Weidner
Influence of Plastic Strain Control on Martensite Evolution and Fatigue Life of Metastable
Austenitic Stainless Steel
Reprinted from: *Metals* **2022**, *12*, 1222, doi:10.3390/met12071222 33

Marek Smaga, Annika Boemke, Dietmar Eifler and Tilmann Beck
Very High Cycle Fatigue Behavior of Austenitic Stainless Steels with Different Surface
Morphologies
Reprinted from: *Metals* **2022**, *12*, 1877, doi:10.3390/met12111877 47

Ulrich Krupp and Alexander Giertler
Surface or Internal Fatigue Crack Initiation during VHCF of Tempered Martensitic and Bainitic
Steels: Microstructure and Frequency/Strain Rate Dependency
Reprinted from: *Metals* **2022**, *12*, 1815, doi:10.3390/met12111815 59

Steven Schellert, Julian Müller, Arne Ohrndorf, Bronslava Gorr, Benjamin Butz and
Hans-Jürgen Christ
Characterization of the Isothermal and Thermomechanical Fatigue Behavior of a Duplex Steel
Considering the Alloy Microstructure
Reprinted from: *Metals* **2022**, *12*, 1161, doi:10.3390/met12071161 73

Amin Khayatzadeh, Stefan Guth and Martin Heilmaier
Comparison of the Internal Fatigue Crack Initiation and Propagation Behavior of a Quenched
and Tempered Steel with and without a Thermomechanical Treatment
Reprinted from: *Metals* **2022**, *12*, 995, doi:10.3390/met12060995 91

Ulrike Karr, Bernd M. Schönbauer, Yusuke Sandaiji and Herwig Mayer
Effects of Non-Metallic Inclusions and Mean Stress on Axial and Torsion Very High Cycle
Fatigue of SWOSC-V Spring Steel
Reprinted from: *Metals* **2022**, *12*, 1113, doi:10.3390/met12071113 103

Anna Wildeis, Hans-Jürgen Christ and Robert Brandt
Influence of Residual Stresses on the Crack Initiation and Short Crack Propagation in
a Martensitic Spring Steel
Reprinted from: *Metals* **2022**, *12*, 1085, doi:10.3390/met12071085 121

David Görzen, Pascal Ostermayer, Patrick Lehner, Bastian Blinn, Dietmar Eifler and Tilmann Beck
A New Approach to Estimate the Fatigue Limit of Steels Based on Conventional and Cyclic Indentation Testing
Reprinted from: *Metals* **2022**, *12*, 1066, doi:10.3390/met12071066 137

Sebastian Barton, Maximilian K.-B. Weiss and Hans Jürgen Maier
In-Situ Characterization of Microstructural Changes in Alloy 718 during High-Temperature Low-Cycle Fatigue
Reprinted from: *Metals* **2022**, *12*, 1871, doi:10.3390/met12111871 151

Sebastian Schöne, Sebastian Schettler, Martin Kuczyk, Martin Zawischa and Martina Zimmermann
VHCF Behavior of Inconel 718 in Different Heat Treatment Conditions in a Hot Air Environment
Reprinted from: *Metals* **2022**, *12*, 1062, doi:10.3390/met12071062 165

Selina Körber, Silas Wolff-Goodrich, Rainer Völkl and Uwe Glatzel
Effect of Wall Thickness and Surface Conditions on Creep Behavior of a Single-Crystal Ni-Based Superalloy
Reprinted from: *Metals* **2022**, *12*, 1081, doi:10.3390/met12071081 177

Ivo Šulák, Karel Hrbáček and Karel Obrtlík
The Effect of Temperature and Phase Shift on the Thermomechanical Fatigue of Nickel-Based Superalloy
Reprinted from: *Metals* **2022**, *12*, 993, doi:10.3390/met12060993 189

Aditya Srinivasan Tirunilai, Klaus-Peter Weiss, Jens Freudenberger, Martin Heilmaier and Alexander Kauffmann
Revealing the Role of Cross Slip for Serrated Plastic Deformation in Concentrated Solid Solutions at Cryogenic Temperatures
Reprinted from: *Metals* **2022**, *12*, 514, doi:10.3390/met12030514 201

Konstantina D. Karantza and Dimitrios E. Manolakos
Crashworthiness Analysis of Square Aluminum Tubes Subjected to Oblique Impact: Experimental and Numerical Study on the Initial Contact Effect
Reprinted from: *Metals* **2022**, *12*, 1862, doi:10.3390/met12111862 207

Maxwell Hein, David Kokalj, Nelson Filipe Lopes Dias, Dominic Stangier, Hilke Oltmanns, Sudipta Pramanik, et al.
Low Cycle Fatigue Performance of Additively Processed and Heat-Treated Ti-6Al-7Nb Alloy for Biomedical Applications
Reprinted from: *Metals* **2022**, *12*, 122, doi:10.3390/met12010122 229

About the Editors

Martin Heilmaier

Prof. Martin Heilmaier is currently full professor at the Institute for Applied Materials at Karlsruhe Institute of Technology (KIT). He is an academic expert on high-temperature structural materials development and testing. His research is mainly targeting integrated materials development guided by thermodynamic modelling and phase predictions, leading to materials synthesis via various methods (arc melting, powder metallurgy as well as additive manufacturing) and ensuing scale-bridging microstructural characterization techniques. His expertise is complemented by several decades of experience in mechanical and oxidation testing at a wide span of temperatures and environmental conditions. Since starting his scientific career in 1996, Martin Heilmaier has currently published over 270 papers in high-ranking international materials and manufacturing journals. They have been cited more than 6600 times presently with an h-index of 43. Because of his internationally known expertise in the field of intermetallic compounds, he has also served as editor of the Elsevier journal Intermetallics for the past 9 years and since 2005 as a key reader for Metallurgical and Materials Transactions A.

Martina Zimmermann

Martina Zimmermann is currently full professor at the Institute of Materials Science at the Faculty of Mechanical Science and Engineering at the TU Dresden. She is also division manager of the competence field Materials Characterization and Testing at the Fraunhofer Institute of Materials and Beam Technology (IWS) in Dresden. She is an academic expert in the field of mechanical behavior primarily of metallic materials. Her research focuses in particular on the fatigue behavior of materials and the underlying microstructure related damage mechanisms. Among others, she investigates the influence of material processing such as laser joining or cutting as well as laser assisted additive manufacturing on the structural reliability of component parts. Joining the materials science and engineering community only after her PhD in 2001 in the field of the fatigue behavior of weldings, she has since published well over 200 papers in peer-reviewed journals and conference proceedings. In 2014 Martina Zimmermann has been honoured with the Galileo Prize (awarded by the German Society for Materials Science (DGM), the German Association for Materials Research and Testing (DVM) and the Steel Institute VDEh) for her outstanding achievements in the field of materials testing. She also acted as organizer of international conferences such as the 7. Very High Cycle Fatigue Conference in 2017 and the Materials Science and Engineering Conference in Darmstadt in 2022.

Preface

The mechanical behavior of metals has been the subject of extensive research work ever since metals were applied in structural applications. However, the need for a comprehensive understanding of process history – (micro)structure – property correlations has not become less important throughout the years. On the contrary, with an increasing demand for a sustainable utilization of materials, their strength potentials under static as well as cyclic loading and the more so in combination with harsh environment conditions have to be fully rationalized. The aim of this Special Issue is to address recent new results and findings on the deformation behavior of single- and multi-phase (particle strengthened) metallic materials, including metal matrix composites, particularly research addressing underlying microstructural mechanisms and consequences on material design and application. The Special Issue contains 17 articles written by well-known experts in the field of fatigue and fracture and related topics. Within the range of contributions latest results also refer to creep and thermomechanical behavior of metallic materials. This Special Issue highlights the present and future significance of an experimental assessment of the mechanical behavior of metallic materials and its consequences for material design and application and will therefore be useful for researchers and engineers alike.

Martin Heilmaier and Martina Zimmermann
Editors

Article

Influence of a Pronounced Pre-Deformation on the Attachment of Melt Droplets and the Fatigue Behavior of Laser-Cut AISI 304

André T. Zeuner [1,*], Thomas Wanski [1], Sebastian Schettler [2], Jonas Fell [3], Andreas Wetzig [2], Robert Kühne [2], Sarah C. L. Fischer [4] and Martina Zimmermann [1,2]

[1] Institute of Materials Science, Technische Universität Dresden, 01069 Dresden, Germany
[2] Fraunhofer Institute for Material and Beam Technology IWS, 01277 Dresden, Germany
[3] Department of Materials Science and Engineering, Saarland University, 66123 Saarbrucken, Germany
[4] Fraunhofer Institute for Nondestructive Testing IZFP, 66123 Saarbrucken, Germany
* Correspondence: andre_till.zeuner@tu-dresden.de

Abstract: Laser cutting is a suitable manufacturing method for generating complex geometries for sheet metal components. However, their cyclic load capacity is reduced compared to, for example, milled components. This is due to the influence of the laser-cut edge, whose characteristic features act as crack initiation sites, especially resolidified material in the form of burr and melt droplets. Since sheet metal components are often formed into their final geometry after cutting, another important factor influencing fatigue behavior is the effect of the forming process on the laser-cut edge. In particular, the effect of high degrees of deformation has not yet been researched in detail. Accordingly, sheets of AISI 304 were processed by laser cutting and pre-deformed. In the process, α′-martensite content was set to be comparable despite different degrees of deformation. It was found that deformation to high elongations caused a large part of the melt adhesions to fall off, but those still attaching were partially detached and thus formed an initial notch for crack initiation. This significantly lowered the fatigue strength.

Keywords: laser cutting; fatigue behavior; melt adhesion; resolidified material; detachment; notch; micro-computed tomography analysis

1. Introduction

Laser beam cutting is a highly flexible and precise process capable of cutting a wide range of industrially relevant materials. The cut is realized by the energy of the laser beam, whereby molten material is expelled by process gas. This manufacturing process offers a number of advantages. Since there is no tool-to-workpiece contact, tool wear is avoided. In addition, no special tools are required for different component geometries or materials, which saves retooling time and reduces the required storage capacity. Laser cutting is, in particular, beneficial for the processing of materials that are difficult to mechanically process, such as titanium alloys or stainless steel. Despite the exceptional mechanical properties of these materials they are easily cut by laser beam cutting due to contactless processing [1,2].

The metastable austenitic stainless steel AISI 304 is a very commonly used sheet metal to which the scenario described above applies. AISI 304 is used in different industries such as chemistry, construction, energy, food, medicine, mechanical engineering, or mobility. For the last two fields of application, cyclic loading is common, for example, in the case of the exhaust gas system of automobiles. The wide field of application results from the advantageous properties of high strength, high ductility, and good corrosion resistance of AISI 304. The latter can be attributed to high chromium content and the face-centered cubic crystal lattice [3–5].

The mechanical behavior of AISI 304 is linked to a complex microstructural evolution which is called martensitic transformation. The martensitic transformation can be thermal, stress, or deformation induced, whereby deformation-induced martensitic transformation has the highest significance for industrial application. This is the transformation of face-centered cubic γ-austenite into body-centered cubic α′-martensite under plastic deformation. It may incorporate the intermediate step of forming hexagonal close-packed ε-martensite [6–11]. For the mechanical properties, α′-martensite is particularly important because it has higher strength and hardness than γ-austenite [12–14].

Martensitic transformation is one factor why machining AISI 304 is challenging, which is a factor why machining by laser can be economically attractive [15]. However, studies on laser-cut materials showed that the fatigue limit is significantly reduced compared to milled or polished material; punched material achieves similar fatigue limits [16–18]. The reason for this is the notch effect of the re-solidified material, which is formed by the solidification of molten material on the melt ejection side of the sheet metal during laser cutting. Removing melt droplets slightly increases fatigue strength. However, pores in the re-solidified material adhering to the cutting edge act as crack initiation sites. Therefore, fatigue strength is still inferior compared to a classical mechanical subtractive machining process [19].

In a previous study, the influence of laser-cutting parameters and a subsequent pre-deformation on the fatigue behavior of laser-cut AISI 304 was investigated [20]. It was shown that depending on laser-cutting parameters, a pre-deformation can cause the detachment of melt droplets. This had a direct influence on the fatigue strength achieved. Laser-cutting parameters that realized melt droplets with a high tendency to detach during subsequent deformation resulted in higher fatigue strength. This is an important factor for the application of laser-cut sheet metal components, which subsequently have to be formed to the final component geometry. However, this detachment effect was only investigated in fatigue tests for amounts of pre-deformation below 30% strain. A comparison of different amounts of deformation showed that a large proportion of the melt droplets had already detached at up to 20% strain, while no significant additional detachment effect was achieved for deformation of up to 50% during tensile testing. The result of this investigation is depicted in Figure 1 for two batches. At this point, the question arises as to how the melt adhesions behave under further plastic deformation significantly above the investigated deformations of less than 30%. After all, significantly higher forming degrees can be achieved in industrial forming steps.

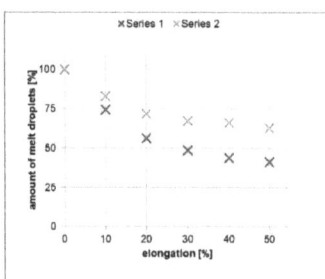

Figure 1. Detachment behavior of melt droplets caused by pre-deformation [20].

In summary, the pre-deformation of a laser-cut AISI 304 has several effects on fatigue behavior. On the one hand, work hardening as well as deformation-induced α′-martensite transformation leads to a strengthening of the material, which leads to an increase in cyclic load-bearing capacity. On the other hand, melt droplets can detach from the workpiece due to weak attachment. This reduces the notch effect and also increases fatigue strength. However, a second mentioned was only investigated in fatigue tests for pre-strains below 30%.

Since degrees of deformation significantly above 30% strain can be achieved in many applications of formed sheet metal components, the objective of this study was to investigate the effect of high grades of pre-deformation on laser-cut AISI 304 material and its influence on fatigue behavior. As the attachment of melt droplets has been found to be an important factor for deformation-induced detachment as well as fatigue behavior, it was particularly important to investigate whether and to what extent the attachment to the base material changes in the case of high degrees of deformation.

2. Material and Methods

2.1. Specimen Preparation

The fatigue behavior of laser-cut AISI 304 is sensitive to a number of influencing variables, some of which have already been described in the first section. Accordingly, it was important that all specimens were manufactured from the same sheet metal, received the same pre-treatment, and were cut on the same laser-cutting machine with the same cutting parameters. The chemical composition and mechanical properties of the investigated 2 mm thick AISI 304 steel sheet are shown in Tables 1 and 2, respectively.

Table 1. Chemical composition of AISI 304 provided by the manufacturer.

Element	C	S	P	Mn	Si	Cr	Ni	N	Fe
wt-%	0.016	0.001	0.036	1.90	0.36	18.2	8.1	0.01	Bal.

Table 2. Mechanical properties of AISI 304 provided by the manufacturer.

Yield Strength/MPa	Tensile Strength/MPa	Elongation at Break/%
275	620	54

Another measure was the heat treatment of the sheet material before laser cutting to ensure homogeneous phase distributions and properties. Therefore, the material was heat treated at 1363 K for 1 h and cooled fast in moving air for complete austenitization and recrystallization, which eliminates the rolling texture. Thus, influences from the rolling process could be excluded in the following investigations. Subsequently, a laser-cutting machine with a TruDisk 5001 disk laser of the company Trumpf (Ditzingen, Germany) was used. All specimens were produced with the same processing parameters, which are summarized in Table 3. Here, f_{coll} stands for collimation length, f_{foc} for focal length, d_{nozzle} for nozzle diameter, P_L for laser power, v_f for feed rate, d_z for focal position, d_s for stand-off distance, p_{gas} for gas pressure, λ_L for laser wavelength, M^2 for beam quality factor, and d_f for focus diameter. Nitrogen was used as a process gas.

Table 3. Laser-cutting parameters.

f_{coll}/mm	f_{foc}/mm	d_{nozzle}/mm	P_L/kW	v_f/m·min^{-1}	d_z/mm	d_{ns}/mm	p_{gas}/bar	λ_L/µm	M^2	d_f/µm
100	150	2.3	3	16.5	0	0.8	11	1.03	13.0	192

2.2. Examination of Melt Droplets by Micro-Computed Tomography

The central question of how the bonding of melt droplets to the base material changes at high degrees of deformation was investigated in detail. For this reason, high-resolution micro-computed tomography (µCT) was chosen as the method of investigation so that 3D information could be generated. Due to the high density of the material investigated, the material volume must be small for this type of investigation. Accordingly, Figure 2 shows the used specimen geometry with a bridge volume of 2 mm^3, which was deformed to a local strain of 44% in the bridge close to the ultimate tensile strength of the specimens. The deformed bridge was cut out of the specimens and examined by µCT. Non-deformed specimens were examined for comparison.

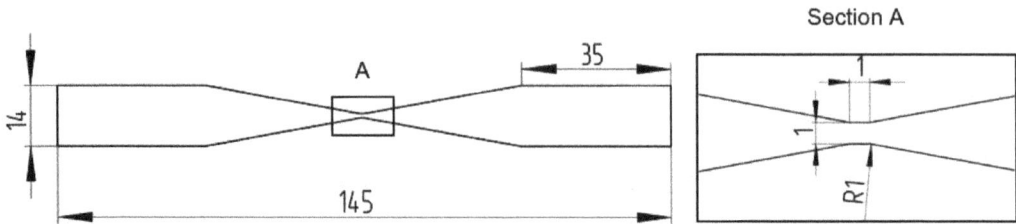

Figure 2. Specimen geometry for pre-deformation and µCT examinations made of 2 mm thick AISI 304 steel sheet with given dimensions in mm. Test region is shown magnified.

Micro-computed tomography was performed with a custom-built X-ray system at Fraunhofer IZFP, Saarbrücken, equipped with a micro-focus tube and a 4 k pixel detector. For the µCT measurement, 1600 projections over 360° were acquired with a tube voltage of 220 kV and a tube power of 100 Watt. A voxel size of approximately 2 µm could be reached. Volume reconstruction was carried out simultaneously with a custom reconstruction algorithm. Sectional planes showing the connection to the base material were then examined from this.

2.3. Fatigue Testing

The effect of large degrees of deformation of laser-cut sheets examined by the µCT studies on fatigue behavior was another focus of the investigations presented. For fatigue testing, the specimen geometry in Figure 3 was used. For this purpose, a series of specimens with a deformation of 28% strain was investigated, as was done in a previous study [20]. A second series of specimens underwent a deformation of 44% strain, as did the specimens in the µCT study.

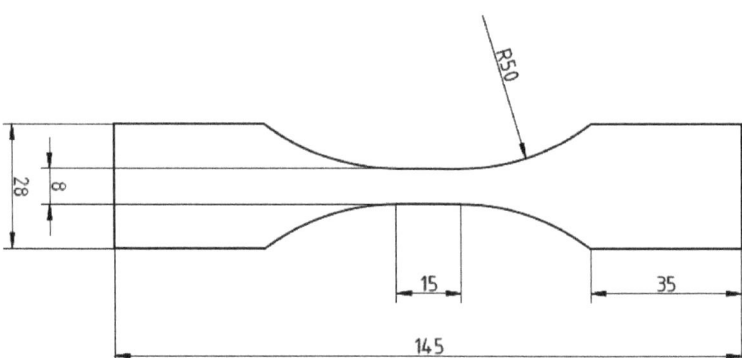

Figure 3. Specimen geometry used for tensile and fatigue testing made of 2 mm thick AISI 304 steel sheet with given dimensions in mm.

However, in order for these two series of specimens with different amounts of strain to be comparable, it was necessary to ensure that similar α'-martensite content was present in both series of specimens after pre-deformation. Otherwise, the specimen series with the higher α'-martensite content would already exhibit a higher cyclic load capacity due to the strengthening effect of α'-martensite. Accordingly, tensile tests to examine the influence of the deformation parameters strain, strain rate, and temperature on the deformation-induced martensitic transformation behavior were carried out for the given chemical composition and specimen geometry first. Therefore, a tensile testing machine from the company ZwickRoell (Ulm, Germany) and a climate chamber from the company mytron (Heilbad

Heiligenstadt, Germany) were used. In addition, a Feritscope from the company Fischer was used to measure the transformed α'-martensite content during the test. The calibration was carried out according to [21]. Using this test setup, tensile tests were performed at laboratory conditions (298 K), 273 K, 258 K, and 243 K. At each temperature, tests were performed at strain rates of 10 mm/min, 1 mm/min, and 0.1 mm/min achieved in the bridge of the specimens, respectively. Three samples were tested for each deformation state of temperature and strain rate. The results of this preliminary investigation are shown in Figure 4. Here the α'-martensite volume fraction M_V is plotted against temperature and stress.

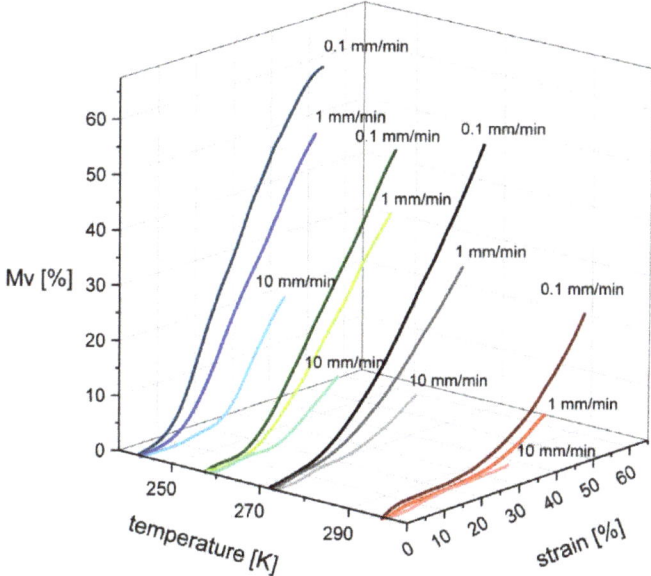

Figure 4. Relationship among ambient temperature, strain, strain rate, and α'-martensite volume fraction (Mv) for the chemical composition of AISI 304.

Based on tensile test results, two deformation regimes were selected that resulted in comparable α'-martensite contents, see Table 4. At both conditions, 15 specimens each were manufactured to investigate fatigue behavior. The first series of specimens were deformed at 298 K up to 44% and the second series of specimens were deformed at 273 K up to 28%, and they are accordingly referred to as 298K-44% and 273K-28%, respectively, in the remainder of this paper.

Table 4. Designation of specimen series, ambient temperature during pre-deformation, strain, strain rate, and number of specimens used to characterize fatigue strength.

Series Name	Temperature/K	Strain/%	Strain Rate/mm/min	α'-Content/%	No. of Samples
298K-44%	298	44	1	10 ± 2	15
273K-28%	273	28	1	10 ± 2	15

To investigate fatigue behavior, a Gigaforte testing machine from the company Russenberger Prüfmaschinen (Neuhausen am Rheinfall, Switzerland) was used. This machine realizes test frequencies of about 1 kHz. AISI 304 heats adiabatically under cyclic loading and therefore has to be cooled sufficiently [22,23]. Pre-deformed specimens already showed significantly reduced adiabatic heating compared to non-deformed AISI 304. Additionally, cooling by means of a vortex tube was used and a stress ratio R of 0.5 was selected. The lower stress amplitudes compared to R of −1 or 0.1 resulted in a further significant reduction in adiabatic heating at high test frequencies. The combination of these actions ensured that the specimens did not exceed a temperature of 303 K during fatigue testing, which was monitored with a pyrometer. The tests were carried out to at least 10^8 load cycles in order to be counted as a run-out. The procedure was in accordance with DIN 50100 (2016) [24], which describes a method to statistically determine fatigue strength by selecting the stress amplitude depending on the result of the previously tested sample (break or run out). This method allows an evaluation of the stress amplitude with a 50% failure probability, which is used as a characteristic value for the discussion of fatigue strength. A change in resonance frequency Δf of 2 Hz was applied as a criterion to stop fatigue testing, indicating technical crack initiation. Subsequently, fracture surface analyses were performed using a scanning electron microscope (SEM) JSM-6610LV of the company JEOL (Tokyo, Japan) to determine the location of crack initiation.

3. Results and Discussion

3.1. Effect of High Pre-Strains on the Bonding of Melt Droplets to the Base Material

Since the pre-deformation was performed after cutting, the laser-cut edge was also deformed. The deformed laser-cut edges were examined by light microscopy and are shown exemplarily in Figure 5 for the conditions as cut (a) and pre-deformed to 28% strain (b). Comparing the two figures, the detachment phenomenon, which has been described in past work [20], due to plastic pre-deformation was again evident at the lower sheet edge.

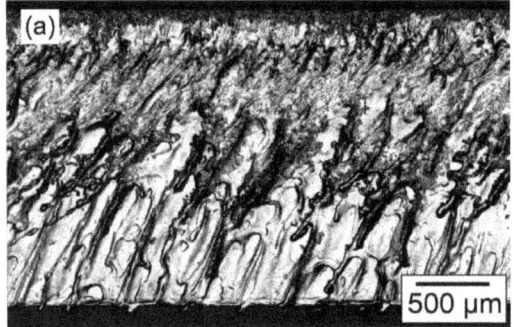 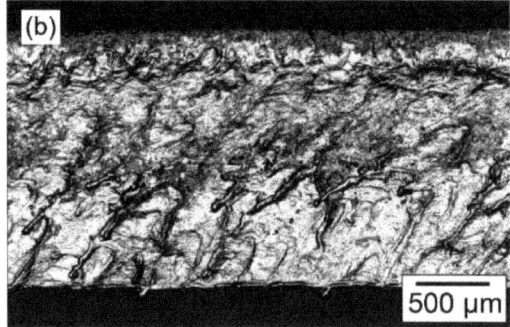

Figure 5. Top view of the laser-cut edge. (**a**) As cut, (**b**) Pre-deformed (ε = 28%).

Figures 6 and 7 show the results of the μCT examination. Here, slices of exemplary reconstructed melt droplets are shown for the two conditions as-cut and pre-strained to 44% highlighting the bonding condition on different observation planes, respectively. The images are to be understood in such a way that the labels above the images represent the depth of observation. In this case, surface means that the surface of the laser-cut edge—to the underside of which the melt droplet is attached—was observed. In the case of the label 36 μm, the observation plane was shifted 36 μm from the surface of the laser-cut edge into the sheet material. The melt droplet of the non-deformed sample in Figure 6 showed strong adherence to the base material, i.e., complete attachment to the base material was observed for the entire melt droplet volume. Opposite to the non-deformed sample, weak adherence of the melt droplet of the pre-deformed specimen can be clearly seen in Figure 7. Only at the surface was the droplet bonded to the base material. At a depth of 36 μm, the reconstructed slice already showed a large gap. This gap increased with greater observation depth to the surface of the laser-cut edge. At 88 μm, the melt droplet did not show a bond to the base material at all. This means that the melt droplet adhered only to the outer edge of the sheet and otherwise no longer had any bond to the base material. These results indicate that the partial detachment created a notch between the melt droplet and the base material, which could be detrimental to the fatigue strength of the material at high levels of pre-deformation.

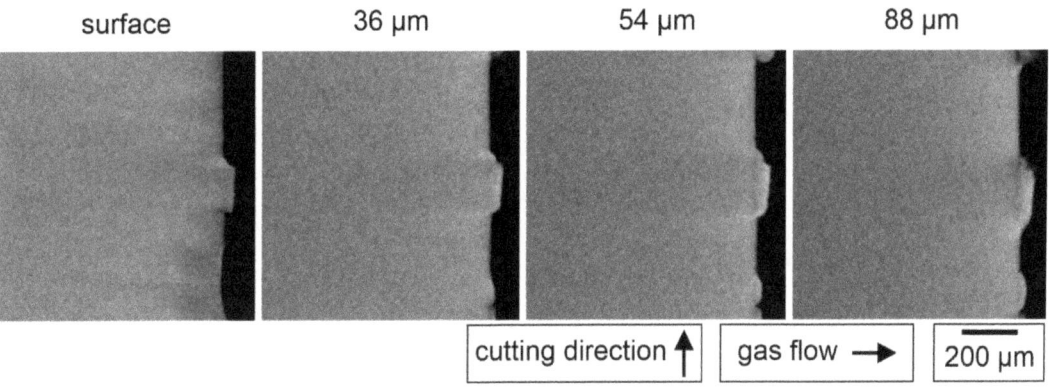

Figure 6. Selection of observation planes of the same re-solidified material droplet in the as-cut condition at different observation depths starting from the surface of the laser-cut edge.

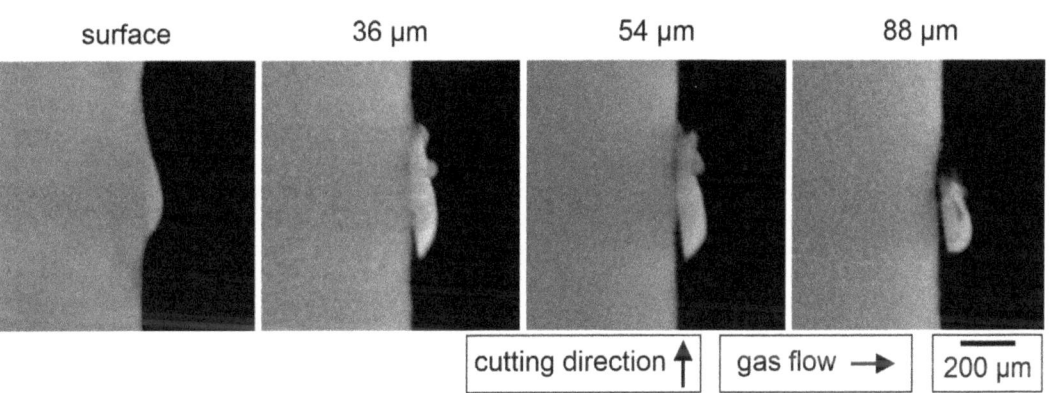

Figure 7. Selection of observation planes of the same re-solidified material droplet in the pre-deformed condition (44%) at different observation depths starting from the surface of the laser-cut edge.

3.2. Effect of High Pre-Strains on Fatigue Behavior

The results of fatigue testing are shown in Figure 8, where stress is plotted over the number of load cycles in a S-N-diagram. It can be seen that the fatigue behavior of the tested specimen series differed significantly from each other. The highest stress levels were reached by the sample series 273K-28% (i.e., pre-deformed at 273 K and a strain of 28%). The fatigue strength for 50% failure probability at 10^8 load cycles was 170 MPa according to DIN 50100.

On the contrary, sample series 298K-44% only showed a fatigue strength for a 50% failure probability of 145 MPa. Additionally, the measured values for load cycles until failure scattered considerably more. The failure of a specimen was even observed shortly before reaching 10^8 load cycles.

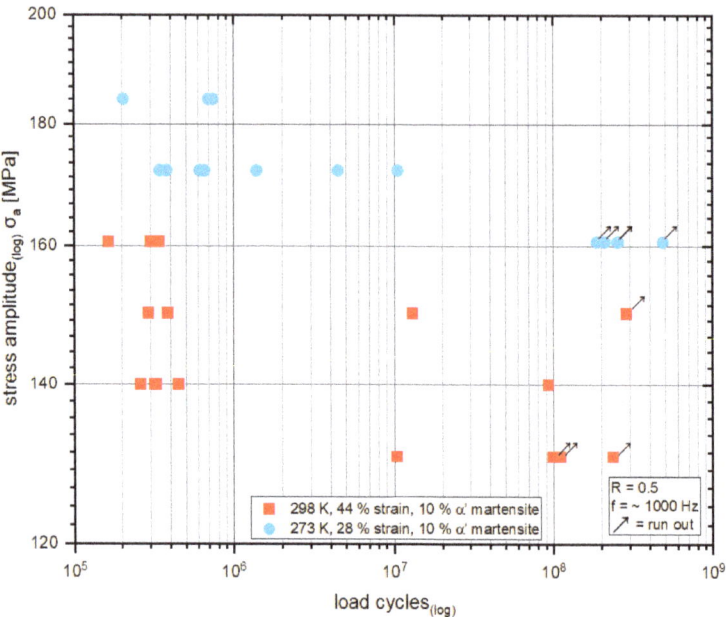

Figure 8. S-N diagram of fatigue tests on laser-cut AISI 304.

Following the fatigue tests, the fracture surfaces of all specimens were examined by SEM. It was observed consistently for all specimens that the location of crack initiation was at the melt ejection site of the laser-cut edge where molten material had re-solidified. In the case of the 298K-44% sample series, it was observed that the crack initiation site showed a characteristic feature with a smooth surface. This characteristic feature is shown in Figure 9, where in the case of (a), the fracture surface is shown from the top view. Figure 9b shows the specimen rotated by 45°. In this view, the sharp interface is visible between the smooth surface and the fatigue crack that propagated from this interface into the specimen volume. In the case of the specimen series 273K-28%, this observation was not made. For this sample series, the location of crack initiation is shown representatively in Figure 9c,d in the same way as for the 298K-44% sample series. It can be observed that the fatigue crack started in the outermost corner of the re-solidified material. A specific feature that served as crack initiation as was the case for 298K-44% is not identifiable here.

Through the previous investigations, it was hypothesized that the resolidified material was strongly deformed at the laser-cut edge, melt adhesions (partially) detached, and this led to a severe, crack-like notch for introducing premature fatigue crack growth. To support this argument, 44% of pre-deformed but non-fatigued specimens were re-examined by SEM. It was observed that the melt adhesions have a melt inflow, which remained at the laser-cut edge. This can be seen in Figure 10. This melt inflow showed a smooth surface similar to the crack-initiating feature in Figure 9a,b. Hence, it can be assumed that fatigue crack propagation indeed started at a notch-like partial detachment of a melt adhesion. Contrary to Figure 10, the melt adhesion itself is not visible in Figure 9, as it fell off either due to pre-deformation or during fatigue loading.

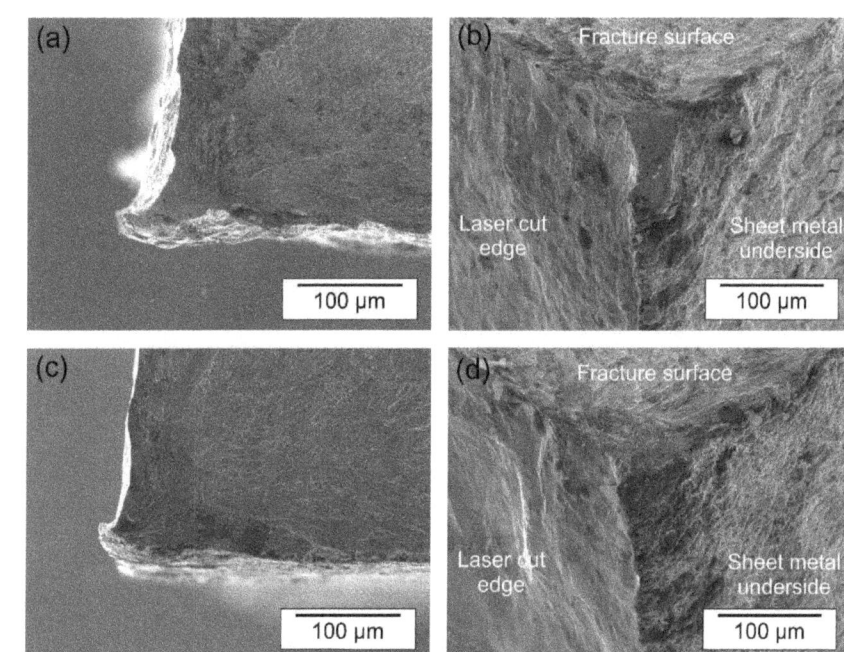

Figure 9. Selection of fracture surfaces captured by SEM. (**a**) 298K-44%, top view, (**b**) 298K-44%, 45°-perspective view, (**c**) 273K-28%, top view, (**d**) 273K-28%, 45°-perspective view.

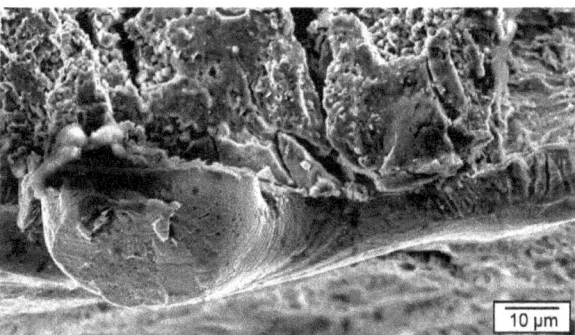

Figure 10. SEM detail image of a melt adhesion showing a melt inflow with a smooth surface, similar to the crack initiation site of the fracture surface for a sample pre-deformed to 44% elongation and subsequently fatigued.

In summary, both series 298K-44% and 273K-28% were manufactured from the same sheet, laser cut with the same parameters, and ultimately pre-deformed in a way that they had comparable α'-martensite content in the base material. Their main difference can be found in the amount of pre-strain introduced. In the end, they showed clear differences in their fatigue behavior. When pre-forming, a significantly large number of melt droplets fell off the laser-cut material within the first 20% elongation, as was shown in previous work [20]. After that, the number of melt droplets decreased marginally. As soon as high elongations in a range close to the tensile strength were reached, it was observed by means of μCT, that the remaining melt droplets no longer detached completely, but partially. Due to only partial bonding to the base material, a crack-like notch formed, see Figure 7, which served as a crack initiation site during the cyclic loading in the fatigue test. This assumption is supported by the smooth features of the fracture surfaces in Figure 9a,b as well as Figure 10, which could not be found in Figure 9c,d. For a better understanding of the significance of the partial detachment of the melt adhesion, the particular mesoscopic features of the pre-damage introduced through the combination of laser cutting and subsequent pre-deformation are schematically drawn in Figure 11.

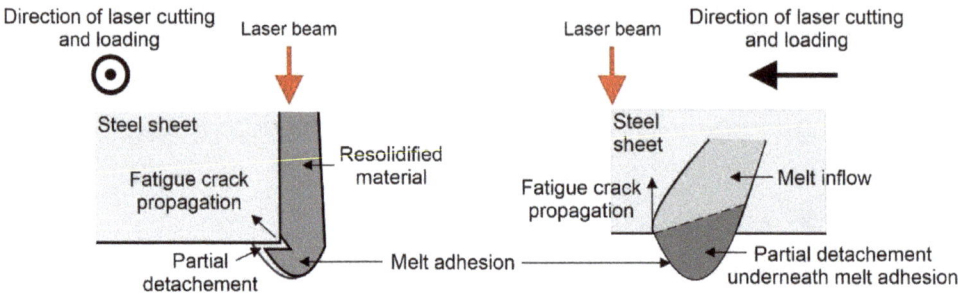

Figure 11. Schematic drawing for partial detachment of melt adhesions on strongly deformed laser-cut edges.

The fatigue test results in Figure 8 showed that this pre-damage due to laser cutting and pre-forming to high degrees of deformation can lead to significantly reduced fatigue strength and very late failures close to 10^8 load cycles. However, both series have in common that the crack initiation took place in the lower outermost corner of the laser-cut edge at the re-solidified material, i.e., melt droplets. The significant difference in pre-strain in the base material did not change this situation. This underlines the central importance of the re-solidified material for the fatigue behavior of laser-cut materials. In this case, it was discovered that the partial detachment of melt droplets and the associated notch formation leads to crack initiation under cyclic loading and hence, has a significant influence on the cyclic strength. Through a careful selection of the laser-cutting parameters alone, the cyclic strength of AISI 304 can already be influenced, as could be shown in [20]. A subsequent pre-deformation—as would be typical for a car body application—will have an additional effect on the fatigue strength. For laser-cut sheet metals, this pre-deformation effect cannot solely be explained by the two strengthening mechanisms, namely strain-hardening and deformation-induced α′ martensite formation. These purely microstructure-related mechanisms have already been investigated comprehensively in the past, see e.g., [25–28]. In the study presented, the higher pre-deformation should have resulted in a beneficial effect due to an increased strain-hardening since the α′ martensite volume fraction was kept constant. However, the higher pre-deformation, in fact, caused a pronounced decrease in the cyclic strength. Fractographic analyses and µCT examinations showed that the effect of dross formation and its partial detachment excel the microstructural strengthening effects on fatigue behavior. From this, it can be deduced that in the case of laser cutting and subsequent forming of AISI 304, a reliable prediction of the fatigue strength strongly depends on a comprehensive understanding of the process–history interactions and their effect both on microstructural changes as well as mesoscopic notch effects.

In previously typical applications for laser cutting, a "clean and accurate" cutting edge is considered as one of the decisive evaluation criteria. Interdependencies between remaining re-solidified material after processing and the thermal impact on the materials' microstructure are of minor importance. However, in the case of laser-cut parts subjected to cyclic mechanical loading, this no longer holds. As is well known from former very high-cycle fatigue research, the influence of isolated micro-notches plays a decisive role in fatigue behavior with an increasing number of cycles. Transferring this observation to the particularly rugged surface relief at the laser-cutting edge, crack initiation will most likely occur at a single, most detrimental, and therefore, failure-relevant micro-notch. The stochastic distribution of the mesoscopic geometrical characteristics of the rugged surface relief of the laser-cut surface is practically not determinable. Nonetheless, the results presented allow for an evaluation of the interaction between the material's microstructural and laser-cut surface characteristics in dependence on the process–history. It could be shown that the effect of dross formation and its partial detachment excel the microstructural strengthening effects.

4. Conclusions

By means of μCT investigations on laser-cut and highly pre-formed specimens, it could be shown that this process–history leads to the partial detachment of melt droplets, which causes notch formation on a mesoscopic scale. Fatigue tests up to 10^8 loading cycles were conducted, since in this particular regime localized discontinuities define the fatigue behavior. Two deformation states (28 and 44% pre-strain) were prepared under different deformation conditions but in a way that similar α'-martensite volume fractions were obtained. The fatigue tests were paired with fracture surface analysis. The following conclusions can be drawn from the investigations:

- Re-solidified material (most often in the form of melt droplets) at the melt ejection site of the laser-cut edge is the dominating fatigue crack initiation site independent of the degree of pre-deformation introduced.
- Melt droplets with good bonding to the base material can show partial detachment at high degrees of deformation.
- Partial detachment of melt droplets leads to mesoscopic notch formation, which causes a significant decrease in fatigue strength and can lead to late failures even at load cycles well beyond the classical "durability limit" (formerly defined at 2 mio. cycles).

Author Contributions: Conceptualization, A.T.Z., S.C.L.F. and M.Z.; Methodology, A.T.Z., S.S. and S.C.L.F.; Validation, A.T.Z., S.C.L.F. and M.Z.; Formal Analysis, A.T.Z., R.K., S.S. and S.C.L.F.; Investigation, A.T.Z., T.W., S.S., J.F., R.K. and S.C.L.F.; Writing—Original Draft Preparation, A.T.Z.; Writing—Review and Editing, S.C.L.F. and M.Z.; Visualization, A.T.Z., T.W. and J.F.; Supervision, A.W., S.C.L.F. and M.Z.; Project Administration, A.W., S.C.L.F. and M.Z.; Funding Acquisition, A.W. and M.Z. All authors have read and agreed to the published version of the manuscript.

Funding: Funded by Deutsche Forschungsgemeinschaft (DFG, German Research Foundation): ZI 1006/15-1—Project-ID 413507401. This work was partially supported by the Fraunhofer Internal Programs under Grant No. Attract 025-601314 awarded to S.C.L. Fischer.

Data Availability Statement: The data presented in this study are available on request from the corresponding author.

Conflicts of Interest: The authors declare no conflict of interest.

References

1. Ion, J. Laser Processing of Engineering Materials. In *Principles, Procedure and Industrial Application*, 1st ed.; Hill, J., Ed.; Elsevier: Amsterdam, The Netherlands, 2005.
2. Sharma, A.; Yadava, V. Experimental analysis of Nd-YAG laser cutting of sheet materials—A review. *Opt. Laser Technol.* **2018**, *98*, 264–280. [CrossRef]
3. Lo, K.H.; Shek, C.H.; Lai, J. Recent developments in stainless steels. *Mater. Sci. Eng. R Rep.* **2009**, *65*, 39–104.
4. Davison, R.M.; Laurin, T.R.; Redmond, J.D.; Watanabe, H.; Semchyshen, M. A Review of Worldwide Developments in Stainless Steels. *Materials & Design* **1986**, *7*, 111–119.
5. Davis, J.R. *Stainless Steels*; ASM International: Materials Park, OH, USA, 2010.
6. Olson, G.B.; Cohen, M. A general mechanism of martensitic nucleation: Part I. General concepts and the FCC → HCP transformation. *Metall. Trans. A* **1976**, *7*, 1897–1904.
7. Olson, G.B.; Cohen, M. A general mechanism of martensitic nucleation: Part II. FCC → BCC and other martensitic transformations. *Metall. Trans. A* **1976**, *7*, 1905–1914.
8. Meyrick, G.; Powell, G.W. Phase Transformations in Metals and Alloys. *Annu. Rev. Mater. Sci.* **1973**, *3*, 327–362. [CrossRef]

9. Schramm, R.E.; Reed, R.P. Stacking fault energies of seven commercial austenitic stainless steels. *Metall. Trans. A* **1975**, *6*, 1345–1351. [CrossRef]
10. Angel, T. Formation of Martensite in Austenitic Stainless Steels. Effects of Deformation, Temperature, and Composition. *J. Iron Steel Inst.* **1954**, *177*, 165–174.
11. Shen, Y.F.; Li, X.X.; Sun, X.; Wang, Y.D.; Zuo, L. Twinning and martensite in a 304 austenitic stainless steel. *Mater. Sci. Eng. A* **2012**, *552*, 514–522. [CrossRef]
12. Müller-Bollenhagen, C.; Zimmermann, M.; Christ, H.-J. Very high cycle fatigue behaviour of austenitic stainless steel and the effect of strain-induced martensite. *Int. J. Fatigue* **2010**, *32*, 936–942. [CrossRef]
13. Smaga, M.; Walther, F.; Eifler, D. Deformation-induced martensitic transformation in metastable austenitic steels. *Mater. Sci. Eng. A* **2008**, *483–484*, 394–397. [CrossRef]
14. Krupp, U.; Christ, H.-J.; Lezuo, P.; Maier, H.; Teteruk, R. Influence of carbon concentration on martensitic transformation in metastable austenitic steels under cyclic loading conditions. *Mater. Sci. Eng. A* **2001**, *319–321*, 527–530. [CrossRef]
15. Jadhav, A.; Kumar, S. Laser cutting of AISI 304 material: An experimental investigation on surface roughness. *Adv. Mater. Process. Technol.* **2019**, *5*, 429–437. [CrossRef]
16. Geiger, M.; Bergmann, H.W.; Nuss, R. Laser Cutting Of Steel Sheets. In *Proceedings Volume 1022, Laser Assisted Processing*; Laude, L.D., Rauscher, G.K., Eds.; International Congress on Optical Science and Engineering: Hamburg, Germany, 1988.
17. Meurling, F.; Melander, A.; Linder, J.; Larsson, M. The influence of mechanical and laser cutting on the fatigue strengths of carbon and stainless sheet steels. *Scand. J. Metall.* **2001**, *30*, 309–319. [CrossRef]
18. Mäntyjärvi, K.; Väisänen, A.; Karjalainen, J.A. Cutting method influence on the fatigue resistance of ultra-high-strength steel. *Int. J. Mater. Form.* **2009**, *2*, 547–550. [CrossRef]
19. Pessoa, D.; Grigorescu, A.; Herwig, P.; Wetzig, A.; Zimmermann, M. Influence of Notch Effects Created by Laser Cutting Process on Fatigue Behavior of Metastable Austenitic Stainless Steel. *Procedia Eng.* **2016**, *160*, 175–182. [CrossRef]
20. Wanski, T.; Zeuner, A.T.; Schöne, S.; Herwig, P.; Mahrle, A.; Wetzig, A.; Zimmermann, M. Investigation of the influence of a two-step process chain consisting of laser cutting and subsequent forming on the fatigue behavior of AISI 304. *Int. J. Fatigue* **2022**, *159*, 106779. [CrossRef]
21. Talonen, J.; Aspegren, P.; Hänninen, H. Comparison of different methods for measuring strain induced α-martensite content in austenitic steels. *Mater. Sci. Technol.* **2004**, *20*, 1506–1512. [CrossRef]
22. Das, A.; Sivaprasad, S.; Ghosh, M.; Chakraborti, P.C.; Tarafder, S. Morphologies and characteristics of deformation induced martensite during tensile deformation of 304 LN stainless steel. *Mater. Sci. Eng. A* **2008**, *486*, 283–286. [CrossRef]
23. Hecker, S.S.; Stout, M.G.; Staudhammer, K.P.; Smith, J.L. Effects of Strain State and Strain Rate on Deformation-Induced Transformation in 304 Stainless Steel: Part I. Magnetic Measurements and Mechanical Behavior. *Metall. Trans. A* **1982**, *13*, 619–626. [CrossRef]
24. DIN 50100:2016-12, Fatigue Testing—Performance and Evaluation of Cyclic Tests with Constant Load Amplitude for Metallic Material Specimens and Components. Available online: https://dx.doi.org/10.31030/2580844 (accessed on 20 December 2021).
25. Grigorescu, A.; Hilgendorff, P.M.; Zimmermann, M.; Fritzen, C.P.; Christ, H.J. Effect of Geometry and Distribution of Inclusions on the VHCF Properties of a Metastable Austenitic Stainless Steel. *Adv. Mater. Res.* **2014**, *891–892*, 440–445. [CrossRef]
26. Hilgendorff, P.-M.; Grigorescu, A.C.; Zimmermann, M.; Fritzen, C.-P.; Christ, H.-J. Cyclic deformation behavior of austenitic Cr–Ni-steels in the VHCF regime: Part II—Microstructure-sensitive simulation. *Int. J. Fatigue* **2016**, *93*, 261–271. [CrossRef]
27. Bayerlein, M.; Christ, H.-J.; Mughrabi, H. Plasticity-induced martensitic transformation during cyclic deformation of AISI 304L stainless steel. *Mater. Sci. Eng. A* **1989**, *114*, L11–L16. [CrossRef]
28. Smaga, M.; Boemke, A.; Daniel, T.; Skorupski, R.; Sorich, A.; Beck, T. Fatigue Behavior of Metastable Austenitic Stainless Steels in LCF, HCF and VHCF Regimes at Ambient and Elevated Temperatures. *Metals* **2019**, *9*, 704. [CrossRef]

Disclaimer/Publisher's Note: The statements, opinions and data contained in all publications are solely those of the individual author(s) and contributor(s) and not of MDPI and/or the editor(s). MDPI and/or the editor(s) disclaim responsibility for any injury to people or property resulting from any ideas, methods, instructions or products referred to in the content.

Article

Influence of Surface Mechanical Attrition Treatment (SMAT) on Microstructure, Tensile and Low-Cycle Fatigue Behavior of Additively Manufactured Stainless Steel 316L

Thomas Wegener [1,*,†], Tao Wu [1,2,3,*,†], Fei Sun [4], Chong Wang [5], Jian Lu [6] and Thomas Niendorf [1]

1. Institut für Werkstofftechnik, Metallische Werkstoffe, Universität Kassel, Mönchebergstraße 3, 34125 Kassel, Germany
2. Research Center Carbon Fibers Saxony (RCCF), Technische Universität Dresden, Breitscheidstraße 78, 01237 Dresden, Germany
3. Fraunhofer Institute for Ceramic Technologies and Systems (IKTS), Winterbergstraße 28, 01277 Dresden, Germany
4. Department of Materials Design Innovation Engineering, Graduate School of Engineering, Nagoya University, Furo-cho, Chikusa-ku, Nagoya 464-8603, Japan
5. Institute of Solid Mechanics, Beihang University, 37 Xueyuan Road, Beijing 100083, China
6. Department of Mechanical Engineering, City University of Hong Kong, 83 Tat Chee Ave, Hongkong 518057, China
* Correspondence: t.wegener@uni-kassel.de (T.W.); tao.wu1@mailbox.tu-dresden.de (T.W.)
† These authors contributed equally to this work.

Citation: Wegener, T.; Wu, T.; Sun, F.; Wang, C.; Lu, J.; Niendorf, T. Influence of Surface Mechanical Attrition Treatment (SMAT) on Microstructure, Tensile and Low-Cycle Fatigue Behavior of Additively Manufactured Stainless Steel 316L. *Metals* 2022, *12*, 1425. https://doi.org/10.3390/met12091425

Academic Editor: Pavel Krakhmalev

Received: 11 August 2022
Accepted: 22 August 2022
Published: 29 August 2022

Publisher's Note: MDPI stays neutral with regard to jurisdictional claims in published maps and institutional affiliations.

Copyright: © 2022 by the authors. Licensee MDPI, Basel, Switzerland. This article is an open access article distributed under the terms and conditions of the Creative Commons Attribution (CC BY) license (https://creativecommons.org/licenses/by/4.0/).

Abstract: Direct Energy Deposition (DED), as one common type of additive manufacturing, is capable of fabricating metallic components close to net-shape with complex geometry. Surface mechanical attrition treatment (SMAT) is an advanced surface treatment technology which is able to yield a nanostructured surface layer characterized by compressive residual stresses and work hardening, thereby improving the fatigue performances of metallic specimens. In the present study, stainless steel 316L specimens were fabricated by DED and subsequently surface treated by SMAT. Both uniaxial tensile tests and uniaxial tension-compression low-cycle fatigue tests were conducted for as-built and SMAT processed specimens. The microstructure of both conditions was characterized by roughness and hardness measurements, scanning electron microscopy and transmission electron microscopy. After SMAT, nanocrystallites and microtwins were found in the top surface layer. These microstructural features contribute to superior properties of the treated surfaces. Finally, it can be concluded that the mechanical performance of additively manufactured steel under static and fatigue loading can be improved by the SMAT process.

Keywords: additive manufacturing; direct energy deposition; surface treatment; stainless steel; microstructure; low-cycle fatigue

1. Introduction

Additive manufacturing (AM), also referred to as 3D printing, is a novel manufacturing process that is capable of fabricating near-net-shape parts with complex geometries in a layer-by-layer manner, directly from the 3D model data, without any molds or tools [1–3]. In a laser-based AM metal process, e.g., laser-based powder bed fusion (PBF-LB) and direct energy deposition (DED), a rasterized laser beam melts the metal powder in a pattern that progressively fills the volume of the designed CAD model and eventually fabricates the metallic part. Selective laser melting (SLM), as one common PBF-LB technique, utilizes the laser to layer-by-layer melt the deposited power bed on the initial substrate plate [4].

DED, also referred to as laser metal deposition (LMD) and laser engineered net shaping (LENS), has been applied for various applications in medical, aerospace, automotive, oil, gas and space industries [5–7]. The DED process can be characterized by the adopted laser powder, laser type, powder delivery method and feedback control. In the DED process, parts and components are fabricated by focusing a laser beam with high volume energy on the deposition substrate, where the metallic powders are simultaneously delivered and injected by inert gas [6]. The laser and inert gas melt the surface of the previously deposited layer and deliver powder, respectively, to create metallurgical bonding layer by layer in the Z-direction. DED can be used to customize the repair of pre-existing parts, generate components with geometries beyond the capabilities of conventional manufacturing techniques and reduce material wastage significantly by directly depositing the materials with fine precision [8,9]. Furthermore, functionally graded material can be fabricated by adjusting the multi-channel powder feeder or combining powder and wire deposition [10].

The austenitic stainless steel 316L (SS 316L), also known as 1.4404 or X2CrNiMo17-12-2, is used in a wide variety of applications, ranging from nuclear to chemical, petrochemical, marine and offshore oil-related fields, due to its outstanding toughness, ductility and resistance to corrosion [7]. In recent years, there has been an apparently increasing interest in the fabrication of SS 316L by using DED due to its intrinsic advantages mentioned above. Moreover, decreasing prices for metal powder allows DED to substitute other fabrication processes, such as diffusion bonding [11]. A large number of studies have been published investigating the influences of several factors, including process parameters (laser type, laser power, scanning strategy, time interval, preheating temperature), building direction, thickness of the components and powder recycling. Furthermore, surface roughness, residual stress, microstructure and mechanical performance of the additively manufactured SS 316L specimen under quasi-static and cyclic loadings were in focus of those studies [5,7,12–17].

The microstructures of metallic materials determine their macroscopic mechanical properties. The DED process is characterized by a fast heating and cooling rate, which results in a high solidification rate. Thus, often highly directional columnar structures are established, these being characterized by internal substructures, i.e., microstructure refinement [18]. A solidification map showing the effects of temperature gradient and growth rate on the morphology and size of the resulting microstructure can be found in [19,20]. Regarding the whole volume of a deposited component, as a consequence of the complex heat transfer during the DED process, it was reported that the columnar structures are strongly affected by the maximum thermal gradient, where in the last deposited layers, microstructures can be slightly different from the remaining bulk [14].

It is widely agreed that SS 316L fabricated by AM processes is characterized by excellent mechanical properties under monotonic loading, i.e., good ductility and a high yield strength (YS), which is significantly higher compared to specimens that are fabricated by conventional manufacturing techniques, such as hot forging and casting [5,21]. It is believed that the high YS is partly because of the high dislocation density in the AM processed steel. The factors that contribute to the final mechanical properties are the reduced grain and dendrite sizes, the presence of residual δ ferrite and the already mentioned presence of a dense dislocation network (eventually forming substructures of submicron size). These factors can also be used to rationalize the relatively low ductility values sometimes reported for deposited parts. The strengthening effect of the refined microstructure can be correlated with the well-known Hall–Petch equation that associates the material grain size (taking into account the substructures, i.e., dislocation cells) and the yield stress. Detailed comparisons of the microstructure and the mechanical performance among additively manufactured specimens, including DED and SLM, and the ones fabricated by traditional manufacturing methods, such as forging and casting, can be found in [14,17,18].

After manufacturing, additively manufactured metallic components are commonly subjected to post-process heat treatments in order to allow for the elimination of residual stresses and the homogenization of the microstructure. It was found that after heat treatment, the YS and ultimate strength of DED SS 316L were reduced due to the decrease in the ferrite content and the decrease in dislocation density [21,22]. Shot peening has been applied for improving the surface hardness, introducing compressive surface residual stresses and refining grain size close to the surface, such that the fatigue response of additively manufactured SS 316L specimens could be improved [23].

Surface mechanical attrition treatment (SMAT) is an emerging post-treatment technology among various peening processes, such as ultrasonic peening, laser shot peening and grit blasting, which is capable of inducing a nanostructured layer on the surface of a metallic component and is commonly used in order to improve the fatigue resistance [24–26]. SMAT has been recognized as a technique for upgrading the microstructures and properties of materials by generating a gradient-structured layer on the material surface without tampering with the local chemical compositions and near-surface compressive residual stresses.

In the past few years, numerous studies experimentally revealed that SMAT has a very positive effect on the fatigue behavior of different materials such as steel [27], titanium [28] and magnesium [29] alloys. The AM groups have drawn attention to this surface treatment technique and also applied it to additively manufactured components. Yan et al. [30] have applied SMAT for the modification of the surface layer of Ti6Al4V fabricated by PBF-LB to improve the fatigue performance. It was found that the specimen after SMAT exhibited significantly higher fatigue strength as compared to the non-treated counterparts in both low- and high-cycle fatigue regimes. Sun et al. [31] applied SMAT to PBF-LB-processed stainless steel specimens. The authors showed that the process-induced surface roughness was reduced by up to 96%, such that a surface finish similar to that produced by surface grinding could be achieved. In [32], SMAT was performed on SS 316L parts produced by PBF-LB. It was found that the SMAT treatment can reduce the surface roughness by a factor of 10 and increase the microhardness in the layer beneath the treated surface by up to 45%. The authors further showed that SMAT could also transform the initial near-surface tensile residual stresses present in AM parts into compressive ones and eventually enhance the mechanical properties of the PBF-LB-processed parts [32]. Such studies already demonstrated the potential of SMAT as a post-treatment to improve surface quality and increase the strength while retaining good ductility of PBF-LB manufactured parts.

As the introduction clearly outlines, DED has become an attractive AM technique in numerous fields. However, since many components manufactured by DED are used under very complex loading regimes, often including cyclic loads, the performance under fatigue loading needs to be studied. In this context, mechanical surface treatment processes are often used to improve the fatigue properties of metallic materials. In order to close prevalent research gaps, the present study focused on a unique combination of DED and SMAT. Promising results with respect to the behavior under cyclic loading were expected. To substantiate this expectation, SS 316L components were manufactured using DED. After fabrication, miniature specimens being cut from the components by electro discharge machining (EDM) were surface treated by SMAT. The mechanical performances of as-built and SMAT processed specimens were characterized by monotonic tensile and strain-controlled low-cycle fatigue (LCF) tests. A comparison of the resulting cyclic stress amplitude, half-life hysteresis loops and cyclic hardening/softening was performed for the untreated and SMAT processed specimens. Surface hardness was measured and microstructural features were characterized using electron backscatter diffraction (EBSD) and transmission electron microscopy (TEM) analysis, respectively. The effect of SMAT on microstructural features and the mechanical performance of the DED 316L specimens under static and cyclic loading was comprehensively studied. The macroscopic properties were rationalized in detail based on the microstructure information.

2. Material and Characterization Methods

2.1. Specimen Manufacturing

The SS 316L parts were fabricated using the DED technology at the Dalian University of Technology, China. The system consisted of a Kuka six-axes robot (ZH 30/60III, KUKA, Augsburg, Germany), a Laser Line diode laser generator (LMD 4000-100, Laserline GmbH, Mülheim-Kärlich, Germany) with 4000 W maximum power, a Precitec laser cladding head (YC52, Precitec KG, Gaggenau, Germany) with four coaxial nozzles and a Raychem metal powder feeder (RC-PGF-D, Raychem RPG Pvt Ltd., Mumbai, India). The powder of SS 316L with a size range of 45-150 μm and a spherical shape, produced by Höganäs (Höganäs, Sweden), was laser deposited on a SS 316L substrate with the dimensions of 130 mm × 40 mm × 15 mm. The nominal chemical composition (in wt.%) of the powder used is given in Table 1. Ar with 99.99% purity was used as a carrier and shielding gas in all processes at flow rates of 400 and 600 L/h. The process parameters adopted are given as follows: laser peak power of 1000 W, scanning speed of 6 mm/s, powder feed rate of 6 mm/s, T-pulse of 25 ms and z-increment of 0.2 mm between two layers for all deposited parts. The dimensions of the components were 60 mm × 10 mm × 40 mm. A bi-directional scanning strategy without rotation between consecutive layers was used. For microstructural and mechanical characterization and SMAT surface treatment, flat dog-bone shaped specimens with nominal gauge section dimensions of 8 mm × 3 mm × 2.0 mm were cut from the initial DED parts by electro-discharge machining (EDM). Further information on the dimensions of the specimen and the actual scanning strategy can be found in Figure 1.

Table 1. Chemical composition of the SS 316L powder used in the present work (wt.%).

Fe	Ni	Cr	Mn	Mo	Nb	Ti	Al	Cu	C	Si
Bal.	12.5	17.1	1.6	2.5	–	–	–	–	0.02	0.7

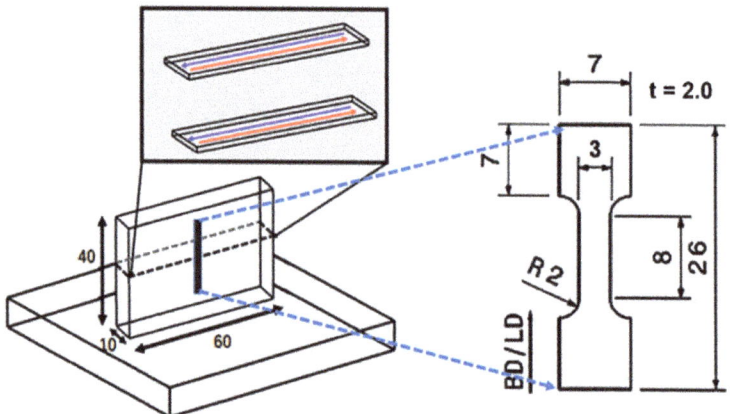

Figure 1. Schematic giving detailed information on the building direction and scanning strategy in the DED process and the dimensions of the specimens (dimensions in mm) cut for SMAT processing.

2.2. SMAT

Upon EDM, some of the DED manufactured SS 316L specimens were subjected to SMAT. No grinding or polishing was conducted between these process steps. The surface treatment was applied to two surfaces of the dog-bone shaped specimens cut from the initial DED cuboids (cf. Figure 1). Figure 2 shows a schematic diagram of the SMAT system. As can be deduced from the schematic, the major part of the specimen's surface, except for 1 mm of the clamping section on each side, was processed by SMAT. In the SMAT process, a chamber was attached to the vibrator, and spherical balls (304 stainless steel + ZrO2) with a diameter of 3 mm were placed inside the sealed chamber. During the SMAT process, the specimen secured to the top of the chamber was impacted by the spherical balls driven by the vibrations of the chamber attached to the vibrator. The same procedure was separately performed on the two large surfaces of each specimen. During SMAT processing of one side of the surface, the opposite side was completely fixed (cf. Figure 2). Due to the size of the spherical balls used, partial SMAT co-processing of the side surfaces of the specimen can be excluded. The vibration frequency and amplitude were electromagnetically controlled by a signal generator and an amplifier. The repeated impacts of spherical balls at a high speed and a high frequency can cause indentation and plastic deformation on the treated surface of a specimen. In the present work, the AM processed specimens were treated for 1 h at a vibration frequency of 20 Hz. After SMAT processing, no further grinding or polishing was conducted prior to mechanical testing. This also held true for the as-built counterpart condition.

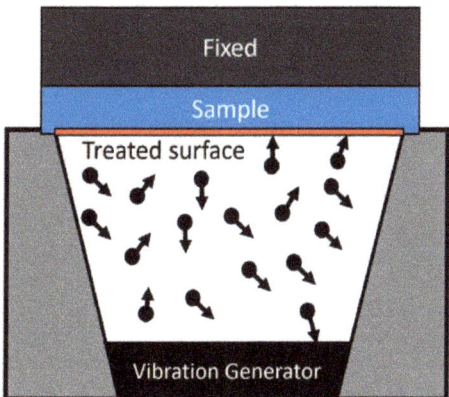

Figure 2. Schematic highlighting the most important features of the SMAT system.

2.3. Mechanical and Microstructure Characterization

Surface roughness measurements were conducted using a Mitutoyo SJ-210 (Mitutoyo, Kawasaki, Japan). Five readings within a specimen length of 0.8 mm were recorded on the large surfaces of the specimen directly in the gauge length. The roughness values R_a reported represent average values calculated from all readings. Characterization of mechanical properties comprised hardness investigations as well as tensile and fatigue tests. Vickers hardness testing was carried out using a Struers DuraScan-70 system (Struers, Copenhagen, Denmark) employing a load of 4.9 N. A screw-driven MTS criterion load frame with a maximum load capacity of 20 kN was used to perform uniaxial tensile tests under displacement control with a constant crosshead speed of 2 mm/min. Strain measurement was conducted using an MTS miniature extensometer with a gauge length of 5 mm directly attached to the surface of the specimen. The same extensometer was used for strain control during the LCF tests, performed in fully reversed push–pull loading ($R_\varepsilon = -1$). A constant strain rate of 6×10^{-3} s^{-1} and total strain amplitudes of $\Delta\varepsilon_t/2 = \pm 0.2\%$, $\Delta\varepsilon_t/2 = \pm 0.35\%$ and $\Delta\varepsilon_t/2 = \pm 0.5\%$ were used. The fatigue investigations were performed on a digitally controlled servo-hydraulic load frame with a maximum force capacity of 16 kN. In total, three tensile tests and two fatigue tests were performed for each condition.

Microstructure analysis of as-built DED specimens—i.e., investigation of grain size and grain morphology—was conducted using a Zeiss ULTRA GEMINI high-resolution SEM (Carl Zeiss AG, Oberkochen, Germany) operating at an acceleration voltage of 20 kV. The SEM was equipped with an electron back-scatter diffraction (EBSD) unit and a backscattered electron detector. For EBSD measurements, all specimens were mechanical ground down to 5 μm grit size using silicon carbide (SiC) paper, and vibration polished for 16 h using a conventional oxide polishing suspension (OPS) with a grain size of 0.04 μm. The measurements were performed using two different magnifications (100× and 500×) and step sizes (1 and 0.2 μm). The EBSD data analysis was performed with the Bruker Esprit 2.3 software (Bruker, Billerica, MA, USA).

Detailed microstructure analysis of as-built and SMAT processed specimen was additionally conducted by TEM using a JEM 2100PLUS system (JEOL, Freising, Germany) operated at 200 kV. TEM foils were selected from different local regions, i.e., close to the surface and in the center of the specimen. The TEM foils were prepared by focused ion beam (FIB) using a FEI Helios NanoLab 600i DualBeam SEM/FIB system (FEI, Hillsboro, OR, USA) operated at 30 kV.

3. Results and Discussion

3.1. Microstructural Analysis of As-Built Specimen by SEM

In the DED process, the thermal history during processing plays a crucial role in establishing specific microstructural features, e.g., morphology and grain size. In general, the microstructure is determined by the solid–liquid interface velocity R, the thermal gradient G and the alloy composition. In particular, the solidification morphology parameter (G/R) and the cooling rate level G×R define the solidification mode, and therefore, the microstructure morphology and its inherent dimensions. In general, as-built DED specimens were characterized by an obvious solidification pattern on the macroscopic scale and cellular substructures on the microscopic scale [12]. In conventionally processed but rapidly cooled austenitic stainless steel, depending on its chemical composition, two different microstructural constituents can be found: austenite γ and the so-called ferrite δ [18].

Figure 3a shows a representative EBSD inverse pole figure (IPF) map. The measurement depicted was performed in the clamping section of a SS 316L specimen (cf. Figure 1). The grain orientations of the EBSD micrograph are plotted with respect to the build direction (BD). The microstructure exhibits large columnar grains with a length of up to 300 µm and width of about 100 µm, respectively. No obvious melt pool boundaries are noticeable. DED can be considered as a directional solidification process characterized by a high temperature gradient and rapid cooling rate. Thus, the grains are mainly orientated in the building direction; however, they are characterized by a slight shift towards the laser travel direction (LTD). This observation was similarly made in previous investigations on DED SS 316L [33] and can be explained by the local conductive heat transfer, i.e., the direction of heat flux. It can further be seen that the SS 316L specimen in the as-built condition was characterized by slightly increased texture intensity in BD. The IPF map is characterized by a comparatively increased fraction of {101}-oriented grains. In these cases, the grains grow parallel to the thermal gradient and direction of heat flux, with a growth rate, which is strictly related to the scan speed used during the building process. The slightly increased texture intensity is additionally highlighted by the IPF displayed in Figure 3e, extracted from the EBSD data of the micrograph shown in Figure 3a. A similar texture evolution was also reported by Andreau et al. [34] for PBF-LB-processed 316L parts. Most importantly, it has to be noted that the process parameters, including laser powder, scan speed, and scanning strategy, significantly influence the size and shape of the melt pool, which in turn affects the resulting texture and grain morphology [3]. However, further in-depth texture investigations, including X-ray diffraction analysis, across the entire part built, have to be conducted in order to analyze the texture's evolution in more detail. In literature, solidification maps have been successfully used for assessing the effects of temperature gradient and growth rate on the morphologies and sizes of microstructural features [19,20]. Figure 3b shows a magnified view of the marked white dashed rectangle in the IPF map in Figure 3a. From the variations in the color within the grains, the presence of subgrains and lattice distortions can be deduced. This is in line with the observations made by Belsvik et al. [33] with respect to the microstructure evolution in DED-processed SS 316L-Si. In addition, some non-indexed areas can be noticed in the magnified IPF map in Figure 3b. From the corresponding image quality map in Figure 3c, these areas can be identified as distinct boundaries inside the grains. Figure 3d shows a phase map of a magnified area of the image quality map in Figure 3c. From this map, the boundaries can be identified as phase-boundaries. In the study of Belsvik et al. [33], a volume fraction of about 2.5% δ-ferrite on the interdendritic and subgrain boundaries was determined by EBSD analysis. In another study [35] investigating the stress corrosion cracking susceptibility of 304L substrate and 308L weld metal exposed to a salt spray, the authors showed that in a conventional fusion weld δ-ferrite can be formed intra- and inter-granularly. The presence of δ-ferrite in DED-processed SS 316L was additionally reported by Saboori et al. [14] based on X-ray diffraction data. From the phase map shown, minor amounts of ferrite can be observed. Generally, the presence and amount of δ-ferrite formed in SS 316L were discussed by Bedmar et al. [18]. In their study reporting on a comparison of different AM methods for 316L stainless steel, a Schaeffler diagram was used to predict the phases based on the chromium and nickel equivalent when the cooling was not in full equilibrium, as is the case in AM and conventional welding. For SS 316L, austenite with a fraction of 5–10% of δ-ferrite was predicted. According to the findings of the EBSD measurement shown in Figure 3d, a volume fraction of about 2% was observed for the DED SS 316L condition of the present study. This in line with the results from Belsvik et at. [33] reporting on a volume fraction of about 2.5% δ-ferrite determined by EBSD. According to [33], differences compared to the theoretical value may be related to poor statistics of the quantitative analysis. As stated in [18], in welding, the presence of a 5–10% δ-ferrite phase improves the behavior of austenitic steels. Moreover, values above 10% lead to reductions in ductility, toughness and corrosion resistance whereas values below 5% can cause solidification cracking. As a result, the presence of 2% δ-ferrite could degrade the properties of the DED SS 316L specimens. However, traces of solidification cracking were not found in the present work.

3.2. Microstructural Features on the Nano-Scale

Figure 4 shows the microstructural evolution within specific regions close to the top surface layer for a specimen which was subject to SMAT processing. Figure 4a shows a gradient structure (in terms of the grain size) in the probed region close to the top surface (marked as layer 1 in the schematic depicted in Figure 6). Figure 4b shows the specific microstructure of the area marked by a red rectangle in Figure 4a, whereas Figure 4c shows the corresponding SAED pattern revealing the presence of a polycrystalline structure (single spots are labelled for clarity). It can be clearly seen that nanograins formed in the region close to the top surface layer ($\approx$0.025–0.05 mm), consistently with the TEM bright field image in Figure 4b. This observation is in line with the literature reporting on SMAT processed SS 316L. According to Tao et al. [36], strain-induced grain refinement and martensite transformation can take place in the top surface layer during SMAT. The grain sizes of the nanocrytallites formed were reported to be in the range of 8 to 60 nm with a mean value of about 30 nm, which correspond well with the results shown in Figure 4b. The authors further reported that the grain refinement in SS 316L after SMAT processing can be attributed to the formation of planar dislocation arrays and mechanical twins; a grain subdivision by mechanical twins and martensite transformation; or the formation of nanocrystallites [36]. Generally, as a result of mechanical surface treatment processes, the elastic-plastic deformation leads to severe local shearing of the surface layer, including activation of dynamic recrystallisation, and thus to grain refinement down to the nanometer range [37,38]. With increasing distance from the specimen surface ($\approx$ 0.1–0.15 mm), e.g., within layer 2 in the schematic in Figure 6, differences in the microstructure became obvious by TEM nano-scale analysis. Figure 4d,e shows bright-field and dark-field TEM images of a nanotwinned microstructure, respectively. Figure 4f shows the corresponding selected area electron diffraction (SAED) pattern, clearly revealing the presence of deformation-induced twins. The austenite matrix and the nano-scaled deformation twins are indexed in the diffraction pattern. In accordance with these observations, Tao et al. [36] reported that at a distance from the treated surface of 30–40 μm, lamellar structures with a width of approximately less than 100 nm were formed, which can be attributed to the evolution of nano-sized mechanical twins. Twinning in SS 316L as a result of surface treatment processes was also shown for other processes. Agrawal et al. [39] reported multiple twin systems resulting in the formation of a dense crisscrossed twin structure by using the novel vaporizing foil actuator (VFA) technique. Wang et al. [40] studied the microstructural evolution and the mechanical behavior of 316L parts after surface treatments by ultrasonic impact peening (UIP) and laser shock peening (LSP). The authors reported that mechanical twinning was almost completely absent in the specimen processed by UIP, whereas twinning was frequently observed in specimens treated by LSP. It was concluded that the magnitude of peak pressure determined the transition from a dislocation-dominated mechanism (≈680 MPa for UIP) to a twinning-dominated mechanism (≈2200 MPa for LSP). Thus, it can be assumed that the peak pressure induced by the SMAT treatment in the current study was high enough to induce twinning in the surface treated region.

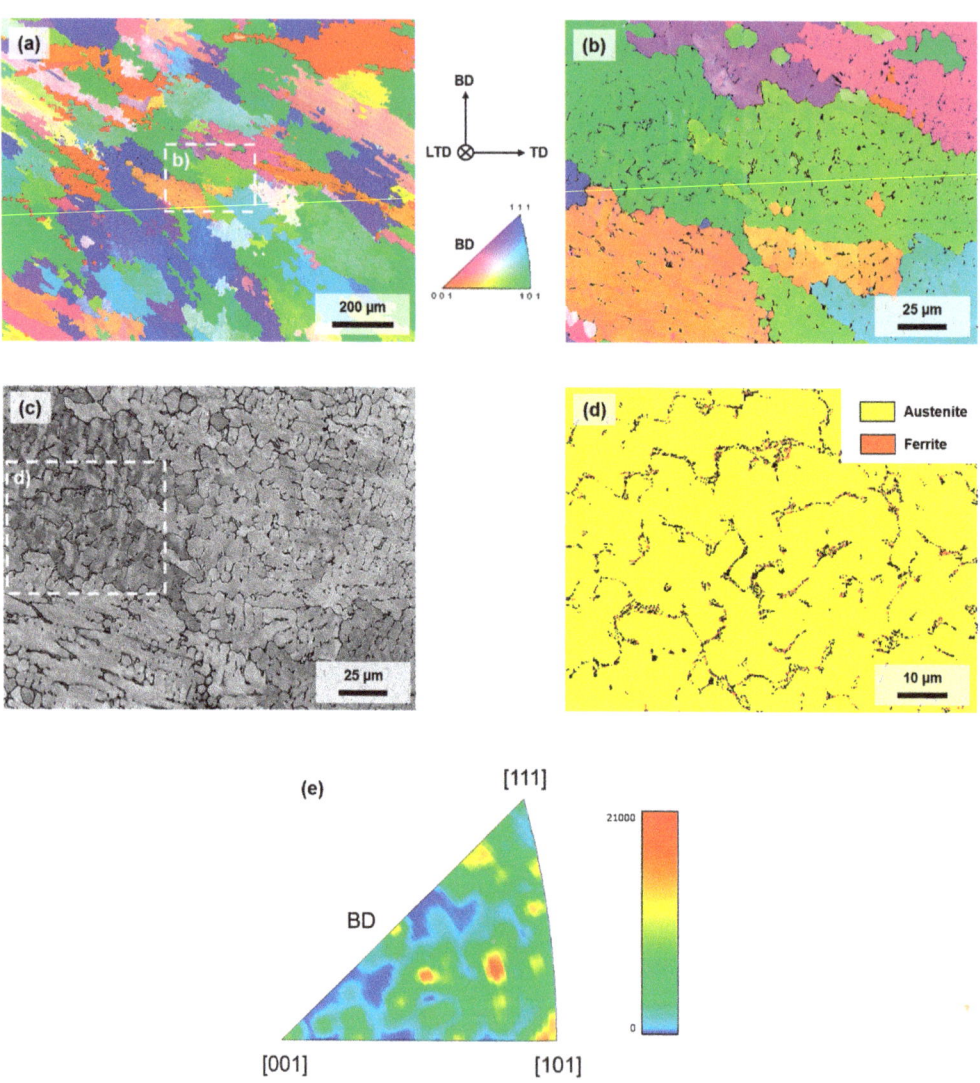

Figure 3. (**a**) EBSD inverse pole figure (IPF) map of an as-built DED SS 316L specimen. (**b**) A magnified view of the marked white dashed rectangle in (**a**). (**c**) Image quality map highlighting the grain structures in addition to the IPF map in (**b**). (**d**) EBSD phase map of the marked white dashed rectangle in (**c**). The grain orientations in (**a,b**) are plotted with respect to the build direction (BD). (**e**) Inverse pole figure calculated from EBSD data of the micrograph shown in (**a**) revealing the micro-texture of the DED SS 316L. Data are plotted with respect to the BD.

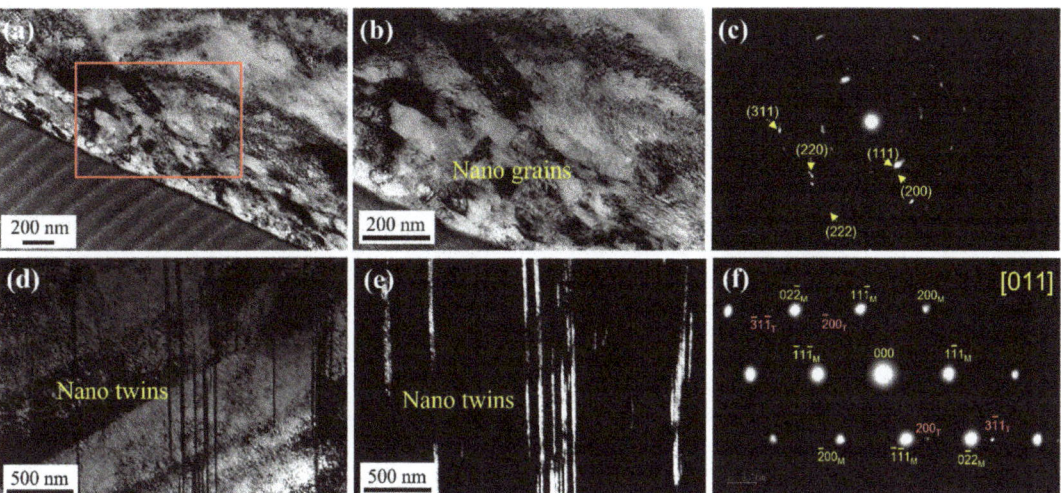

Figure 4. TEM micrographs of regions adjacent to the top surface layer of the SMAT processed condition of the SS 316L specimen and magnified view of a selected locations: (**a**) bright-field image of a nano-grained layer; (**b**) higher magnification micrograph of nanograins in the region marked by the red rectangle in (**a**); (**c**) SAED of nanograins; (**d**) bright-field image showing nanotwins and slip bands, (**e**) dark-field image highlighting the nanotwins; (**f**) SAED of the nanotwins.

In addition to the formation of nanocrystallites and deformation-induced twins, deformation-induced martensite was found in the surface treated layer in some studies reporting on mechanically surface-treated SS 316L. Jayalakshmi et al. [41] reported on a gradient nanostructured layer after severe shot peening. The initial hot-rolled austenitic microstructure, being characterized by grain sizes in the range of 40–80 µm, was refined to a dislocation cell-type structure with deformation-induced martensitic structures having cell sizes in the range of 100–140 nm. The authors showed that the martensite, assessed by TEM investigations, was nucleated at multiple locations in the austenite matrix. However, with the measurement methods used in the present study, no martensitic transformation could be observed. The diffraction spots of the SAED pattern shown in Figure 4f, associated with Figure 4d,e, clearly correspond to the austenite phase, additionally revealing twin reflections. The lath-like microstructure appearance in Figure 4d might thus be attributed to the formation of shear bands. In a study reporting on the microstructural evolution of SS 316L subjected to SMAT treatment, Bahl et al. [42] showed that microbands and shear bands were formed as a result of SMAT processing. The authors further concluded that deformation twinning and dynamic recrystallization within the shear bands are responsible for nanocrystallization.

With a further increase in the specimen depth analyzed, i.e., at a larger distance to the treated surface, Figure 5a,b shows TEM bright-field images of the local microstructure (in a distance from the surface of approximately 1.0–1.2 mm corresponding to layer 3 in Figure 6). From both micrographs, it can be deduced that within the investigated region, dislocation cell walls (marked by the yellow arrows) were formed during the SMAT processing. As will be pointed out by the results of hardness investigations in Section 3.3 (cf. Figure 7), the area under investigation in Figure 5 corresponded to the non-affected as-built DED microstructure. This specific microstructure was in good agreement with data available in literature. In their study investigating the origin of dislocation structures in an additively manufactured austenitic SS 316L, Bertsch et al. [43] reported that the highest dislocation densities were found in DED 3D and PBF-LB parts, manifested in the formation of dislocation cells approximately 300–450 nm in diameter. The authors further showed that dislocation structures in AM originated as a consequence of thermal distortions during processing, which were primarily dictated by constraints surrounding the melt pool and thermal cycling, respectively.

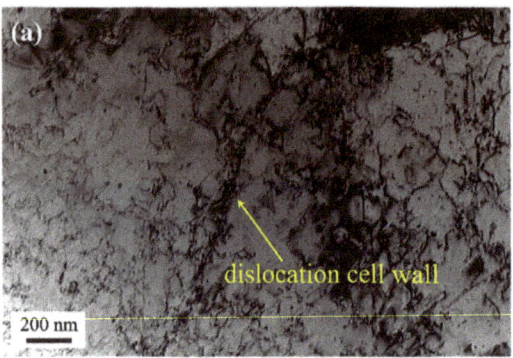 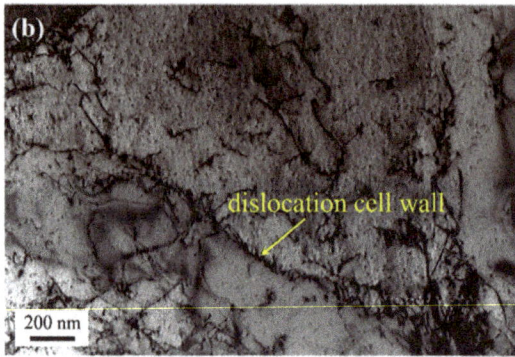

Figure 5. (a,b) Representative bright-field TEM micrographs of the center part of a SMAT processed SS 316L specimen.

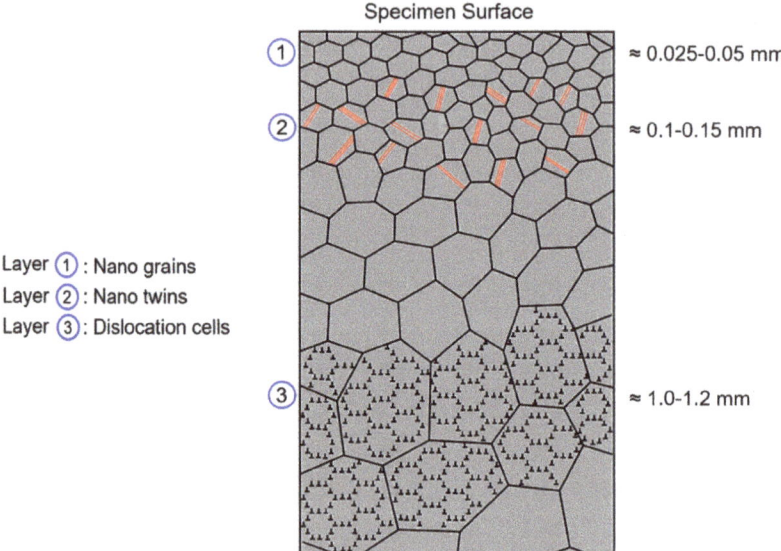

Figure 6. Schematic of the cross-sectional microstructure of the SMAT processed SS 316L specimen consisting of a nanograin layer adjacent to the top surface, a nanotwin layer below the nanograin layer and a layer characterized by dislocation cells in the central region.

The results obtained by TEM nano-scale analysis are summarized in the schematic of the cross-sectional microstructure of a SMAT processed SS 316L specimen displayed in Figure 6. As is obvious, the specimen was characterized by various types of microstructures after surface treatment, being explicitly characterized by the gradually increasing grain size with an increasing distance to the surface. In the immediate vicinity of the surface, i.e., a depth of ≈0.025–0.05 mm, nanocrystallites were found, whereas with increasing distance to the surface (≈0.1–0.15 mm), mechanically induced nanotwins with a mean thickness of approximately 100 nm were observed within the deformed grains. As reported in [34], such a gradient microstructure including randomly-oriented equiaxed nanocrystallites in the top surface layer and mechanical nano-sized twins below this layer were also found for Inconel

600 specimens following SMAT surface treatment. In the central region (≈1.0–1.2 mm), i.e., in the non-surface treated bulk material, the microstructure was characterized by high density of dislocation cells formed within the coarse grains directly stemming from the DED process.

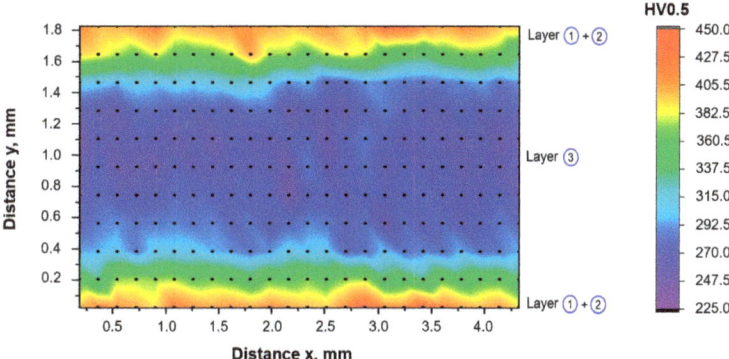

Figure 7. Hardness map showing the hardness distribution in the cross-section of a SMAT processed SS 316L specimen. Each black dot represents the position of a single measurement. The different microstructural layers characterized by TEM measurements (cf. Figure 6) are also marked.

3.3. Roughness, Hardness and Mechanical Response under Tensile Loading

Depending on the initial surface appearance before the SMAT treatment, the roughness profile can either be improved or deteriorated [32,44]. In order to analyze the effect of SMAT, surface roughness measurements were carried out on the large surfaces of specimens from both conditions considered. While the as-built condition after EDM was characterized by an average roughness R_a of 3.84 µm, smoothing of the surface can be observed after the SMAT treatment, revealing an average roughness value of 1.40 µm. This is in line with the investigations of Portella et al. [32] reporting that the surface roughness of as-built SLM parts was considerably reduced by a SMAT post-treatment.

In order to analyze the effective depth of a specimen being influenced by the SMAT surface treatment, Vickers hardness measurements were carried out on a cross-section of a specimen cut from the clamping area. In total, 264 individual measurements were carried out. The results obtained are illustrated in the hardness map in Figure 7. The position of each measurement point is represented by a black dot. Additionally, 10 measurements were carried out in the clamping section of a non-surface treated specimen in order to determine the mean value of the hardness after the DED process. As a result of these measurements (results not shown in Figure 7), the medium hardness of a specimen without SMAT was found to be approximately 237.7 HV0.5, whereas the maximum and minimum values were determined to be 227 and 247 HV0.5, respectively. A similar hardness can also be derived for the central area of the cross-section of a SMAT processed specimen from the hardness map in Figure 7. In contrast, an increase in hardness can be seen for the near-surface layer of each SMAT processed surface. In fact, in the direct vicinity of the treated surface the hardness was increased by a factor of about two, reaching maximum values of up to 450 HV0.5. Taking the results of the nano-scaled microstructural investigations into account, the increased hardness in direct vicinity of the surface can be attributed to the formation of the nanocrystallites, and thus to the intense level of grain refinement in this area. This observation can be explained based on the well-known Hall–Petch relation, according to which the increase in hardness or strength is inversely proportional to the square-root of the mean grain diameter [45]. Although an increase in grain sizes was observed with a concomitant increase in the distance to the surface, this area was still characterized by a higher mean hardness of approximately 375 HV0.5 as compared to the central region. This observation can be attributed to the formation of the mechanically induced nano-sized twins in this area. The formation of these twins by the SMAT treatment led to the so-called dynamic Hall–Petch effect, i.e., an increasing density of interfaces within the grains eventually, promoting the increase in hardness [46,47]. The hardness map further demonstrates that the SMAT treatment is characterized by an effective depth of approximately 0.5 mm (on both sides of the specimen), as this is the range, the hardness converges to the values obtained for a non-surface treated specimen before. Taking the nominal specimen thickness of 2.0 mm into account, it can be concluded that almost 50% of the cross-sections of the specimens considered in present work were strengthened by the SMAT process.

Figure 8 highlights the behavior under quasi-static tensile loading by depicting representative stress–strain curves of the as-built and SMAT processed SS 316L specimens. The mechanical response of the as-built specimen was characterized by well-defined elastic–plastic behavior, a YS of approximately 300 MPa, an ultimate tensile strength (UTS) of 715 MPa and an elongation at fracture of 19%. With respect to the SMAT processed specimen, the mechanical performance and quasi-static tensile strength were characterized by the following values: YS of 375 MPa, UTS of 800 MPa and elongation at fracture of approximately 6.5%. It can be derived that the value of the YS of the SMAT processed specimen is 25% higher than that of the as-built counterpart. Compared to SS 316L counterparts produced by conventional methods (casting and forging) [48], the as-built DED-processed specimen considered in the present study was characterized by increased YS and UTS values. This fact can be attributed to the unique microstructure of the alloy generated by the AM process. As a consequence of the process's inherent rapid cooling rates, a finer microstructure compared to conventional counterparts with subgrain structures evolves. The formation of δ-ferrite in the DED SS 316L (in the intergranular regions) further strengthens the soft austenitic matrix. Concomitantly, a reduction in the ductility could be observed in comparison with the cast and forged counterparts [48]. In addition, dislocation cells were detected in the non-surface treated microstructure by nano-scale microstructure analysis. The improved mechanical strength of additively manufactured metals has been attributed to the formation of these dislocation cell structures with solute micro-segregation [49,50]. In contrast, the increase in the strength after SMAT processing can be attributed to the volume of material affected by the surface treatment. As evidenced by the hardness map, almost 50% of the cross-section was strengthened by SMAT, explaining the increased strength and more pronounced brittleness. An increase in the strength under monotonic tensile loading and a simultaneous reduction in ductility, have often been reported for other austenitic steels after mechanical surface treatment [46,51].

3.4. Low-Cycle Fatigue Analysis

In this section, the LCF properties of the DED SS 316L are presented and analyzed for the as-built and SMAT processed (applied to the two large surfaces of each specimen) conditions. Figure 9 shows the cyclic deformation responses (CDRs) for the conditions considered, i.e., as-built and SMAT. The LCF tests were conducted at various total strain amplitudes—i.e., Figure 9a, $\Delta\varepsilon_t/2 = \pm 0.2\%$, Figure 9b; $\Delta\varepsilon_t/2 = \pm 0.35\%$; and Figure 9c, $\Delta\varepsilon_t/2 = \pm 0.5\%$. As mentioned in Section 2, two tests were conducted for each condition in order to analyze the reproducibility and scatter behavior of the LCF response. As, with respect to resulting stress amplitudes and number of cycles to failure, no pronounced scatter was detected, only one curve for each condition and loading condition is shown for the sake of clarity. In order to avoid buckling of the miniature specimen, the load was increased stepwise during the first cycles. As a result, the prescribed strain amplitude was reached after approximately 10–50 cycles, depending on the actual strain amplitude. Thus, these initial cycles were not taken into account for evaluation. From the CDRs depicted, it can be deduced that, irrespective of the condition, a higher imposed strain amplitude resulted in higher cyclic stress amplitudes and decreased fatigue life. Furthermore, for a given total strain amplitude, the resulting cyclic stress amplitudes of the SMAT processed specimens were slightly increased compared to those of their untreated counterparts. The hardening/softening behavior, which is characterized by an increase or decrease in the corresponding stress response throughout the fatigue tests, can also be derived from the graphs shown in Figure 9. The CDRs of both conditions are characterized by similar courses for all total strain amplitudes considered. After reaching the prescribed strain amplitude, a slight initial saturation stage is followed by cyclic softening being more pronounced at higher total strain amplitudes and for the SMAT processed specimens in direct comparison to their as-built counterparts. The cyclic softening behavior presumably results from the rearrangement of dislocations and a decrease in the overall dislocation density, as similarly shown in previous studies [52–54]. As revealed by the TEM measurements, fast-cooling and the repeated cyclic heating during the DED processing promoted the formation of dislocation cell walls (cf. Figure 5) in the as-built microstructure. Obviously, the more pronounced softening during cyclic loading, as deduced from the CDRs of the SMAT condition, caused an even higher (local) dislocation density after surface treatment (cf. layer structure shown in Figure 6). Regarding number of cycles to failure for the two conditions considered, a change in the properties can be deduced from the CDRs depicted in Figure 8. While fatigue life at the lowest total strain amplitude of $\Delta\varepsilon_t/2 = \pm 0.2\%$, i.e., about 110,000 cycles for the SMAT condition, is more than 25% higher than for the as-built counterpart ($\approx$80,000 cycles), this trend changes with increasing strain amplitude. For the medium total strain amplitude, i.e., $\Delta\varepsilon_t/2 = \pm 0.35\%$, the as-built condition is already characterized by an increased number of cycles to failure of approximately 40%, whereas for the highest total strain amplitude of $\Delta\varepsilon_t/2 = \pm 0.6\%$, the fatigue life is, with $\approx$3.400 cycles, more than two times higher than that of the SMAT counterpart ($\approx$1.400 cycles). These observations can be explained based on the characteristics of the surface treated conditions. In line with several other surface treatment processes, such as shot peening and deep rolling, SMAT is well known for establishing nanocrystalline near-surface layers characterized by high compressive residual stresses and increased hardness due to work hardening. Depending on the initial surface appearance before SMAT treatment, the roughness profile can either be improved or deteriorated. In addition, deformation-induced twinning or phase transformation can be triggered as a result of the high plastic deformation [32,44,55]. According to [56], the cyclic deformation responses of surface-strengthened specimens are characterized by lower plastic strain amplitudes over the number of cycles due to the residual stresses present and the effective strengthening of the near-surface layer. In order to achieve a permanent improvement in the general performance imposed by these characteristics, the stability of the near-surface layer properties is of decisive importance. However, stability can be detrimentally influenced by thermal and mechanical (quasi-static and/or cyclic) stresses and strains [57]. Figure 10 depicts half-life hysteresis loops for both conditions and all total strain amplitudes considered. The area of a hysteresis loop represents the energy dissipation per cycle and is directly related to the plastic strain amplitude, i.e., half of the maximum width of the hysteresis curve. An increased energy dissipation per cycle can be linked to a more intense dislocation activity, eventually promoting premature failure [53,58]. As can be deduced from Figure 10a for the lowest total strain amplitude of $\Delta\varepsilon_t/2 = \pm 0.2\%$, the SMAT condition's outcomes are characterized by a narrower half-life hysteresis loop, i.e., a lower plastic strain amplitude due to characteristics of the near-surface layer, as described above, resulting in a higher number of cycles to failure as compared to the as-built counterpart. As the hysteresis loop of the SMAT samples is almost fully closed, revealing only a very low contribution of plastic strain, relatively high stability of the near-surface layer properties could be expected. After increasing the total strain amplitude, the appearance of the half-life hysteresis loops changes. For the medium total strain amplitude, the differences in the plastic strain amplitudes of both conditions start to decrease, revealing only a slightly wider opened hysteresis for the as-built condition, but the hysteresis loops become almost equal for the highest total strain amplitude of $\Delta\varepsilon_t/2 = \pm 0.5\%$ (cf. Figure 10b,c). As a common feature of both total strain amplitudes, all half-life hysteresis loops are characterized by a pronounced contribution of plastic strain irrespective of the condition considered. From these results it can be concluded that with increasing total strain amplitude, and thus increasing plastic deformation, the near-surface layer properties become instable and lose their effect. In general, at high plastic strain ranges, i.e., in the LCF regime, materials of higher ductility have a higher crack-initiation resistance, and thus, superior fatigue life. On the contrary, at low plastic strain ranges, i.e., in the high-cycle fatigue (HCF) regime, materials of higher tensile strength are characterized by higher crack-initiation and growth resistance. In conclusion, the fatigue resistance in different regimes comprises a tradeoff among strength and ductility in dependence of the plastic strain evolution [59]. Thus, the improved fatigue life of the as-built DED SS 316L condition at medium and high total strain amplitudes, deduced from the CDRs in Figure 9, can be rationalized by higher ductility compared to the SMAT counterparts (cf. Figure 8) in combination with the degrading near-surface layer properties of the surface treated SMAT condition. In contrast, the stability of the properties induced by SMAT and the generally improved strength lead to improved cyclic properties at low total strain amplitudes. Thus, the results of the present study already indicate that a significant improvement in the cyclic properties in the HCF regime, known to be characterized by crack initiation and failure mainly due to elastic deformation [60,61], can be expected for the SMAT DED SS 316L.

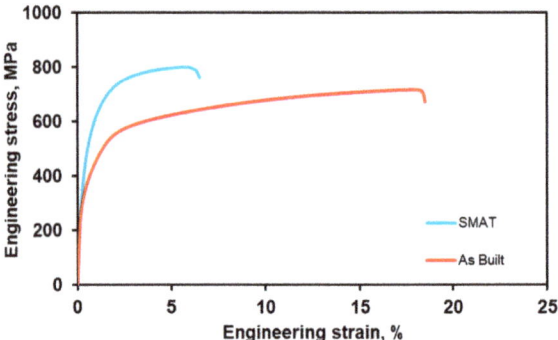

Figure 8. Representative tensile stress–strain curves of as-built and SMAT processed SS 316L specimens.

3.5. Fracture Surface Analysis

After fatigue testing, fracture surface analysis was carried out for all specimens tested. The SEM micrographs obtained for the highest total strain amplitude, i.e., $\Delta\varepsilon_t/2 = \pm 0.5\%$, are shown in Figure 11. As, irrespective of the condition, similar characteristics were revealed for specimens fatigued at all strain levels, the depicted fracture surfaces can be considered as representative. Thus, for sake of brevity, micrographs for only one total strain amplitude are shown. From the fracture surfaces presented, well-known features for fatigue tested specimen can be deduced. Besides areas of fatigue crack initiation (cf. Figure 11b,d) and propagation characterized by submicron fatigue striations, the fracture surfaces of both conditions are characterized by an overload final fracture region with a ductile, dimple-like structure (cf. Figure 11c,f). In line with the number of cycles to failure being increased by more than a factor of two for the as-built condition compared to the SMAT counterpart, an increase in the share of the area characterized by stable fatigue crack propagation can be observed. In addition, the fracture surface of the as-built condition locally seemed to be plastically deformed to a higher extent. This fact can be explained by the higher ductility, as already deduced from the results of the tensile tests (cf. Figure 8). As common feature of both conditions, fatigue crack initiation was always located on the side surface of the specimen, marked by the white arrows in Figure 11a,d. In general, a typical characteristic known from the literature related to mechanical surface treatment processes such as shot peening and deep rolling, is a shift of the crack initiation points away from surface areas towards internal defects. This is a result of the near-surface compressive residual stresses [27,51,62,63]. This has previously been shown in other fatigue studies on SMAT processed materials as well [30,64]. However, as detailed in Section 2.2, SMAT was only performed on the two large surfaces of each specimen in order to gain initial insights into the impacts of this emerging surface treatment process with respect to microstructural evolution and effective depth. Since fatigue cracks for the SMAT condition in the present work always initiated from the surface of the specimen, even for the lowest total strain amplitude of $\Delta\varepsilon_t/2 = \pm 0.2\%$ (not shown for sake of brevity), it can be assumed that the non-treated side surfaces represented the weakest link under fatigue loading. As a result, the conditions in focus in the current study, i.e., as-built and SMAT processing, result in similar crack initiation behavior. Due to the increased brittleness of the SMAT processed samples (cf. Figure 8), a higher crack growth rate can be assumed, eventually leading to premature failure of the specimen compared to the as-built counterpart, at least at medium and high total strain amplitudes characterized by high plastic strains. The positive influence of SMAT treatment at the low strain amplitude can be rationalized based on a surface-core model (in accordance with the well-known Masing model [65,66]). Due to the significant volume fraction of the deformed surface layers, the composite body of surface and core only suffered decreased plastic strain amplitudes (cf. Figure 10). After exceeding the YS of the unaffected core to a much higher degree in the case of the higher strain amplitudes, the local plastic strain for the surface layer was also increased. From the results presented, it can therefore be concluded that for an effective impact of a SMAT treatment, either rotationally symmetrical components should be used or a flat specimen should be processed on all surfaces. Otherwise, non-treated sides can be considered as weak points under cyclic loading, especially at high plastic strains.

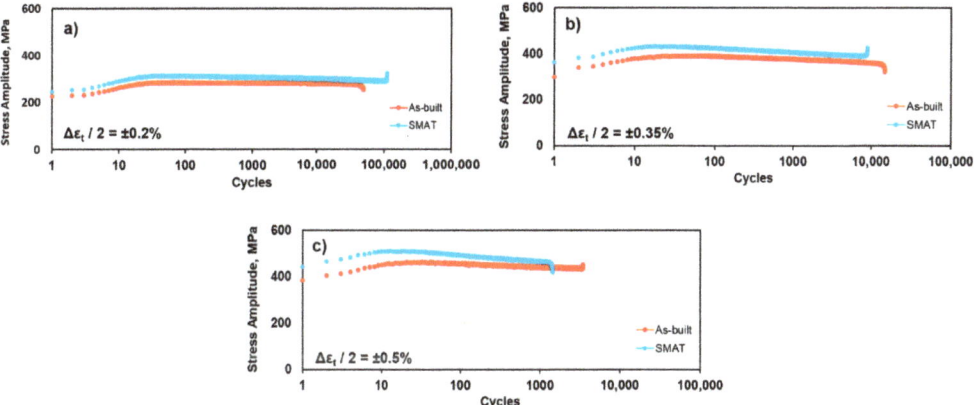

Figure 9. Cyclic stress response at room temperature at total strain amplitudes of (**a**) $\Delta\varepsilon_t/2 = \pm 0.2\%$, (**b**) $\Delta\varepsilon_t/2 = \pm 0.35\%$ and (**c**) $\Delta\varepsilon_t/2 = \pm 0.5\%$ for the DED SS 316L in as-built and SMAT processed condition.

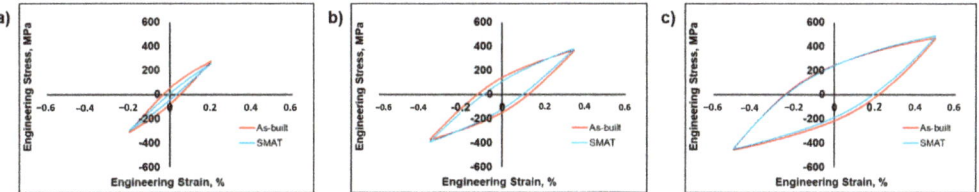

Figure 10. Half-life hysteresis loops for the DED SS 316L in as-built and SMAT processed conditions for total strain amplitudes of (**a**) $\Delta\varepsilon_t/2 = \pm 0.2\%$, (**b**) $\Delta\varepsilon_t/2 = \pm 0.35\%$ and (**c**) $\Delta\varepsilon_t/2 = \pm 0.5\%$.

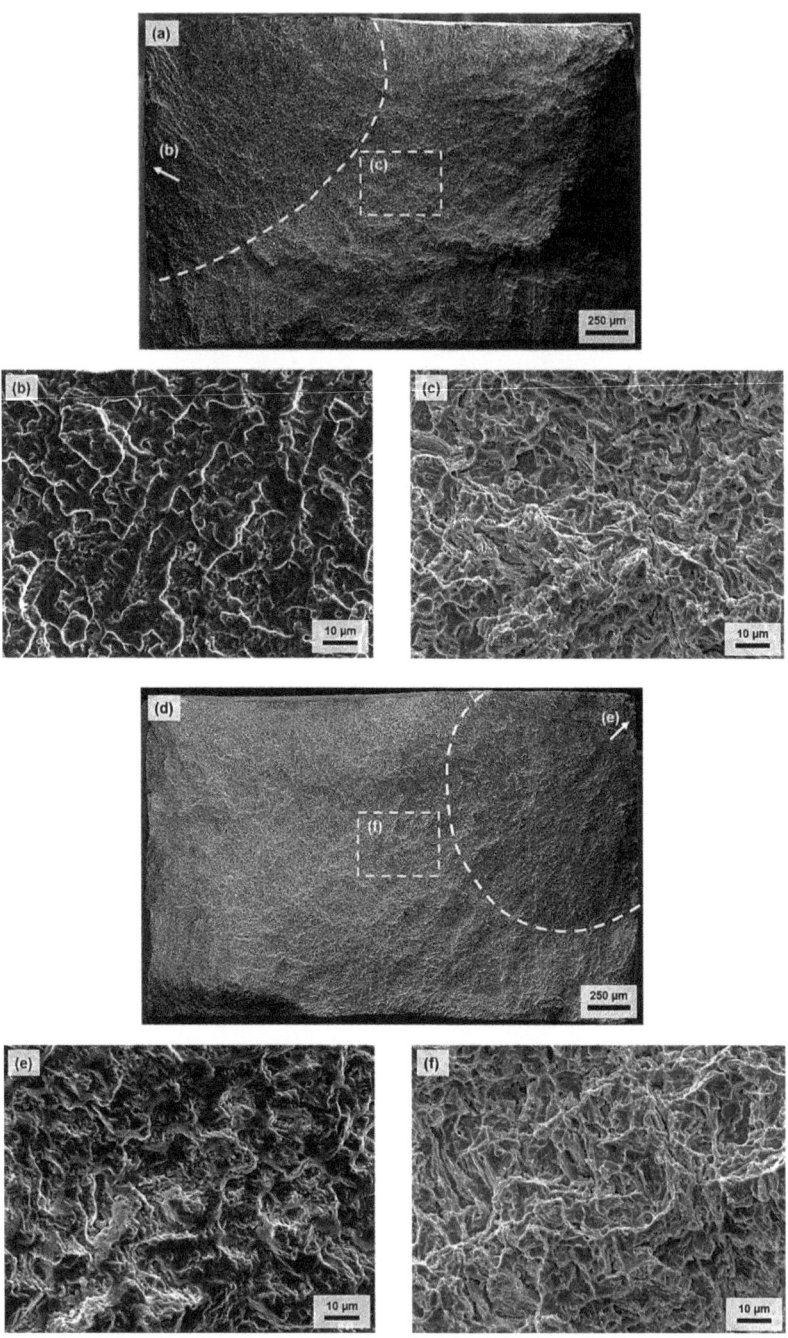

Figure 11. SEM micrographs of fracture surfaces after fatigue testing at a total strain amplitude of $\Delta\varepsilon_t/2 = \pm 0.5\%$ for the DED SS 316L in (**a**–**c**) as-built and (**d**–**f**) SMAT processed conditions. The subimages (**b**,**c**) and (**e**,**f**) show magnified views of the areas marked in (**a**,**d**) with a white arrow and white dashed rectangle, respectively.

4. Conclusions

In the present work, SS 316L was processed by DED and subsequently subjected to SMAT. Besides characterization of hardness and tensile tests, LCF tests with different strain amplitudes were performed focusing on both the as-built and SMAT processed conditions. Fracture surfaces and microstructure were thoroughly characterized by SEM and TEM. The effects of the SMAT processing on the microstructural and mechanical properties of 316L parts were finally evaluated. The following conclusions can be drawn from the results presented:

- The as-built DED microstructure consisted of large columnar grains nearly orientated along the building direction featuring a slight ⟨101⟩ texture. EBSD analysis revealed a small volume fraction of δ-ferrite at interdendritic and subgrain boundaries. TEM studies revealed the formation of dislocation cells promoted by rapid cooling rates and intrinsic heat treatment.
- As a result of the SMAT treatment, the formation of nanograins and nanotwins in the near-surface area was revealed by TEM. Moreover, results obtained by hardness mappings revealed that almost 50% of the cross-sections of surface treated specimens were strengthened by the SMAT process.
- Monotonic tensile loading revealed ductile material behavior of the DED SS 316L as-built condition, being characterized by YS, UTS and elongation at fracture of 300 MPa, 715 MPa and 19%, respectively. After SMAT treatment, the strength of the material was increased (YS and UTS of 375 and 800 MPa) due to the surface strengthening, alongside a concomitant loss of ductility (elongation at fracture of 6.5%).
- The cyclic deformation response of both as-built and SMAT conditions is characterized by slight cyclic softening. For the lowest total strain amplitude considered, fatigue properties of the DED SS 316L were improved as a result of the SMAT treatment. At higher strain amplitudes, the as-built condition was characterized by superior fatigue properties.
- Irrespective of the condition considered, post fatigue fractography revealed crack initiation solely in the direct vicinity of the side surfaces. Sub-surface crack initiation known from mechanical surface treatment processes could not be observed for the SMAT condition at the strain levels considered. Due to the surface treatment only being carried out on two large sides of the surface, the side surfaces represented the weakest link under cyclic loading. As a result, the conditions are characterized by similar crack initiation behavior. The increased brittleness of the SMAT condition therefore leads to an increased crack growth, eventually leading to inferior fatigue properties at high plastic strains.

Author Contributions: Conceptualization, T.W. (Thomas Wegener) and T.W. (Tao Wu); Methodology, T.W. (Thomas Wegener) and T.W. (Tao Wu); Investigation, T.W. (Thomas Wegener), T.W. (Tao Wu), F.S. and C.W.; Writing—Original Draft Preparation, T.W. (Thomas Wegener) and T.W. (Tao Wu); Writing—Review and Editing, T.W. (Thomas Wegener), T.W. (Tao Wu), J.L. and T.N.; Visualization, T.W. (Thomas Wegener) and T.W. (Tao Wu); Supervision, T.N. and J.L.; Project Administration, T.N. and J.L.; Funding Acquisition, T.N. and J.L. All authors have read and agreed to the published version of the manuscript.

Funding: Joint Research Group Program between Kassel University and Shenzhen Research Institute, City University of Hong Kong funded by Sino-German Center for Research Promotion (SGC) with project number GZ 1581. A part of this study was supported by the Ministry of Education, Culture, Sports, Science and Technology through a Grant-in-Aid for EarlyCareer Scientists 20K15057 (2020–2022).

Data Availability Statement: The data presented in this study are available upon request from the corresponding author.

Acknowledgments: The authors would like to thank Rolf Diederich and Rainer Hunke for technical support regarding the maintenance and operation of the laboratory equipment, and Marcel Krochmal for specimen preparation. Additionally, the authors appreciate Dewei Deng for providing stainless steel 316L specimens fabricated by direct energy deposition.

Conflicts of Interest: The authors declare no conflict of interest. The funding agency had no influence on the design of the study, the collection, analysis or interpretation of data, the writing of the manuscript; or the decision to publish the results.

References

1. Abdulhameed, O.; Al-Ahmari, A.; Ameen, W.; Mian, S.H. Additive manufacturing: Challenges, trends, and applications. *Adv. Mech. Eng.* **2019**, *11*, 168781401882288. [CrossRef]
2. Wegener, T.; Koopmann, J.; Richter, J.; Krooß, P.; Niendorf, T. CuCrZr processed by laser powder bed fusion—Processability and influence of heat treatment on electrical conductivity, microstructure and mechanical properties. *Fatigue Fract. Eng. Mat. Struct.* **2021**, *44*, 2570–2590. [CrossRef]
3. Zerbst, U.; Bruno, G.; Buffiere, J.-Y.; Wegener, T.; Niendorf, T.; Wu, T.; Zhang, X.; Kashaev, N.; Meneghetti, G.; Hrabe, N.; et al. Damage tolerant design of additively manufactured metallic components subjected to cyclic loading: State of the art and challenges. *Prog. Mater. Sci.* **2021**, *121*, 100786. [CrossRef] [PubMed]
4. Liverani, E.; Toschi, S.; Ceschini, L.; Fortunato, A. Effect of selective laser melting (SLM) process parameters on microstructure and mechanical properties of 316L austenitic stainless steel. *J. Mater. Process. Technol.* **2017**, *249*, 255–263. [CrossRef]
5. Wang, X.; Deng, D.; Yi, H.; Xu, H.; Yang, S.; Zhang, H. Influences of pulse laser parameters on properties of AISI316L stainless steel thin-walled part by laser material deposition. *Opt. Laser Technol.* **2017**, *92*, 5–14. [CrossRef]
6. Sibisi, P.N.; Popoola, A.P.I.; Arthur, N.K.K.; Pityana, S.L. Review on direct metal laser deposition manufacturing technology for the Ti-6Al-4V alloy. *Int. J. Adv. Manuf. Technol.* **2020**, *107*, 1163–1178. [CrossRef]
7. Guo, P.; Zou, B.; Huang, C.; Gao, H. Study on microstructure, mechanical properties and machinability of efficiently additive manufactured AISI 316L stainless steel by high-power direct laser deposition. *J. Mater. Process. Technol.* **2017**, *240*, 12–22. [CrossRef]
8. Zhang, K.; Wang, S.; Liu, W.; Shang, X. Characterization of stainless steel parts by Laser Metal Deposition Shaping. *Mater. Des.* **2014**, *55*, 104–119. [CrossRef]
9. Sun, G.F.; Shen, X.T.; Wang, Z.D.; Zhan, M.J.; Yao, S.; Zhou, R.; Ni, Z.H. Laser metal deposition as repair technology for 316L stainless steel: Influence of feeding powder compositions on microstructure and mechanical properties. *Opt. Laser Technol.* **2019**, *109*, 71–83. [CrossRef]
10. Syed, W.U.H.; Pinkerton, A.J.; Liu, Z.; Li, L. Coincident wire and powder deposition by laser to form compositionally graded material. *Surf. Coat. Technol.* **2007**, *201*, 7083–7091. [CrossRef]
11. AlHazaa, A.; Haneklaus, N. Diffusion Bonding and Transient Liquid Phase (TLP) Bonding of Type 304 and 316 Austenitic Stainless Steel—A Review of Similar and Dissimilar Material Joints. *Metals* **2020**, *10*, 613. [CrossRef]
12. Wang, X.; Deng, D.; Qi, M.; Zhang, H. Influences of deposition strategies and oblique angle on properties of AISI316L stainless steel oblique thin-walled part by direct laser fabrication. *Opt. Laser Technol.* **2016**, *80*, 138–144. [CrossRef]
13. Zheng, B.; Haley, J.C.; Yang, N.; Yee, J.; Terrassa, K.W.; Zhou, Y.; Lavernia, E.J.; Schoenung, J.M. On the evolution of microstructure and defect control in 316L SS components fabricated via directed energy deposition. *Mater. Sci. Eng. A* **2019**, *764*, 138243. [CrossRef]
14. Saboori, A.; Aversa, A.; Bosio, F.; Bassini, E.; Librera, E.; Chirico, M.; de Biamino, S.; Ugues, D.; Fino, P.; Lombardi, M. An investigation on the effect of powder recycling on the microstructure and mechanical properties of AISI 316L produced by Directed Energy Deposition. *Mater. Sci. Eng. A* **2019**, *766*, 138360. [CrossRef]
15. Thompson, S.M.; Bian, L.; Shamsaei, N.; Yadollahi, A. An overview of Direct Laser Deposition for additive manufacturing; Part I: Transport phenomena, modeling and diagnostics. *Addit. Manuf.* **2015**, *8*, 36–62. [CrossRef]
16. Shamsaei, N.; Yadollahi, A.; Bian, L.; Thompson, S.M. An overview of Direct Laser Deposition for additive manufacturing; Part II: Mechanical behavior, process parameter optimization and control. *Addit. Manuf.* **2015**, *8*, 12–35. [CrossRef]
17. Saboori, A.; Aversa, A.; Marchese, G.; Biamino, S.; Lombardi, M.; Fino, P. Microstructure and Mechanical Properties of AISI 316L Produced by Directed Energy Deposition-Based Additive Manufacturing: A Review. *Appl. Sci.* **2020**, *10*, 3310. [CrossRef]
18. Bedmar, J.; Riquelme, A.; Rodrigo, P.; Torres, B.; Rams, J. Comparison of Different Additive Manufacturing Methods for 316L Stainless Steel. *Materials* **2021**, *14*, 6504. [CrossRef]
19. DebRoy, T.; Wei, H.L.; Zuback, J.S.; Mukherjee, T.; Elmer, J.W.; Milewski, J.O.; Beese, A.M.; Wilson-Heid, A.; De, A.; Zhang, W. Additive manufacturing of metallic components–Process, structure and properties. *Prog. Mater. Sci.* **2018**, *92*, 112–224. [CrossRef]
20. Hunt, J.D. Steady state columnar and equiaxed growth of dendrites and eutectic. *Mater. Sci. Eng.* **1984**, *65*, 75–83. [CrossRef]
21. Morrow, B.M.; Lienert, T.J.; Knapp, C.M.; Sutton, J.O.; Brand, M.J.; Pacheco, R.M.; Livescu, V.; Carpenter, J.S.; Gray, G.T. Impact of Defects in Powder Feedstock Materials on Microstructure of 304L and 316L Stainless Steel Produced by Additive Manufacturing. *Met. Mat. Trans. A* **2018**, *49*, 3637–3650. [CrossRef]

22. Shin, W.-S.; Son, B.; Song, W.; Sohn, H.; Jang, H.; Kim, Y.-J.; Park, C. Heat treatment effect on the microstructure, mechanical properties, and wear behaviors of stainless steel 316L prepared via selective laser melting. *Mater. Sci. Eng. A* **2021**, *806*, 140805. [CrossRef]
23. Bagherifard, S.; Slawik, S.; Fernández-Pariente, I.; Pauly, C.; Mücklich, F.; Guagliano, M. Nanoscale surface modification of AISI 316L stainless steel by severe shot peening. *Mater. Des.* **2016**, *102*, 68–77. [CrossRef]
24. Olugbade, T.O.; Lu, J. Literature review on the mechanical properties of materials after surface mechanical attrition treatment (SMAT). *Nano Mater. Sci.* **2020**, *2*, 3–31. [CrossRef]
25. Roland, T.; Retraint, D.; Lu, K.; Lu, J. Fatigue life improvement through surface nanostructuring of stainless steel by means of surface mechanical attrition treatment. *Scr. Mater.* **2006**, *54*, 1949–1954. [CrossRef]
26. Huang, H.W.; Wang, Z.B.; Lu, J.; Lu, K. Fatigue behaviors of AISI 316L stainless steel with a gradient nanostructured surface layer. *Acta Mater.* **2015**, *87*, 150–160. [CrossRef]
27. Roland, T.; Retraint, D.; Lu, K.; Lu, J. Enhanced mechanical behavior of a nanocrystallised stainless steel and its thermal stability. *Mater. Sci. Eng. A* **2007**, *445–446*, 281–288. [CrossRef]
28. Maurel, P.; Weiss, L.; Bocher, P.; Grosdidier, T. Effects of SMAT at cryogenic and room temperatures on the kink band and martensite formations with associated fatigue resistance in a β-metastable titanium alloy. *Mater. Sci. Eng. A* **2021**, *803*, 140618. [CrossRef]
29. Shi, X.Y.; Liu, Y.; Li, D.J.; Chen, B.; Zeng, X.Q.; Lu, J.; Ding, W.J. Microstructure evolution and mechanical properties of an Mg–Gd alloy subjected to surface mechanical attrition treatment. *Mater. Sci. Eng. A* **2015**, *630*, 146–154. [CrossRef]
30. Yan, X.; Yin, S.; Chen, C.; Jenkins, R.; Lupoi, R.; Bolot, R.; Ma, W.; Kuang, M.; Liao, H.; Lu, J.; et al. Fatigue strength improvement of selective laser melted Ti6Al4V using ultrasonic surface mechanical attrition. *Mater. Res. Lett.* **2019**, *7*, 327–333. [CrossRef]
31. Sun, Y.; Bailey, R.; Moroz, A. Surface finish and properties enhancement of selective laser melted 316L stainless steel by surface mechanical attrition treatment. *Surf. Coat. Technol.* **2019**, *378*, 124993. [CrossRef]
32. Portella, Q.; Chemkhi, M.; Retraint, D. Influence of Surface Mechanical Attrition Treatment (SMAT) post-treatment on microstructural, mechanical and tensile behaviour of additive manufactured AISI 316L. *Mater. Charact.* **2020**, *167*, 110463. [CrossRef]
33. Belsvik, M.A.; Tucho, W.M.; Hansen, V. Microstructural studies of direct-laser-deposited stainless steel 316L-Si on 316L base material. *SN Appl. Sci.* **2020**, *2*, 1967. [CrossRef]
34. Andreau, O.; Koutiri, I.; Peyre, P.; Penot, J.-D.; Saintier, N.; Pessard, E.; Terris, T.; de Dupuy, C.; Baudin, T. Texture control of 316L parts by modulation of the melt pool morphology in selective laser melting. *J. Mater. Process. Technol.* **2019**, *264*, 21–31. [CrossRef]
35. Hsu, C.-H.; Chen, T.-C.; Huang, R.-T.; Tsay, L.-W. Stress Corrosion Cracking Susceptibility of 304L Substrate and 308L Weld Metal Exposed to a Salt Spray. *Materials* **2017**, *10*, 187. [CrossRef]
36. Tao, N.R.; Lu, J.; Lu, K. Surface Nanocrystallization by Surface Mechanical Attrition Treatment. *MSF* **2008**, *579*, 91–108. [CrossRef]
37. Pyun, Y.S.; Suh, C.M.; Yamaguchi, T.; Im, J.S.; Kim, J.H.; Amanov, A.; Park, J.H. Fatigue characteristics of SAE52100 steel via ultrasonic nanocrystal surface modification technology. *J. Nanosci. Nanotechnol.* **2012**, *12*, 6089–6095. [CrossRef]
38. Xie, J.; Zhu, Y.; Huang, Y.; Bai, C.; Ye, X. Microstructure Characteristics of 30CrMnSiNi2A Steel After Ultrasound-Aided Deep Rolling. *J. Mater. Eng. Perform.* **2013**, *22*, 1642–1648. [CrossRef]
39. Agrawal, A.K.; Singh, A.; Vivek, A.; Hansen, S.; Daehn, G. Extreme twinning and hardening of 316L from a scalable impact process. *Mater. Lett.* **2018**, *225*, 50–53. [CrossRef]
40. Wang, Z.D.; Sun, G.F.; Lu, Y.; Chen, M.Z.; Bi, K.D.; Ni, Z.H. Microstructural characterization and mechanical behavior of ultrasonic impact peened and laser shock peened AISI 316L stainless steel. *Surf. Coat. Technol.* **2020**, *385*, 125403. [CrossRef]
41. Jayalakshmi, M.; Huilgol, P.; Bhat, B.R.; Udaya Bhat, K. Insights into formation of gradient nanostructured (GNS) layer and deformation induced martensite in AISI 316 stainless steel subjected to severe shot peening. *Surf. Coat. Technol.* **2018**, *344*, 295–302. [CrossRef]
42. Bahl, S.; Suwas, S.; Ungàr, T.; Chatterjee, K. Elucidating microstructural evolution and strengthening mechanisms in nanocrystalline surface induced by surface mechanical attrition treatment of stainless steel. *Acta Mater.* **2017**, *122*, 138–151. [CrossRef]
43. Bertsch, K.M.; Meric de Bellefon, G.; Kuehl, B.; Thoma, D.J. Origin of dislocation structures in an additively manufactured austenitic stainless steel 316L. *Acta Mater.* **2020**, *199*, 19–33. [CrossRef]
44. Arifvianto, B.; Suyitno; Mahardika, M.; Dewo, P.; Iswanto, P.T.; Salim, U.A. Effect of surface mechanical attrition treatment (SMAT) on microhardness, surface roughness and wettability of AISI 316L. *Mater. Chem. Phys.* **2011**, *125*, 418–426. [CrossRef]
45. Naik, S.N.; Walley, S.M. The Hall–Petch and inverse Hall–Petch relations and the hardness of nanocrystalline metals. *J. Mater. Sci.* **2020**, *55*, 2661–2681. [CrossRef]
46. Oevermann, T.; Wegener, T.; Liehr, A.; Hübner, L.; Niendorf, T. Evolution of residual stress, microstructure and cyclic performance of the equiatomic high-entropy alloy CoCrFeMnNi after deep rolling. *Int. J. Fatigue* **2021**, *153*, 106513. [CrossRef]
47. Butz, A.; Zapara, M.; Helm, D. Modellierung von hochfesten und hochduktilen Blechwerkstoffen aus TWIP-Stahl. *wt Werkstattstech. Online* **2015**, *105*, 47–48. [CrossRef]
48. Yadollahi, A.; Shamsaei, N.; Thompson, S.M.; Seely, D.W. Effects of process time interval and heat treatment on the mechanical and microstructural properties of direct laser deposited 316L stainless steel. *Mater. Sci. Eng. A* **2015**, *644*, 171–183. [CrossRef]
49. Wang, Y.M.; Voisin, T.; McKeown, J.T.; Ye, J.; Calta, N.P.; Li, Z.; Zeng, Z.; Zhang, Y.; Chen, W.; Roehling, T.T.; et al. Additively manufactured hierarchical stainless steels with high strength and ductility. *Nat. Mater.* **2018**, *17*, 63–71. [CrossRef]

50. Saeidi, K.; Gao, X.; Zhong, Y.; Shen, Z.J. Hardened austenite steel with columnar sub-grain structure formed by laser melting. *Mater. Sci. Eng. A* **2015**, *625*, 221–229. [CrossRef]
51. Oevermann, T.; Wegener, T.; Niendorf, T. On the Evolution of Residual Stresses, Microstructure and Cyclic Performance of High-Manganese Austenitic TWIP-Steel after Deep Rolling. *Metals* **2019**, *9*, 825. [CrossRef]
52. Niendorf, T.; Lotze, C.; Canadinc, D.; Frehn, A.; Maier, H.J. The role of monotonic pre-deformation on the fatigue performance of a high-manganese austenitic TWIP steel. *Mater. Sci. Eng. A* **2009**, *499*, 518–524. [CrossRef]
53. Lambers, H.-G.; Rüsing, C.J.; Niendorf, T.; Geissler, D.; Freudenberger, J.; Maier, H.J. On the low-cycle fatigue response of pre-strained austenitic Fe61Mn24Ni6.5Cr8.5 alloy showing TWIP effect. *Int. J. Fatigue* **2012**, *40*, 51–60. [CrossRef]
54. Wegener, T.; Haase, C.; Liehr, A.; Niendorf, T. On the influence of $\varkappa$-carbides on the low-cycle fatigue behavior of high-Mn light-weight steels. *Int. J. Fatigue* **2021**, *150*, 106327. [CrossRef]
55. Rajabi, E.; Miresmaeili, R.; Aliofkhazraei, M. The effect of surface mechanical attrition treatment (SMAT) on plastic deformation mechanisms and mechanical properties of austenitic stainless steel 316L. *Mater. Res. Express* **2019**, *6*, 1250g8. [CrossRef]
56. Schulze, V.; Hoffmeister, J. Mechanische Randschichtverfestigungsverfahren–Verfahren. In *Handbuch Wärmebehandeln und Beschichten*; Zoch, H.-W., Spur, G., Eds.; Carl Hanser Verlag GmbH & Co. KG: München, Germany, 2015; pp. 591–612. ISBN 978-3-446-42779-2.
57. Schulze, V. *Modern Mechanical Surface Treatment: States, Stability, Effects*; Wiley-VCH: Weinheim, Germany, 2006; ISBN 978-3-527-31371-6.
58. Tian, Y.Z.; Sun, S.J.; Lin, H.R.; Zhang, Z.F. Fatigue behavior of CoCrFeMnNi high-entropy alloy under fully reversed cyclic deformation. *J. Mater. Sci. Technol.* **2019**, *35*, 334–340. [CrossRef]
59. Manson, S.S.; Hirschberg, M.H. The role of ductility, tensile strength and fracture toughness in fatigue. *J. Frankl. Inst.* **1970**, *290*, 539–548. [CrossRef]
60. Mughrabi, H. Microstructural mechanisms of cyclic deformation, fatigue crack initiation and early crack growth. *Philos. Trans. A Math. Phys. Eng. Sci.* **2015**, *373*, 20140132. [CrossRef]
61. Xin, Q. Durability and reliability in diesel engine system design. In *Diesel Engine System Design*; Elsevier: Amsterdam, The Netherlands, 2013; pp. 113–202. ISBN 9781845697150.
62. Wegener, T.; Krochmal, M.; Oevermann, T.; Niendorf, T. Consequences of Deep Rolling at Elevated Temperature on Near-Surface and Fatigue Properties of High-Manganese TWIP Steel X40MnCrAl19-2. *Appl. Sci.* **2021**, *11*, 10406. [CrossRef]
63. Oguri, K. Fatigue life enhancement of aluminum alloy for aircraft by Fine Particle Shot Peening (FPSP). *J. Mater. Process. Technol.* **2011**, *211*, 1395–1399. [CrossRef]
64. Gao, T.; Sun, Z.; Xue, H.; Retraint, D. Effect of surface mechanical attrition treatment on high cycle and very high cycle fatigue of a 7075-T6 aluminium alloy. *Int. J. Fatigue* **2020**, *139*, 105798. [CrossRef]
65. Masing, G. Zur Heyn'schen Theorie der Verfestigung der Metalle durch verborgen elastische Spannungen. In *Wissenschaftliche Veröffentlichungen aus dem Siemens-Konzern*; Harries, C.D., Ed.; Springer: Berlin/Heidelberg, Germany, 1923; pp. 231–239. ISBN 978-3-642-98848-6.
66. Masing, G. Eigenspannungen in kaltgereckten Metallen. *Z. Für. Tech. Phys.* **1925**, *6*, 569–573.

Article

Influence of Plastic Strain Control on Martensite Evolution and Fatigue Life of Metastable Austenitic Stainless Steel [†]

Matthias Droste, Sebastian Henkel *, Horst Biermann and Anja Weidner

Institute of Materials Engineering, Technische Universität Bergakademie Freiberg, D-09599 Freiberg, Germany; matthiasdroste@web.de (M.D.); biermann@ww.tu-freiberg.de (H.B.); anja.weidner@ww.tu-freiberg.de (A.W.)
* Correspondence: henkel@ww.tu-freiberg.de
† Dedicated to Hans-Jürgen Christ, Universität Siegen, Germany, for his valuable contributions to the understanding of fatigue and fracture of metals and alloys.

Abstract: Metastable austenitic stainless steel was investigated during fatigue tests under strain control with either constant total or constant plastic strain amplitude. Two different material conditions with coarse-grained and ultrafine-grained microstructure were in focus. The influence of plastic strain control of the fatigue test on both the martensitic phase transformation as well as on the fatigue lives is discussed. In addition, an approach for calculating the Coffin–Manson–Basquin parameters to estimate fatigue lives based on strain-controlled tests at constant total strain amplitudes is proposed for materials undergoing a strong secondary hardening due to martensitic phase transformation.

Keywords: fatigue; cyclic plastic strain; martensite

Citation: Droste, M.; Henkel, S.; Biermann, H.; Weidner, A. Influence of Plastic Strain Control on Martensite Evolution and Fatigue Life of Metastable Austenitic Stainless Steel. *Metals* **2022**, *12*, 1222. https://doi.org/10.3390/met12071222

Academic Editors: Martin Heilmaier and Martina Zimmermann

Received: 31 May 2022
Accepted: 15 July 2022
Published: 19 July 2022

Publisher's Note: MDPI stays neutral with regard to jurisdictional claims in published maps and institutional affiliations.

Copyright: © 2022 by the authors. Licensee MDPI, Basel, Switzerland. This article is an open access article distributed under the terms and conditions of the Creative Commons Attribution (CC BY) license (https://creativecommons.org/licenses/by/4.0/).

1. Introduction

The first investigations on the fatigue failure during cyclic loading of metallic materials carried out by Wöhler in the 19th century revealed that the fatigue life is a function of the applied cyclic stress. However, since that time, a great increase in knowledge on different factors influencing the fatigue life was gained. Thus, the influence of cyclic plastic strain on the crack initiation was intensively studied, resulting in the well-known Coffin–Manson approach [1–3], which showed that the fatigue life, particularly in the low-cycle fatigue (LCF) regime, is governed by the cyclic plastic strain. Later on, it was shown by Lukas et al. [4] that this relationship can be extended also to the high-cycle fatigue (HCF) regime.

Nevertheless, the majority of fatigue life investigations are still performed under stress control, or at least under strain control at constant total strain amplitude. However, the performance of stress-controlled or total strain-controlled fatigue tests disregards changes in the cyclic plastic strain caused by cyclic hardening or softening of materials. In particular, for ductile metallic materials in single (copper, e.g., [5,6], nickel [7]) and polycrystalline (AISI 316L [8], nickel [9], copper [6], aluminum [10]) conditions, a broad variety of investigations were performed, demonstrating the influence of cyclic plastic strain on the dislocation arrangement in the microstructure and, therefore, also on the crack initiation and propagation. However, in the case of materials undergoing a strong cyclic hardening, such as metastable austenitic stainless steels, the majority of fatigue life investigations are performed under stress control or under strain control at constant total strain amplitude [11]. This is caused mainly by the fact that these materials yield a significant secondary cyclic hardening caused by the martensitic phase transformation (e.g., [12,13]). This hardening behavior impedes the performance of fatigue tests under strain control at constant plastic strain amplitude since the cyclic stress vs. plastic strain hysteresis loop is changing from cycle to cycle due to the influence of hardening. Moreover, the specimen stiffness as well as the Young's modulus is changing during cyclic deformation due to the phase change from austenite to α′-martensite, making fatigue tests even more complex.

Furthermore, it is well known that the grain size has a significant influence on fatigue lives, as well. Thus, investigations on the fatigue lives of coarse-grained (CG) and ultrafine-grained (UFG) metallic materials have been the focus of scientific interest for long time [14,15]. It is well known and commonly accepted from total strain vs. fatigue life diagrams, particularly for pure materials such as copper [16–18], aluminum [19] or nickel [20], that CG microstructures yield superior fatigue lives in the LCF regime, whereas in the HCF regime, the fatigue lives of UFG microstructures are superior compared to CG material conditions [15]. This behavior is explained by the increased strength and the loss of ductility of UFG microstructures compared to their CG counterparts.

To date, however, only few papers are known [21–30] to focus on the influence of grain refinement on the fatigue lives of metastable austenitic stainless steels. The results of these investigations can be summarized by the following main findings: (i) the initial stress amplitude during strain-controlled cycling is a function of the grain size leading to significantly higher values for grain-refined material conditions; (ii) the cyclic deformation curves of grain-refined states are characterized by an initial softening, followed subsequently by a cyclic hardening, which is related to the cyclic-strain induced α'-martensitic phase transformation; (iii) the stress amplitude reached at the end of fatigue life is directly related to the volume fraction of α'-martensite. Related to the mechanism of martensitic phase transformation in grain-refined material states, it was suggested by Droste et al. [30] that in UFG materials, the transformation does not occur inside deformation bands as in CG reference states. Instead, the transformation starts at grain boundaries and covers larger areas of the grains with dimensions less than 1 µm.

The aim of the present investigation was to demonstrate the influence of both the grain size as well as the mode of strain control during testing on the fatigue lives of metastable austenitic stainless steel.

2. Materials and Methods

A X2CrMnNi16-7-6 steel with two different grain sizes (62 µm and 0.8 µm) was investigated during cyclic loading under constant total and constant plastic strain at different amplitudes. The evolution of the ferromagnetic phase fraction (i.e., α'-martensite) was recorded in situ during the fatigue experiments. The fatigue lives of the total strain and plastic strain-controlled tests were compared. In addition, Coffin–Manson plots were calculated from strain-controlled tests at constant total strain amplitudes using stress amplitudes at different fractions of fatigue life N_f and compared to fatigue lives obtained from plastic strain-controlled fatigue tests. The investigations were corroborated by microstructural investigations using scanning electron microscopy (SEM) showing the grain size and orientation as well as the morphology and distribution of α'-martensite.

2.1. Material Conditions

The material under investigation was a metastable austenitic stainless steel based on low interstitial contents of carbon and nitrogen and high concentrations of alloying elements such as chromium (16 wt %), manganese (6 wt %) and nickel (6 wt %). The material was studied in two different microstructural conditions: (i) coarse-grained (CG) microstructure and (ii) ultrafine-grained (UFG) microstructure.

The CG material was achieved via the powder metallurgical production route. The steel with the chemical composition given in Table 1 was atomized under nitrogen atmosphere, resulting in a steel powder with a particle size d_{50} = 26 µm. Subsequently, the steel powder was pressed into a green body with an applied stress of 60 MPa under laboratory atmosphere at room temperature. This was followed by a hot-pressing process under vacuum at 1250 °C and 30 MPa for 30 min. The heating and cooling rates during the process were 10 K/min and 5 K/min, respectively. The hot-pressed material was manufactured at the Fraunhofer Institute for Ceramic Technologies and Systems (IKTS, Dresden, Germany). The microstructure is characterized by an average grain size of D = 62 µm ± 39 µm containing high volume fraction of annealing twins (see Figure 1a) characterized by a misorientation angle of 60° around ⟨111⟩ axis (see Figure 1e).

Table 1. Chemical composition and mechanical properties of CG and UFG material conditions (Fe balance). In addition, M_s, SFE, area-weighted average grain size D and martensite content after tensile testing are provided.

Steel	C	N	Cr	Mn	Ni	Si	SFE (mJ/m^2)	M_s (°C)	D (µm)	$Rp_{0.5}$ (MPa)	UTS (MPa)	A (%)	Vol % α' (%)
CG	0.02	0.06	16.4	7.1	6.3	0.1	17.9	−28	62	267	697	69	18
UFG	0.03	0.02	15.6	7.1	6.1	0.9	11.9	−21	0.78	923	1063	35	20

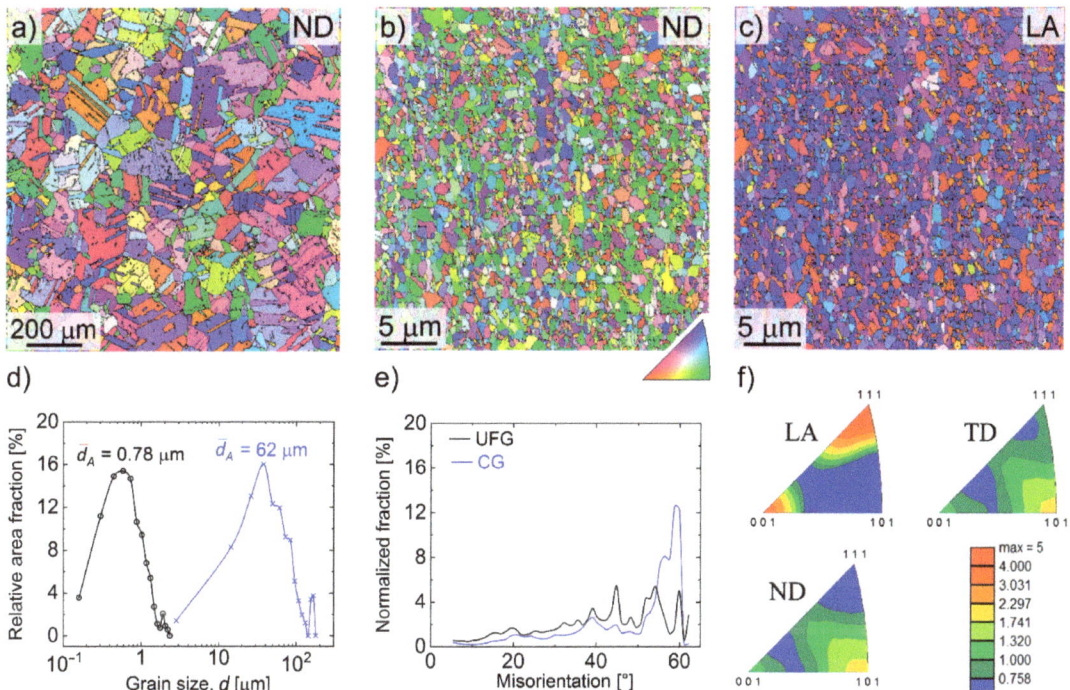

Figure 1. Initial microstructures of CG and UFG materials conditions obtained from EBSD investigations. (a–c) Crystallographic orientation maps of CG (a) and UFG (b,c) conditions. (d) Grain size distributions. (e) Misorientation distributions. (f) Texture of UFG material. ND—normal direction, TD transverse direction, LA—loading axis.

The UFG material was obtained by thermomechanically controlled processing (TMCP) [31]. The initial material was an as-cast steel with the chemical composition given in Table 1. A steel rod with a diameter of 50 mm was cold-formed in a 4-jaw rotary swaging machine at Leibniz-Institut für Festkörper- und Werkstoffforschung Dresden (IFW, Dresden, Germany). Cold forming was performed in 11 passes with cross-sectional changes of 19% each. Between the passes, the bars were cooled down to room temperature to enable the formation of a high fraction of deformation-induced α'-martensite (86 vol %) during swaging. Subsequently, a conventional reversion annealing process was performed in a preheated tube furnace under argon atmosphere at 700 °C for 5 min, followed by water quenching. The microstructure of this material is fully austenitic (Figure 1b,c) and is characterized by an area-weighted average grain size of D = 0.8 µm ± 0.4 µm (see Figure 1d, Table 1). The

EBSD measurements showed a pronounced ⟨001⟩ and ⟨111⟩ texture in load axis (Figure 1f). More details on the process are given elsewhere [30].

Based on the chemical composition, the martensite start temperature M_s as well as the stacking fault energy (SFE) were calculated for both UFG and CG material conditions according to [31] and [32], respectively. In both cases, the M_s is well below room temperature, indicating that no thermal martensite has been formed. However, both material conditions exhibit a high ability to show strain-induced martensitic phase transformation under mechanical loading and also a higher dissociation width of Shockley partial dislocations forming stacking faults due to stacking fault energy (SFE) value below 18 mJ/m². Due to the different heat treatments, further influences might impact the mechanical behavior. Thus, the UFG structure was obtained by ageing at 700 °C, whereas the CG state was set at 1250 °C. Therefore, it cannot be excluded that precipitates may have developed at 700 °C. However, we assume that these should have only minor relevance for the cyclic behavior compared to the grain size difference.

The mechanical properties such as yield strength ($Rp_{0.5}$), ultimate tensile strength (UTS) and elongation at fracture (A) under tensile loading at RT of both material states are included in Table 1, together with the achieved martensite content in the gauge length of specimens after tensile testing.

2.2. Mechanical Testing

Cylindrical specimens were manufactured for both CG and UFG material conditions with a gauge length of 14 mm and a diameter of 6 mm and 5 mm, respectively. The specimen geometries are shown in Figure 2. Different sample geometries were necessary because after completion of the tests on the CG material (specimen in Figure 2a), it was realized that the UFG material was only available in smaller dimensions and the geometry in Figure 2b had to be chosen. The fatigue tests were carried out using two servo-hydraulic testing machines: MTS Landmark 250 with max. load capacity of 250 kN for CG specimens and MTS Landmark 100 for UFG specimens (both MTS Systems Corporation, Eden Prairie, MN, USA). The tests were performed using a clip-on extensometer with a gauge length of 12 mm and a measuring range of ±9%. The fatigue tests were performed for CG and UFG material conditions, both under total strain control in a range of $0.3\% \leq \Delta\varepsilon_t/2 \leq 1.2\%$, as well as under plastic strain control in a range of $0.14\% \leq \Delta\varepsilon_{pl}/2 \leq 1.0\%$. In all tests, a triangular waveform command signal was applied and the total strain rate was set to $\dot{\varepsilon} = 4 \times 10^{-3}\ \text{s}^{-1}$. Plastic strain was evaluated according to procedure described by Sommer et al. [33] regarding the stress dependence of Young's modulus in high strength materials. Sommer et al. suggested an approach for the relationship between stress and elastic strain that accounts for this stress dependence using a quadratic complement to Hooke's law according to Equation (1):

$$\sigma = E_0 \cdot \varepsilon_{el} + z \cdot \varepsilon_{el}^2 \tag{1}$$

where z is constant with z < 0, which should be evaluated using partial unloading during the first loading cycle under total strain control.

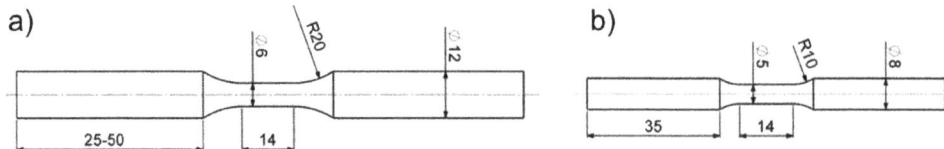

Figure 2. Specimen geometry for fatigue tests of CG (**a**) and UFG (**b**) material conditions. All dimensions are in mm.

The fatigue experiments were corroborated by ferromagnetic phase fraction measurements. Thus, Fischerscope® MMS® PC or Feritscope® FMP30 (both Helmut Fischer GmbH, Sindelfingen, Germany) were used to measure the ferromagnetic phase fraction in situ during the fatigue tests at a frequency of 5 Hz using a ferrite probe of type FGAB1.3-Fe. The detection depth decreases with increasing ferromagnetic phase content and is 3–3.5 mm at 7% ferrite. The ferrite probes were calibrated with the same calibration set based on four ferrite standards. Since this calibration refers to the δ-ferrite content of a sample, the measured values must be multiplied by a correction factor of 1.7 according to Talonen et al. [34] to obtain the content of α'-martensite. This correction factor is valid up to measured values of maximum 55 Fe-%. Since this value is exceeded in some cases of the present work, the uncorrected ferrite probe signal is given in Fe-% throughout the paper. More-over, a roundness correction due to the cylindrical samples or an edge distance correction in the case of the flat samples is omitted. Instead, the measured lengths of the samples were scanned with the ferrite probe after the end of the test in order to calculate a mean α'-martensite content. Subsequently, the test data were corrected to this mean value.

Microstructural investigations were performed using scanning electron microscopy (SEM) on the initial material conditions as well as for selected fatigue experiments under total strain control. Thus, electron-backscattered diffraction (EBSD) measurements were conducted using a field-emission SEM (Mira3, Tescan, Brno, Czech Republic) operated between 20 kV and 25 kV acceleration voltage equipped with an EBSD detector and OIM acquisition/analysis software (both EDAX, TSL, Mahwah, NJ, USA).

3. Results

3.1. Cyclic Deformation Curves

The results of the fatigue tests under total and plastic strain control in terms of cyclic deformation curves are summarized in Figure 3 for CG (Figure 3a) and UFG material (Figure 3b). The plastic strain amplitudes were set according to the initial plastic strain amplitudes obtained in the respective total strain-controlled tests. Thus, pairs of tests resulted, which had initially the same plastic strain amplitudes. It must be noted that the plastic strain amplitudes of the total strain-controlled tests decreased during the tests in the case of pronounced cyclic hardening and increased for cyclic softening, respectively. Furthermore, the results of two further tests at lower (0.235%) and higher (1.0%) plastic strain amplitudes were included. First, significant differences between CG and UFG material become apparent, as expected already from the quasi-static mechanical properties (Table 1). The initial stress amplitudes for the UFG material state are significantly higher due to the small grain size (<1 µm). Moreover, the fatigue life of the UFG state is identical or slightly enhanced compared to the CG condition for tests at high or small strain amplitudes, respectively. No significant differences were observed in the cyclic stress–strain responses obtained under plastic strain control or total strain control. The small discontinuities in cyclic hardening/softening curves obtained under plastic strain control are related to adjustments of the stress vs. plastic strain hysteresis loops caused by the hardening behavior of the material under cyclic loading. Overall, a pronounced secondary hardening is observed in both material conditions at higher strain amplitudes, which is related to the formation of α'-martensite. For further analysis of the cyclic deformation behavior of the investigated steel under total strain control, the reader is referred to our recent papers [27,30,35].

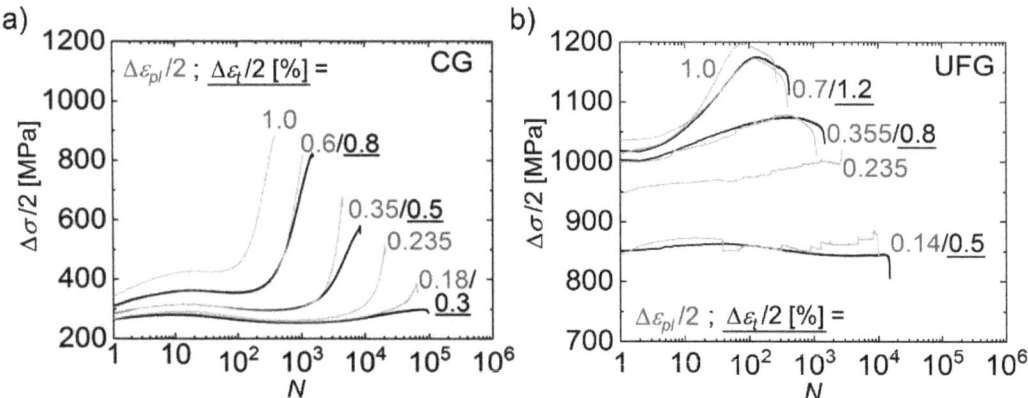

Figure 3. Cyclic hardening/softening curves for (**a**) CG and (**b**) UFG material states. Comparison of total strain-controlled ($\Delta\varepsilon_t/2$) (black curves) and plastic strain-controlled ($\Delta\varepsilon_{pl}/2$) (gray curves) fatigue tests.

3.2. Evolution of α′-Martensite Volume Fraction

The evolution of ferromagnetic phase fractions during fatigue tests under total and plastic strain control is shown for both CG and UFG material states in Figure 4. It can be seen that for both materials and at all strain amplitudes, an incubation period is needed for the onset of the formation of cyclic deformation-induced α′-martensite. Furthermore, it is visible that the incubation period is always significantly shorter for the UFG material compared to CG condition, in particular at high total strain amplitudes (Figure 4a,b). The α′-martensite fractions at the end of fatigue life are more or less comparable at high total strain amplitudes, whereas at small total strain amplitudes, the CG condition exhibits higher martensitic phase fractions. The plastic strain-controlled tests in Figure 4c,d clearly support the statement of significantly shorter incubation periods for the UFG material. Thus, at the highest strain amplitude of $\Delta\varepsilon_t/2 = 1.2\%$, the incubation period is in the range of a few cycles for the UFG material, whereas for the CG material, it is around 100 cycles (compare Figure 4a,b). The phase fractions at the end of fatigue life, on the other hand, do not differ that much between the conditions, with a trend for the UFG material to exhibit a slightly higher volume fraction of α′-martensite. In addition, at small strain amplitudes $\Delta\varepsilon_t/2 < 0.4\%$, the formation of α′-martensite is negligible in both CG and UFG material. However, it must be mentioned that some problems occurred regarding the ferrite probe measurements in the case of the CG material cyclically deformed under total strain control (Figure 4a). For $\Delta\varepsilon_t/2 \leq 0.3\%$, the measurement was not sensitive enough at small α′-martensite fractions. There was some α′-martensite detected along the gauge lengths after the tests. This is why some α′-martensite fractions are only indicated by a cross in the diagram representing the values at the end of fatigue life.

Furthermore, it turned out that the formation of cyclic deformation-induced α′-martensite is more intense and the incubation period is even shortened under plastic strain control (Figure 4c,d). This can be explained in terms of the control mode and the threshold value of the cumulative plastic strain $\lambda_{p,th}$ [35], which is calculated by Equation (2):

$$\lambda_{p,th} = 4 \cdot \sum_{i=1}^{N_f} \Delta\varepsilon_{pl}/2. \tag{2}$$

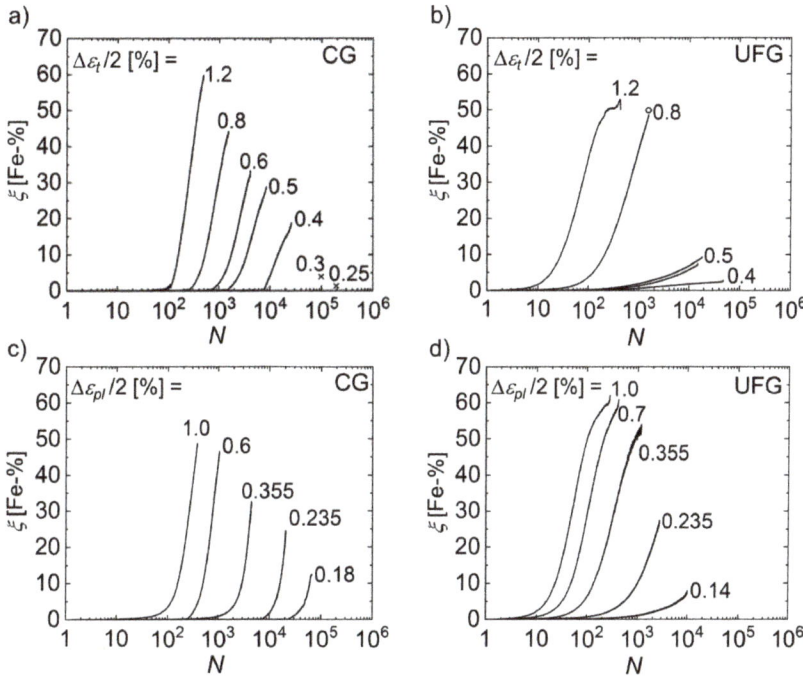

Figure 4. Evolution of ferromagnetic phase fraction in CG (**a,c**) and UFG material (**b,d**) conditions during cyclic loading under (**a,b**) total strain control and (**c,d**) plastic strain control with different amplitudes.

The evolution of the ferromagnetic phase fraction as a function of $\lambda_{p,th}$ is shown in Figure 5 for two selected strain amplitudes of both total strain (Figure 5a) and plastic strain-controlled (Figure 5b) fatigue tests. In general, the cumulated plastic strain needed for the onset of martensitic phase transformation is significantly lower for the UFG condition as well as for the plastic strain-controlled tests. Thus, the formation of α'-martensite starts earlier. Moreover, it is noticeable that at the end of the fatigue life, the α'-martensite content of the UFG material is always higher than for the CG material for the corresponding plastic strain amplitude. Only for small total strain amplitudes was the martensitic transformation of the UFG material significantly lower compared to CG material (compare Figure 5a,b).

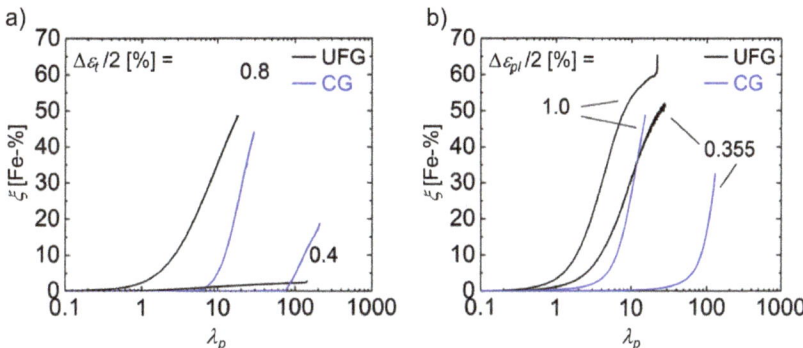

Figure 5. Evolution of ferromagnetic phase fraction as a function of cumulative plastic strain for UFG and CG material conditions during total strain (**a**) and plastic strain (**b**) controlled fatigue tests.

These results allow the conclusion that the UFG condition, despite its small grain size, has a higher tendency to form α'-martensite than the CG counterpart. This contradicts experimental findings on the destabilization of metastable austenitic steels during monotonic loading, in which decreasing grain size typically inhibits and retards α'-martensite formation [36–40]. Dissenting observations have mostly been attributed to the reduction of austenite stability due to the binding of carbon and nitrogen in precipitates [24,25] or to deformed microstructure components left over from fabrication [41]. In contrast, the UFG condition of the present study exhibits a completely recrystallized, homogeneous microstructure and a low proportion of interstitial elements due to the alloy design. In addition to the high stress amplitudes as a result of the small grain size, the strong texture with $\langle 001 \rangle$ and $\langle 111 \rangle$ lattice directions parallel to the loading axis (compare Figure 1f) contributes decisively to a pronounced phase transformation. These crystallographic orientations are preferred for large dissociation width of Shockley partial dislocations—grains with $\langle 111 \rangle$ orientation under tensile load and grains with $\langle 001 \rangle$ orientation under compressive loads facilitate the formation of α'-martensite [42,43]. Moreover, the chemical driving force for martensitic phase transformation expressed by martensite start temperature M_s (compare Table 1) is slightly increased for the UFG condition compared to coarse-grained counterpart. Finally, the SFE of UFG condition is lower compared to the CG material (see Table 1) due to the lower N content resulting in a higher tendency of the formation of extended stacking faults, which are a pre-cursor for α'-martensite formation.

3.3. Microstructure

In addition to the in situ ferromagnetic measurements, microstructural investigations were conducted for a selected total strain amplitude on CG and UFG material conditions. Figure 6 shows the results of EBSD measurements unraveling the martensitic phase transformation in CG und UFG material states at different total strain amplitudes. Figure 6a shows the phase map of CG material cyclically deformed at $\Delta\varepsilon_t/2 = 0.3\%$. Figure 6b,c are related to UFG material cycled at $\Delta\varepsilon_t/2 = 0.4\%$ and $\Delta\varepsilon_t/2 = 0.5\%$, respectively.

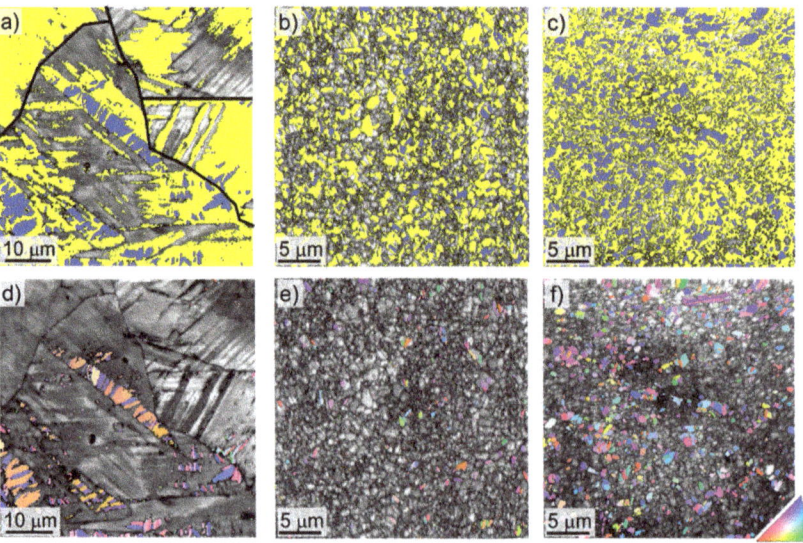

Figure 6. Results of EBSD measurements on CG (**a,d**) and UFG (**b,c,e,f**) material conditions at different total strain amplitudes of $\Delta\varepsilon_t/2 = 0.3\%$ (**a,d**), $\Delta\varepsilon_t/2 = 0.4\%$ (**b,e**) and $\Delta\varepsilon_t/2 = 0.5\%$ (**c,f**). Color code of phase maps (**a–c**): gray—austenite, yellow—ε-martensite, blue—α'-martensite. Crystallographic orientation maps of α'-martensite (**d–f**) are shown by inverse pole figure color code of the normal direction.

As expected for CG material condition, the formation of deformation bands along different activated slip systems is clearly visible. Some of these bands are transferred to ε-martensite (yellow) due to the high density of stacking faults and, in addition, α'-martensite grains (blue in Figure 6a) with different crystallographic orientations (compare Figure 6d) are formed in some bands. Figure 6b,c,e,f shows the microstructure of the UFG condition after fatigue tests at $\Delta\varepsilon_t/2 = 0.4\%$ (b,e) and $\Delta\varepsilon_t/2 = 0.5\%$ (c,f). The different martensitic phase transformation of metastable UFG steel becomes apparent. Due to the small grain size (<1 µm), the formation of deformation bands is impeded. Instead of formation of α'-martensite within deformation bands, the small grains transform successively either to ε-martensite or to α'-martensite (see [30]). The volume fraction of α'-martensite increases significantly with the increase in strain amplitude (compare Figure 6b,c), which is in agreement with measurements of ferromagnetic phase fraction.

4. Discussion

The fatigue life with respect to the applied total strain amplitude was analyzed for both CG and UFG material conditions according to the relationship of Basquin and Coffin–Manson in Equation (3):

$$\Delta\varepsilon_t/2 = \frac{\sigma'_f}{E} \cdot \left(2N_f\right)^b + \varepsilon'_f \cdot \left(2N_f\right)^c, \tag{3}$$

where σ'_f represents the fatigue strength coefficient, E is the Young's modulus, b is the fatigue strength exponent, ε'_f is the fatigue ductility coefficient and c is the fatigue ductility exponent. The resulting fatigue life curves for the CG and UFG material states are shown in Figure 7a. The plots unravel two main aspects: (i) identical fatigue lives of UFG and CG material states in the LCF regime, and (ii) superior fatigue lives of UFG state compared to CG counterpart in the HCF regime. In principle, the latter fact of higher durability in the HCF range is typical for UFG materials tested under total strain control [14,15]. The higher fatigue lives at lower total strain amplitudes are caused by the superior strength due to the small grain size compared to the CG condition, which is accompanied by significantly lower plastic strain amplitudes (compare Figure 7b).

In addition, the UFG state has good ductility (compare Table 1) due to its fully recrystallized microstructure and can tolerate a correspondingly high degree of plasticity. In contrast, lower lifetimes of UFG material states compared to their CG counterparts are expected at high strain amplitudes, i.e., in the LCF regime [14,15]. However, in the present case, the fatigue lives of UFG and CG conditions are identical in the LCF regime under total strain-controlled tests, as seen from Figure 7a.

The fatigue lives of the plastic strain-controlled tests on for UFG and CG material conditions are plotted in Figure 7c. The diagram reveals that the fatigue lives at the highest plastic strain amplitude of 1% are nearly identical for both material conditions. However, the fatigue lives of the CG condition are significantly higher at small plastic strain amplitudes compared to the UFG material. At $\Delta\varepsilon_{pl}/2 = 0.15\%$, the difference is approximately one decade. This illustrates the influence of the control mode of the fatigue tests on the lifetime.

During the total strain-controlled tests, the opposite was observed: a higher fatigue life of the UFG condition at low strain amplitudes (compare Figure 7a). The strength of the material was life-determining [2,3]. For the stronger UFG material, the elastic portion of the strain amplitude was larger and the plastic proportion was smaller.

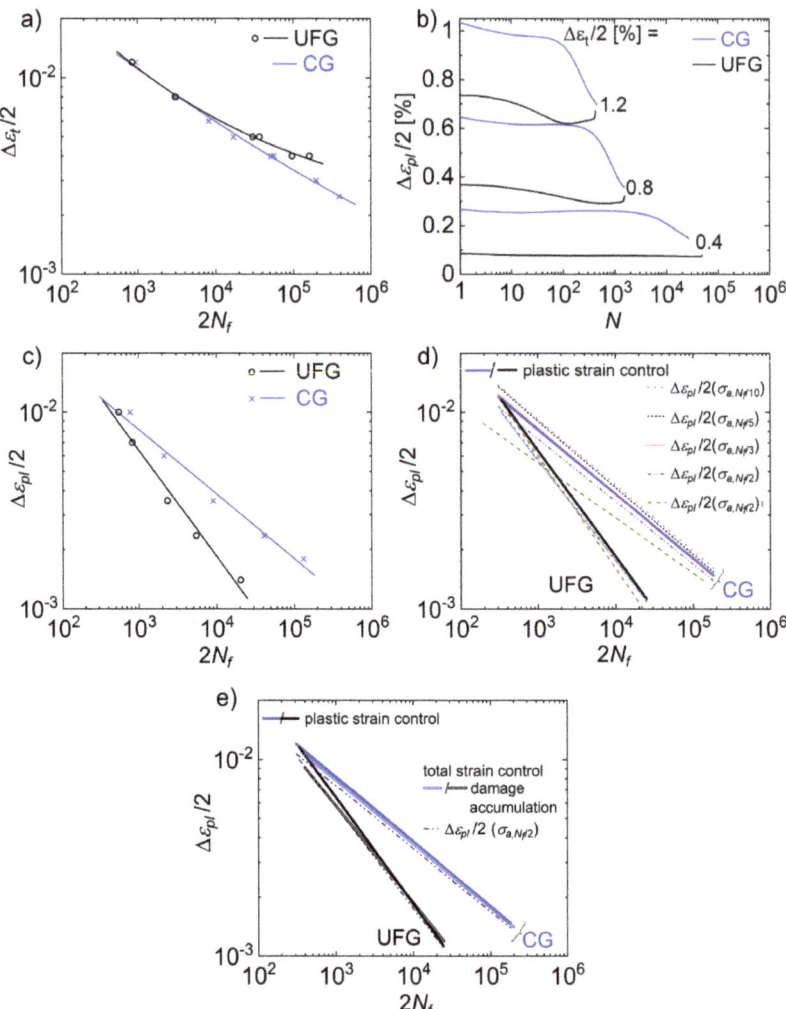

Figure 7. Fatigue life curves of total strain amplitude vs. reversals to failure of UFG and CG conditions (**a**) and evolution of plastic strain amplitude vs. number of cycles (**b**) obtained during total strain-controlled fatigue tests. (**c**) Fatigue lives obtained under plastic strain control for UFG and CG conditions. (**d**) Coffin–Manson plots calculated from total strain-controlled tests using stress amplitudes at different fractions of fatigue life N_f compared to fatigue lives obtained from plastic strain-controlled fatigue tests. (**e**) Coffin–Manson plots calculated from total strain-controlled tests using linear damage accumulation fit compared to fatigue lives obtained from plastic strain-controlled fatigue tests and the normally used comparison point at $N_f/2$.

The fatigue life curves of total strain-controlled tests are calculated according to Equation (1), which is based on the sum of the elastic (Basquin) and plastic (Coffin–Manson) strain contributions. In general, the saturated stress amplitude usually taken at half fatigue life is used to determine the elastic and plastic components and the Basquin/Coffin–Manson parameters (see Equation (1)). However, it is obvious from the cyclic stress–strain response shown in Figure 3 that no saturation plateau is reached for the investigated metastable austenitic steels due to their pronounced secondary cyclic hardening caused by the martensitic phase transformation. Thus, there is no universal rule on which stage of fatigue life should be used for parameter calculation. To prove different approaches, various stress amplitudes such as σ_a at $N_f/10$, $N_f/5$ or $N_f/3$ were used. The results in terms of the Coffin–Manson plots calculated based on the total strain-controlled tests are shown in Figure 7d (thin lines) together with the results from plastic strain-controlled tests, which represent the experimentally revealed Coffin–Manson plots (bold lines). Depending on the calculation approach, different progressions for the Coffin–Manson plots arise. Since both total and plastic strain-controlled tests were performed in this work, different approaches can be compared and their accuracy evaluated. It turned out that for the CG condition, the calculation of parameters ε'_f and c strongly depend on the chosen approach. Thus, markedly different courses for the Coffin–Manson plots arise for different approaches. For example, calculations based on the maximum stress amplitude result in a particularly high elastic fraction of the total strain (Basquin), which in turn causes a particularly low plastic fraction. Consequently, the latter is underestimated and the Coffin–Manson plot predicts a too low fatigue life. The opposite is provided by calculating ε'_f and c using the stress amplitude at $N_f/10$. This is still within the incubation period. Thus, the martensitic phase transformation has not yet set in and the stress amplitudes are comparatively low. Therefore, this approach results in too low elastic and too high plastic fraction. If stress amplitudes from later phases of the cyclic deformation are used, the plastic fraction is reduced as a consequence of the cyclic hardening, and the Coffin–Manson plots are shifted downward. The best agreement of the Coffin–Manson curve with that of the plastic strain-controlled tests was obtained using the stress amplitude at $N_f/3$ in our case. The most common approach to determine the Coffin–Manson parameters based on the stress amplitude at half-fatigue life results in an underestimation of the plastic part due to the strong cyclic hardening of the CG state (compare Figure 7d).

As in the total strain-controlled tests, the plastic strain amplitude changes during the test in each cycle with the cyclic hardening or softening. We suggest in this work an additional approach with a regression calculation based on the linear damage accumulation hypothesis similar to Palmgreen and Miner. In this approach, the damage sum S over all cycles is determined for each test (t) based on a Coffin–Manson approach:

$$S(t) = \sum_{N=1}^{N_f} 2 \left(\frac{\Delta \varepsilon_{pl}(N)}{\varepsilon'_f} \right)^{-\frac{1}{c}} \tag{4}$$

A sum U of the error squares of the deviations of the damage sums of all cycles from 1 each is formed:

$$U = \sum_{t=1}^{t_{max}} (S(t) - 1)^2 \tag{5}$$

The coefficients of this Coffin–Manson equation are then determined with a linear regression calculation in which the sum U is minimized.

Figure 7e shows that the Coffin–Manson curves obtained from the total strain-controlled tests agree better with the measurements at plastic strain control than those determined from plastic strains at half-life. The authors consider this method useful when predicting Coffin–Manson parameters under plastic strain control from total strain-controlled tests, in particular for materials with severe cyclic hardening or softening.

In the case of the UFG condition, the scatter of the Coffin–Manson curves calculated according to different approaches is much smaller than for the CG condition, as seen from Figure 7d. This is caused by the less pronounced cyclic hardening/softening (compare Figure 3b). The comparatively small differences in the stress amplitudes lead to similar Coffin–Manson plots, which agree well with the reference curve from the plastic strain-controlled tests shown in Figure 7d.

5. Summary

The influence of cyclic plastic strain control on the evolution of α'martensite and, hence, on the fatigue life was studied for a metastable austenitic stainless steel with two different grain sizes: (i) coarse-grained (62 μm) and ultrafine-grained (0.8 μm) material. The main results can be summarized as follows:

- The initial stress amplitude is significantly higher for the ultrafine-grained material condition during both total and plastic strain-controlled fatigue tests.
- In both material conditions, a pronounced secondary hardening is observed, which is attributed to the martensitic phase transformation.
- During plastic strain-controlled tests, the cumulated plastic strain needed for the onset of martensitic phase transformation is significantly lower compared to total strain control.
- The UFG condition shows the higher ability for martensitic phase transformation, which is caused by the pronounced texture.
- Based on strain-controlled tests at constant total strain, an approach for the calculation of Coffin–Manson parameters was proposed for materials undergoing a strong secondary hardening due to martensitic phase transformation. The best agreement of the Coffin–Manson curve with that of the plastic strain-controlled test was obtained when calculations were based on the stress amplitude at $N_{f/3}$.
- A linear damage accumulation approach can be used to estimate Coffin–Manson parameters close to those obtained from plastic strain-controlled tests without an arbitrary comparison point.

Author Contributions: Conceptualization, mechanical testing, data analysis, M.D.; methodology, validation, writing, data analysis, S.H.; supervision, project administration, funding acquisition, H.B.; validation, writing—original draft preparation, A.W. All authors have read and agreed to the published version of the manuscript.

Funding: Financial support for this research from Deutsche Forschungsgemeinschaft (DFG, German Research Foundation) for the Collaborative Research Centre "TRIP-Matrix-Composite" (SFB 799, project B3) under Project-ID 54473466 is gratefully acknowledged. The APC was funded by Publication Fund of the TU Bergakademie Freiberg.

Data Availability Statement: Not applicable.

Acknowledgments: The authors wish to thank R. Prang and K. Becker for preparation of the specimens for SEM investigations, Furthermore, J. Freudenberger and D. Seifert (both IFW Dresden) are gratefully acknowledged for conducting the hot forming and rotary swaging. Open Access Funding by the Publication Fund of the TU Bergakademie Freiberg.

Conflicts of Interest: The authors declare no conflict of interest. The funders had no role in the design of the study; in the collection, analyses, or interpretation of data; in the writing of the manuscript, or in the decision to publish the results.

References

1. Coffin, L.F., Jr. Low Cycle Fatigue—A Review. *Appl. Mater. Res.* **1962**, *1*, 129.
2. Manson, S.S.; Halford, G.R. Inst. Metals Monograph and Rept. Set., No 32, 1967, p. 154, cited in: C.E. Feltner and R.W. Landgraf. *J. Basic Eng.* **1971**, *93*, 444.
3. Feltner, C.E.; Landgraf, R.W. Selecting materials to resist low cycle fatigue. *J. Basic Eng.* **1971**, *93*, 444–452. [CrossRef]
4. Lukas, P.; Klesnil, M.; Polak, J. High cycle fatigue life of metals. *Mater. Sci. Eng.* **1974**, *15*, 239–245. [CrossRef]
5. Cheng, A.S.; Laird, C. The high cycle fatigue life of copper single crystals tested under plastic-strain-controlled conditions. *Mater. Sci. Eng.* **1981**, *51*, 55–60. [CrossRef]
6. Wang, R.; Mughrabi, H. Secondary cyclic hardening in fatigued copper monocrystals and polycrystals. *Mater. Sci. Eng.* **1984**, *63*, 147–163. [CrossRef]
7. Blochwitz, C.; Kahle, E. A method for the determination of the cyclic stress-strain curve of persistent slip bands in fatigued single crystals. *Krist. Tech.* **1980**, *16*, 977–986. [CrossRef]
8. Li, Y.; Laird, C. Cyclic response and dislocation structures of AISI 316L stainless steel. Part 2: Polycrystals fatigued at intermediate strain amplitude. *Mater. Sci. Eng. A* **1994**, *186*, 87–103. [CrossRef]

9. Buque, C.; Bretschneider, J.; Schwab, A.; Holste, C. Effect of grain size and deformation temperature on the dislocation structure in cyclically deformed polycrystalline nickel. *Mater. Sci. Eng. A* **2001**, *319–321*, 631–636. [CrossRef]
10. Videm, M.; Ryum, N. Cyclic deformation and fracture of pure aluminium polycrystals. *Mater. Sci. Eng. A* **1996**, *219*, 11–20. [CrossRef]
11. Polak, J. Plastic strain-controlled short crack growth and fatigue life. *Int. J. Fatigue* **2005**, *27*, 1192–1201. [CrossRef]
12. Smaga, M.; Boemke, A.; Daniel, T.; Skorupski, R.; Sorich, A.; Beck, T. Fatigue Behavior of Metastable Austenitic Stainless Steels in LCF, HCF and VHCF Regimes at Ambient and Elevated Temperatures. *Metals* **2019**, *9*, 704. [CrossRef]
13. Lei, Y.; Xu, J.; Wang, Z. Controllable Martensite Transformation and Strain-Controlled Fatigue Behavior of a Gradient Nanostructured Austenite Stainless Steel. *Nanomaterials* **2021**, *11*, 1870. [CrossRef] [PubMed]
14. Mughrabi, H.; Höppel, H.W.; Kautz, M. Fatigue and microstructure of ultrafine grained metals produced by severe plastic deformation. *Scr. Mater.* **2004**, *51*, 807–812. [CrossRef]
15. Mughrabi, H.; Höppel, H.W. Cyclic deformation and fatigue properties of very fine-grained metals and alloys. *Int. J. Fatigue* **2010**, *32*, 1413–1427. [CrossRef]
16. Kunz, L.; Lukáš, P.; Svoboda, M. Fatigue strength, microstructural stability and strain localization in ultrafine-grained copper. *Mater. Sci. Eng. A* **2006**, *424*, 97–104. [CrossRef]
17. Höppel, H.W.; Zhou, Z.M.; Kautz, M.; Mughrabi, H.; Valiev, R.Z. Cyclic deformation behaviour of ultrafine-grained copper produced by equal channel angular pressing. In *Proceedings of the Fatigue 2002*; Blom, A.F., Ed.; Engineering Advisory Services: West Midlands, UK, 2002; Volume 3, pp. 1617–1622.
18. Agnew, S.R.; Weertman, J.R. Cyclic softening of ultrafine grained copper. *Mater. Sci. Eng. A* **1998**, *244*, 145–152. [CrossRef]
19. Patlan, V.; Vinogradov, A.; Higashi, K.; Kitagawa, K. Overview of fatigue properties of fine grain 5056 Al–Mg alloy processed by equal-channel angular pressing. *Mater. Sci. Eng. A* **2001**, *300*, 171–182. [CrossRef]
20. Holste, C. Cyclic plasticity of nickel, from single crystal to submicrocrystalline polycrystals. *Philos. Mag.* **2004**, *84*, 299–315. [CrossRef]
21. Stolarz, J.; Baffie, N.; Magnin, T. Fatigue short crack behaviour in metastable austenitic stainless steels with different grain sizes. *Mater. Sci. Eng. A* **2001**, *319–321*, 521–526. [CrossRef]
22. Vogt, J.B.; Poulon, A.; Brochet, S.; Glez, J.C.; Mithieux, J.D. LCF behaviour of fine grained austenitic stainless steels. In Proceedings of the 6th International Conference on Low Cycle Fatigue (LCF 6), Berlin, Germany, 8–12 September 2008; pp. 97–102.
23. Chlupová, A.; Man, J.; Kuběna, I.; Polák, J.; Karjalainen, L.P. LCF behaviour of ultrafine grained 301LN stainless steel. *Procedia Eng.* **2014**, *74*, 147–150. [CrossRef]
24. Järvenpää, A.; Jaskari, M.; Man, J.; Karjalainen, L.P. Stability of grain-refined re-versed structures in a 301LN austenitic stainless steel under cyclic loading. *Mater. Sci. Eng. A* **2017**, *703*, 280–292. [CrossRef]
25. Järvenpää, A.; Jaskari, M.; Man, J.; Karjalainen, L.P. Austenite stability in reversion-treated structures of a 301LN steel under tensile loading. *Mater. Charact.* **2017**, *127*, 12–26. [CrossRef]
26. Man, J.; Järvenpää, A.; Jaskari, M.; Kuběna, I.; Fintová, S.; Chlupová, A.; Karjalai-nen, L.P.; Polák, J.; Hénaff, G. Cyclic deformation behaviour and stability of grain-re-fined 301LN austenitic stainless structure. *MATEC Web Conf.* **2018**, *165*, 6005. [CrossRef]
27. Droste, M.; Ullrich, C.; Motylenko, M.; Weidner, A.; Fleischer, M.; Freudenberger, J.; Rafaja, D.; Biermann, H. Fatigue behaviour of an ultrafine-grained metastable CrMnNi-steel tested under total strain control. *Int. J. Fatigue* **2018**, *106*, 143–152. [CrossRef]
28. Järvenpää, A.; Jaskari, M.; Kisko, A.; Karjalainen, P. Processing and properties of re-version-treated austenitic stainless steels. *Metals* **2020**, *10*, 281. [CrossRef]
29. Zhang, Z.; Li, A.; Wang, Y.; Lin, Q.; Chen, X. Low-cycle fatigue behavior and life prediction of fine-grained 316LN austenitic stainless steel. *J. Mater. Res.* **2020**, *35*, 3180–3191. [CrossRef]
30. Droste, M.; Järvenpää, A.; Jaskari, M.; Motylenko, M.; Weidner, A.; Karjalainen, P.; Biermann, H. The role of grain size in static and cyclic deformation behavior of a laser reversion annealed metastable austenitic steel. *Fatigue Fract. Eng. Mater. Struct.* **2021**, *44*, 43–62. [CrossRef]
31. Jahn, A.; Kovalev, A.; Weiß, A.; Scheller, P.R. Influence of manganese and nickel on the α' martensite transformation temperatures of high alloyed Cr-Mn-Ni steels. *Steel Res. Int.* **2011**, *82*, 1108–1112. [CrossRef]
32. Ojima, M.; Adachi, Y.; Tomota, Y.; Katada, Y.; Kaneko, Y.; Kuroda, K.; Saka, H. Weak beam TEM study on stacking fault energy of high nitrogen steels. *Steel Res. Int.* **2009**, *80*, 477–481. [CrossRef]
33. Sommer, C.; Christ, H.-J.; Mughrabi, H. Non-linear elastic behaviour of the roller bearing steel SAE 52100 during cyclic loading. *Acta Metall. Mater.* **1991**, *39*, 1177–1187. [CrossRef]
34. Talonen, J.; Aspegren, P.; Hänninen, H. Comparison of different methods for measuring strain induced α'-martensite content in austenitic steels. *Mater. Sci. Technol.* **2004**, *20*, 1506–1512. [CrossRef]
35. Biermann, H.; Glage, A.; Droste, M. Influence of temperature on fatigue-induced martensitic phase transformation in a metastable CrMnNi-steel. *Metall. Mater. Trans. A* **2016**, *47*, 84–94. [CrossRef]
36. Misra, R.D.K.; Challa, V.S.A.; Venkatsurya, P.K.C.; Shen, Y.F.; Somani, M.C.; Karjalainen, L.P. Interplay between grain structure, deformation mechanisms and austenite stability in phase-reversion-induced nanograined/ultrafine-grained austenitic ferrous alloy. *Acta Mater.* **2015**, *84*, 339–348. [CrossRef]

37. Challa, V.; Wan, X.L.; Somani, M.C.; Karjalainen, L.P.; Misra, R.D.K. Strain hardening behavior of phase reversion-induced nanograined/ultrafine-grained (NG/UFG) austenitic stainless steel and relationship with grain size and deformation mechanism. *Mater. Sci. Eng. A* **2014**, *613*, 60–70. [CrossRef]
38. Challa, V.; Wan, X.L.; Somani, M.C.; Karjalainen, L.P.; Misra, R.D.K. Significance of interplay between austenite stability and deformation mechanisms in governing three-stage work hardening behavior of phase-reversion induced nanograined/ultrafine-grained (NG/UFG) stainless steels with high strength-high ductility combination. *Scr. Mater.* **2014**, *86*, 60–63. [CrossRef]
39. Shen, Y.F.; Jia, N.; Wang, Y.D.; Sun, X.; Zuo, L.; Raabe, D. Suppression of twinning and phase transformation in an ultrafine grained 2 GPa strong metastable austenitic steel: Experiment and simulation. *Acta Mater.* **2015**, *97*, 305–315. [CrossRef]
40. Jin, J.E.; Jung, Y.S.; Lee, Y.K. Effect of grain size on the uniform ductility of a bulk ultrafine-grained alloy. *Mater. Sci. Eng. A* **2007**, *449–451*, 786–789. [CrossRef]
41. Kisko, A.; Hamada, A.S.; Talonen, J.; Porter, D.; Karjalainen, L.P. Effects of reversion and recrystallization on microstructure and mechanical properties of Nb-alloyed low-Ni high-Mn austenitic stainless steels. *Mater. Sci. Eng. A* **2016**, *657*, 359–370. [CrossRef]
42. Yang, P.; Xie, Q.; Meng, L.; Ding, H.; Tang, Z. Dependence of deformation twinning on grain orientation in a high manganese steel. *Scr. Mater.* **2006**, *55*, 629–631. [CrossRef]
43. Ullrich, C.; Eckner, R.; Krüger, L.; Martin, S.; Klemm, V.; Rafaja, D. Interplay of microstructure defects in austenitic steel with medium stacking fault energy. *Mater. Sci. Eng. A* **2016**, *649*, 390–399. [CrossRef]

Article

Very High Cycle Fatigue Behavior of Austenitic Stainless Steels with Different Surface Morphologies

Marek Smaga *, Annika Boemke, Dietmar Eifler and Tilmann Beck

Institute of Materials Science and Engineering, TU Kaiserslautern, 67663 Kaiserslautern, Germany
* Correspondence: smaga@mv.uni-kl.de; Tel.: +49-631-205-2762

Abstract: The fatigue behavior of the two austenitic stainless steels AISI 904L and AISI 347 with different surface morphologies, (i) conventionally turned and finally polished, (ii) cryogenic turned using CO_2 snow, as well as (iii) cryogenic turned and finally polished, was investigated using an ultrasonic fatigue testing system up to the very high cycle fatigue regime using an ultrasonic fatigue testing system. The AISI 904L is stable against deformation-induced phase formation while the AISI 347 is in the metastable state and shows martensite formation induced by cryogenic turning as well as mechanical loading. For the detailed characterization of the surface morphology, confocal microscopy, scanning electron microscopy, and X-ray diffraction methods were used. The specimens from stable austenite failed in the high cycle fatigue and very high cycle fatigue regime. Opposed to this, the metastable austenite achieved true fatigue limits up to load cycle $N = 1 \times 10^9$ and failed only in the high cycle fatigue regime. Furthermore, due to surface modification, an increase of fatigue strength of metastable AISI 347 was observed.

Keywords: austenitic stainless steels; metastability; surface morphology; VHCF

Citation: Smaga, M.; Boemke, A.; Eifler, D.; Beck, T. Very High Cycle Fatigue Behavior of Austenitic Stainless Steels with Different Surface Morphologies. *Metals* **2022**, *12*, 1877. https://doi.org/10.3390/met12111877

Academic Editor: Martin Heilmaier

Received: 30 September 2022
Accepted: 31 October 2022
Published: 3 November 2022

Publisher's Note: MDPI stays neutral with regard to jurisdictional claims in published maps and institutional affiliations.

Copyright: © 2022 by the authors. Licensee MDPI, Basel, Switzerland. This article is an open access article distributed under the terms and conditions of the Creative Commons Attribution (CC BY) license (https://creativecommons.org/licenses/by/4.0/).

1. Introduction

It is well known that austenitic stainless steels (ASSs) can exist in stable or metastable state dependent on their chemical composition [1,2]. Hence, the paramagnetic austenite can transform due to plastic deformation to a more stable microstructure, i.e., paramagnetic ε-martensite and/or ferromagnetic α′-martensite [3]. Because during the discovery of ASSs the passivity (and hence "stainlessness") was the main scope of material development and not the metastability as well as its influence on the mechanical and physical behavior, still extensive experimental work is done at this class of materials. Investigations of ASSs up to the very high cycle fatigue (VHCF) regime show different fatigue behavior depending on their level of metastability. Carstensen et al. [4] performed fatigue tests with tubes of stable AISI 904L and determined specimen failure in the VHCF regime with crack initiation at the surface. Therefore, in stable austenitic stainless steels, no classical fatigue limits up to load cycles $N = 10^7$ exist. Comparable results for AISI 316L were found by Lago et al. [5]. A continuous decrease of fatigue strength was detected up to $N = 1 \times 10^9$. Crack initiation occurred at the surface at all specimens. Subsurface fish-eye fracture without inclusion in austenitic stainless steel SUS 316NG was observed in pre-strained specimen by Takahashi et al. [6] as well as surface crack initiation in the VHCF regime. Grigorescu et al. [7,8] investigated additionally to "quasi-stable" AISI 316L the metastable austenitic AISI 304L. For AISI 304L, a true fatigue limit was detected associated with α′-martensite formation, i.e., specimen failure of AISI 304L only occurs below $N = 1 \times 10^6$ load cycles. For AISI 904L, a clear decrease of fatigue strength in VHCF regime occurs with crack initiation at twin boundaries [7].

Since specimen failure in VHCF regime of ASSs occurred only in the case of the stable austenite and crack initiation occurs mostly on the surface [4,5], surface modification seems to be one possibility to increase the fatigue strength, even in the VHCF regime. The modification of surface morphology improves for stable and metastable ASSs the fatigue strength in the low cycle (LCF) and in the high cycle regime (HCF). Consequently, numbers of cycle higher than $N > 10^7$ can be applied at relatively high stress amplitudes without specimen failure. Hence, the crack initiation and fatigue process of ASSs with modified surface morphologies must be characterized. In the literature, different methods were used for surface modification of austenite, e.g., ultrasonic surface modification [9], cryogenic deep rolling [10], or cryogenic turning [11,12]. All these methods lead to an increase of fatigue strength in LCF and/or HCF regime due to the resulting nanostructured layers and, in the case of the metastable ASSs, ε and/or α´-martensite formation [13,14]. Up to now, only few results of fatigue life in VHCF regime for specimens with modified surface from ASSs exist [15]. Therefore, the research presented in this paper focused on the VHCF behavior of austenitic stainless steels with different surface morphologies. Both type ASS, stable AISI 904L and metastable AISI 347, were investigated. The surface modification was produced by cryogenic turning [11,12]. The fatigue tests were performed with an ultrasonic testing system, developed at the authors' institute [16].

2. Materials and Methods

2.1. Austenitic Stainless Steels

The investigated materials were the stable austenitic steel AISI 904L in solution annealed state (T = 1100 °C, t = 30 min, quenched in H_2O) and the metastable austenitic stainless steel AISI 347, also solution annealed (T = 1050 °C, t = 35 min, quenched in He). Both materials consist of a purely austenitic microstructure in the initial state, have no preferred crystallographic orientation, and contain twins from solution annealing (Figure 1). The grain size including twins is 38 µm for the AISI 904L and 17 µm for the AISI 347. Optical micrographs (Figure 1b,d) show, aside from the grain structure, a band-like structure in the axial direction of the specimen, which indicates a heterogeneity in the chemical distribution [17]. To characterize the metastability of both materials, the $M_{s,Eichelmann}$ temperature [18], $M_{d30,Angel}$ temperature [19], and stacking fault energy (SFE) [20] were calculated according to the chemical composition of the investigated materials (Table 1). The values indicate the stability of AISI 904L and the metastability of AISI 347. Consequently, the metastable AISI 347 is prone to deformation-induced transformation from paramagnetic γ-austenite to paramagnetic ε-martensite as well as to ferromagnetic α´-martensite. Mechanical properties determined in tensile tests at ambient temperature estimated according to the DIN 50125 standard as well as ferromagnetic α´-martensite fractions after specimen failure measured by magnetic Feritscope™ signal (ξ) in FE-% [2] are given in Table 2.

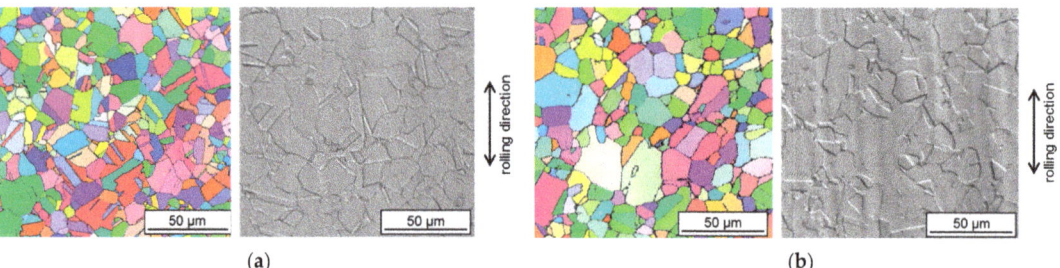

(a) (b)

Figure 1. Initial microstructure of (a) stable AISI 904L and (b) metastable AISI 347. EBSD maps with grain orientation and optical micrographs after V2A etching.

Table 1. Chemical composition of the investigated austenitic stainless steel in wt%.

	C	N	Cr	Ni	Nb	Mo	Cu	Mn
AISI 904L (stable)	0.03	0.06	19.92	24.34	0.03	4.22	1.42	0.95
AISI 347 (metastable)	0.02	0.02	17.19	9.44	0.39	0.23	0.11	1.55

Table 2. Mechanical properties, austenite stability parameters, and α'-martensite fraction (ξ) after specimen failure in tensile tests.

	Young's modulus in GPa	$R_{p0.2}$ in MPa	UTS in MPa	$M_{d30,Angel}$ in °C	$M_{s,Eichelmann}$ in °C	SFE in mJ/m²	ξ in FE-%
AISI 904L (stable)	187	307	631	−220	−1156	54	0
AISI 347 (metastable)	179	225	603	46	-87	26	15

2.2. Experimental Methods

An ultrasonic fatigue testing (USFT) system built at the authors' institute was used for the fatigue tests. The working frequency was ~20 kHz, which was generated by an ultrasonic generator and transformed in a mechanical oscillation by a converter. The specimen was fixed on one hand at a booster which was attached at the converter that amplified the oscillation about factor 2.5. The displacement amplitude at the bottom end of the specimen was measured by a laser vibrometer CLV-2534 from Polytec. The oscillation signal was recorded with a sample rate of 500 kHz. With this experimental setup, the fatigue tests were continuously monitored, and the exact number of cycles could be determined. To perform reliable fatigue tests with the USFT system, an exact pulse shape with short onset and decay times as well as a stable stationary phase must be achieved, a constant displacement stress amplitude must be ensured throughout the complete test, and as the pulse-pause ratio needs to be selected in such a way that excessive specimen heating due to the high-frequency loading is avoided [21]. The temperature in the specimen gauge length was measured by an IR-pyrometer. The temperature change ΔT_{Pulse} is defined as the difference of the measured absolute temperature at the beginning and the end of the pulse. The maximum temperature of the specimen during the tests was limited to 50 °C. The specimens were additionally cooled with compressed air (Figure 2a).

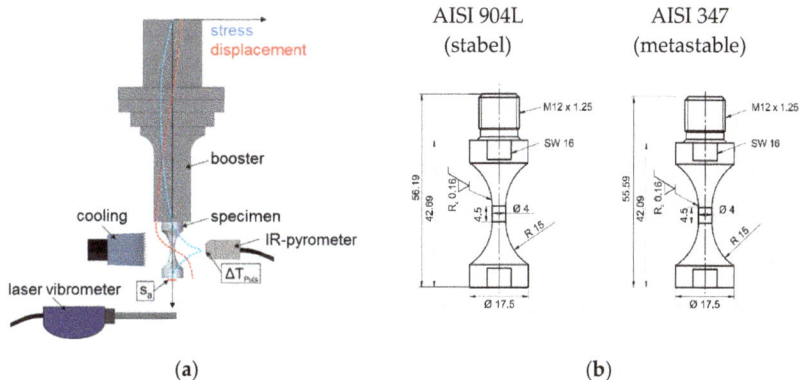

Figure 2. Schematic view of the operating principle of the ultrasonic fatigue testing system with the development of displacement and stress, as well as in situ measurements during the tests (a) and specimen geometry (b).

The geometry of the specimens is given in Figure 2b. The maximum stress amplitude was calculated based on steady state FEM simulations in ABAQUS assuming a linear relationship between the stress amplitude (σ_a) and the displacement amplitude (s_a) at the lower end of the specimen measured by a 1-axis laser vibrometer. A linear correlation factor k between stress amplitude and displacement amplitude of k_{904L} = 23.6 MPa/μm for AISI 904L and k_{347} = 22.7 MPa/μm for AISI 347 was determined. Accordingly, only the stress amplitude is used to describe the fatigue results. Due to the highly transient material behavior of metastable ASSs, the ultrasonic fatigue testing is very challenging. Already a small increase of the α'-martensitic phase leads to a pronounced change of the displacement amplitude as well as the pulse shape. Consequently, to perform VHCF tests on metastable austenite from N = 0 up to the limiting number of cycles, e.g., $N_l = 1 \times 10^9$ at the USFT stysem, a continuous adjustment of the control parameters must be performed, as described in [22].

To characterize the microstructure, metallographic preparation of the sections was performed by grinding up to 1200 grit and electronical polishing. For optical micrographs metallographic sections were etched using V2A solution. The optical micrographs were obtained using a Leica DM 6000 M. Detailed analyses of the crystallographic microstructure were carried out using an SEM/FIB GAIA3 (Tescan s.r.o., Brno, Czech Republic) equipped with an EBSD module "Hikari Plus". The inverse pole figure maps were generated in the normal direction using OIMA software. X-ray diffraction phase analysis was performed using a Bragg Brentano configuration on a PANalytical X'Pert PRO MRD X-ray diffractometer using $CuK_{\alpha 1}$ radiation. The diffractograms were acquired over a range of $2\theta = 40°$ to $100°$ using a step-size of $0.04°$. Phase analyses were performed by using the Rietveld method. To minimize the influence of texture, the diffraction profiles were measured at five different tilt angles in the range $0–40°$ with relative errors in the quantification of the phase contents of about 1–3% for each phase. The same setup was used to perform residual stress measurements on the surface at all considered ablation depths. In this context, all measurements focusing on the (022) γ-austenite plane were measured from $\chi = -45°$ to $+45°$ tilt angle at a step size of $8.2°$. The type I residual stresses determined with the $\sin^2\psi$ method were measured in axial direction. The maximum measurement spot size on the specimen surface used for the phase analyses and the residual stress measurements were, with approx. 1.5 mm × 1.5 mm, significantly smaller than the material areas removed for depth-resolved investigation of the specimen surface layer. These areas were removed by electrolytic ablation using a Struers LectroPol-5 from the specimen surface to a depth of 400 μm between the XRD measurements. The temperature of the cooled electrolyte remained permanently below 20 °C during the ablation process to avoid microstructural changes. Furthermore, an ablation strategy with local removal was applied which, in contrast to a full-surface removal of material layers, has a way smaller influence on the overall residual stresses [23]. The magnetic Feritscope™ measurements were used for detailed characterization of deformation-induced α'-martensite formation. The Feritscope™ magnetic fraction (ξ) is given in volume percent ferrite (FE-%), without converting into α'-martensite content. In literature, a linear correlation between FE-% and α'-martensite content is reported [24].

3. Results

3.1. Surface Morphology

As mentioned above, specimens with different surface morphologies were investigated. Two different manufacturing processes were used for surface modification: conventional turning and cryogenic turning. Specimens from both materials were polished after conventional turning, such that the fatigue behavior could be analyzed in a reference state. These specimens consisting of purely austenitic microstructure in the volume as well as in the near surface area have a smooth surface in the gauge length. The surface morphologies produced by conventional turning and subsequently polishing are called SSL_p (Stable Surface Layer polished) for AISI 904L and ASL_p (Austenitic Surface Layer polished) for AISI 347. The second production process was a cryogenic turning at the Institute for Manufacturing Technology and Production Systems, TU Kaiserslautern, Germany. The specimens were cooled with CO_2 snow using two attached nozzles at the CNC lathe. The chosen cutting speed was v_c = 30 m/min, the depth of cut a_p = 0.2 mm and the two feeds f_1 = 0.15 and f_2 = 0.35 mm/rev (Table 3). More details about the cryogenic turning process were published elsewhere, see e.g., [11,12].

Table 3. Investigated surface morphologies.

AISI 904L	Stable Surface Layer polished	SSL_p
	Stable Surface Layer turned with f = 0.15 mm/rev	SSL_{t015}
	Stable Surface Layer turned with f = 0.35 mm/rev	SSL_{t035}
AISI 347	Austenitic Surface Layer polished	ASL_p
	Martensitic Surface Layer polished	MSL_p
	Martensitic Surface Layer turned with f = 0.15 mm/rev	MSL_{t015}
	Martensitic Surface Layer turned with f = 0.35 mm/rev	MSL_{t035}
	Martensitic Surface Layer of pre-deformed and finial turned with f = 0.35 mm/rev	MSL_{dt035}

To eliminate the relatively high roughness formed by the turning process and to investigate only the influence of the martensitic surface layer on the fatigue behavior, the specimens of the metastable AISI 347 were manufactured by cryogenic turning with lower feed f_1 = 0.15 mm/rev and subsequently polished. This morphology is named MSL_p for Martensitic Surface Layer polished. The mean roughness (R_z) of the surfaces of all investigated morphologies determined with confocal microscopy is given in Figure 3. In general, three levels of the roughness in dependency of the feed rate can be seen (Figure 3a). The polished specimens have a smooth surface with an average roughness value below R_z = 1 µm. The roughness values of the turned specimen is dominated by the grooves resulting from the chosen feed in turning process. Hence, the low feed rate f_1 = 0.15 mm/rev leads to smaller R_z of ~4 µm and higher f_2 = 0.35 mm/rev to higher R_z up to ~11 µm, while the polished surface is standard for fatigue tests, which focus on the determination of fatigue limits of the materials, the turned variants represent typical technical surfaces. Generally, the surface roughness has a large influence of the fatigue properties and typically reduces fatigue life. However, previous results for metastable austenitic steles show that the negative influences of the high surface roughness of fatigue life in the LCF/HCF regime can be compensated by near surface martensite formation during cryogenic turning [11,12], which opens a new perspective to increase the fatigue life of components.

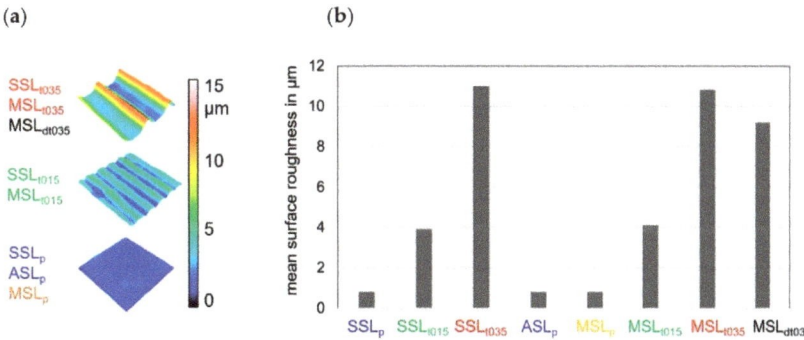

Figure 3. Confocal microscopy images of three representative surface topographies after cryogenic turning with f = 0.35 mm/rev, 0.15 mm/rev, and electrolytical polishing (a) as well as the mean surface roughness R_z for all morphologies (b).

Scanning-ion microscopy (SIM) investigations were performed at FIB cross sections, which were positioned in the gauge length of the VHCF specimens in axial direction in the middle of a turning groove (Figure 4). ASL_p and SSL_p show the microstructure as the initial state (comp. Figure 1), i.e., purely austenitic grains with different grain size. No influence of the production process can be observed. In contrast to that, the cryogenic turned specimens show a nanocrystalline layer (NL) under the surface, with a maximum layer thickness of about 5 µm caused by cryogenic turning. A dependency of the NL thickness on the feed rate cannot be seen. SIM micrographs of the MSL_{dt035} sample indicated aside from the NL a fine microstructure, which is a result of pre-deformation before final cryogenic turning.

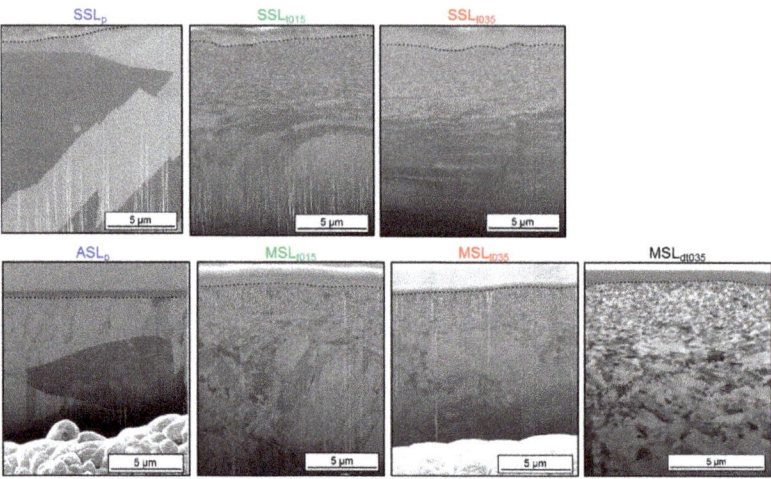

Figure 4. Scanning-ion microscopy images of the near surface microstructure of the investigated morphologies.

Aside from the surface roughness (Figure 3) and near surface microstructure (Figure 4), residual stresses and phase distribution below the specimens' surface influence the fatigue performance. Figure 5 shows the residual stresses (σ_{RS}) and phase distributions of all investigated morphologies. The information for polished specimens (SSL_p, ASL_p, MSL_p) correspond to the state at a distance of about 20–30 µm from the surface. All morphologies show relatively high tensile σ_{RS} at the specimen surface. The maximum σ_{RS} = 750 MPa was measured at MSL_{dt035} and the smallest one on MSL_{t015} with σ_{RS} = 400 MPa. The σ_{RS} reduced to nearly zero at ~50 µm distance from the surface for MSL_{t015} specimens. The residual stresses for MSL_{d035} decreased from σ_{RS} = 750 MPa to 200 MPa at ~100 µm distance from the surface and remained constant. A typical change from tensile to compressive residual stress as described in literature [25] was only observed for MSL_{t035}. Note that, since the specimens with the morphologies ASL_p, SSL_p, SSL_{t015}, and SSL_{t035} consist of a purely austenitic microstructure, the respective phase distributions are not shown in Figure 5. Γ-austenite, α'- and ε-martensite content was measured in the other morphologies of AISI 347 (MSL_{t015}, MSL_{t035}, and MSL_{dt035}). All three morphologies have a maximum content of α'-martensite not directly at the surface but in some distance from the surface: MSL_{t015} at 15 µm with α' = 18 vol.%, MSL_{t035} at 80 µm with α' = 23 vol.% and MSL_{dt035} at 12 µm with 45 vol.%, which corresponds to the fraction caused by pre-deformation. The penetration depth of the X-ray is a few micrometers, so in the first X-ray step, the nanocrystalline structure strongly influences the measurement. However, the strong decrease of α'-martensite in the MSL_{dt035} from 45 vol.% to 23 vol.% indicates a possible back transformation to austenite during the manufacturing process. Only minimal amounts of ε-martensite were found. The MSL_{t015} shows a slightly lower α'-martensite content, which at 98 µm drops to almost 0 vol.%. This coincides with findings in the literature, which show higher α'-martensite proportions with increasing feed in the cryogenic turning [12]. With the disappearance of α'-martensite, small proportions ε-martensite appear of maximum 7 vol.%.

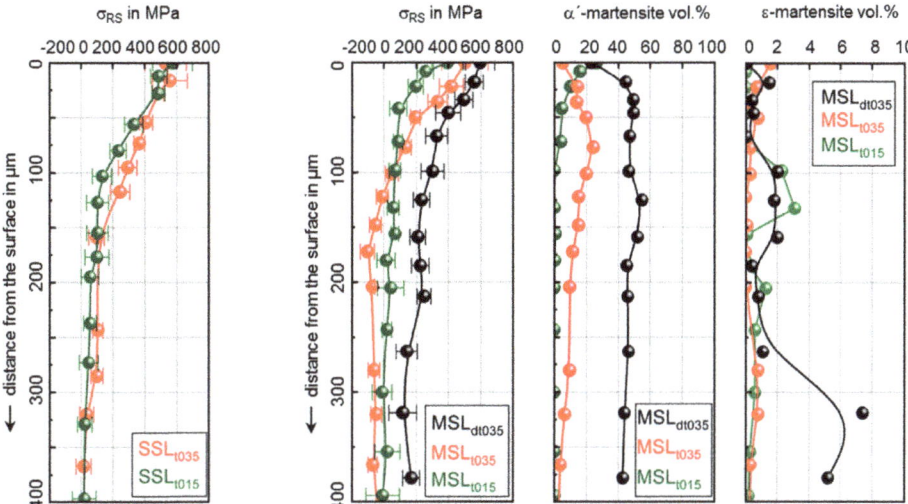

Figure 5. Distribution of residual stresses and phases content until the specimen surface.

3.2. Fatigue Behavior of Stable AISI 904L with Different Surface Morphologies

To study the fatigue behavior of the stable ASS, single step fatigue tests at a load ratio R = −1 were realized at room temperature up to the VHCF range with a maximum number of load cycles of 1×10^9 with the ultrasonic fatigue system with the load frequency of ~20 kHz (see Section 2.2). The S-N diagram of the samples with the three different surface morphologies is shown in Figure 6; run out is marked with an arrow. The specimens with the morphology SSL_p failed in the HCF and the VHCF regime. With decreasing stress amplitude, the number of cycles to failure increased until the ultimate number of cycles is reached. Hence, no classical fatigue limit exits. A similar fatigue behavior was also observed by Carstensen et al. in fatigue tests at thin tubular specimens from AISI 904L [4]. At a stress amplitude of σ_a = 306 MPa, the tested sample with SSL_p reaches the limiting number of cycles without failure. All cryogenically turned specimens exhibit earlier failures than the specimens with SSL_p at the same stress amplitude, where the sample with the morphology SSL_{t015} had the lowest lifetime. Specimen failure occurred only at the specimens' surface due to the high roughness of R_z = 4 µm (SSL_{t015}) and R_z = 11 µm (SSL_{t035}) in HCF regime. The influence of the surface roughness and the high tensile residual stresses seem to be predominant compared to the positive influence of the nanocrystalline structure (Figure 4), which in general reduces the plastic deformation in the specimen [13]. Therefore, no increase of fatigue life was achieved by cryogenic turning of stable ASS. Feritescope™ measurements confirm that no phase transformation occurred during the fatigue tests. Fractographic investigations on the failed specimens with SSL_p show that for $N_f > 1 \times 10^8$, the crack initiation changes from the surface to the volume (Figure 6). Volume defects, such as inclusions, could not be detected. Note that internal cracks can also occur in metallic materials without the presence of inclusions and, e.g., stress concentrations at the grain boundaries can have a crack-initiating effect, especially in the VHCF regime [26,27]. Takahashi and Ogawa also detected on a prestressed sample in compression an internal crack with a fish-eye structure without any internal defect [6]. However, in most cases when examining ASSs up to the VHCF range, surface cracks or cracks originating in the immediate vicinity of the surface are observed [4,5,28]. Bright areas are visible around the start of the crack at a distance of a few 100 µm on which there are extrusions (Figure 6). Probably during micro-crack growth, two halves of the sample do not have contact with each other and in this area, the extrusions developed at the surface fracture.

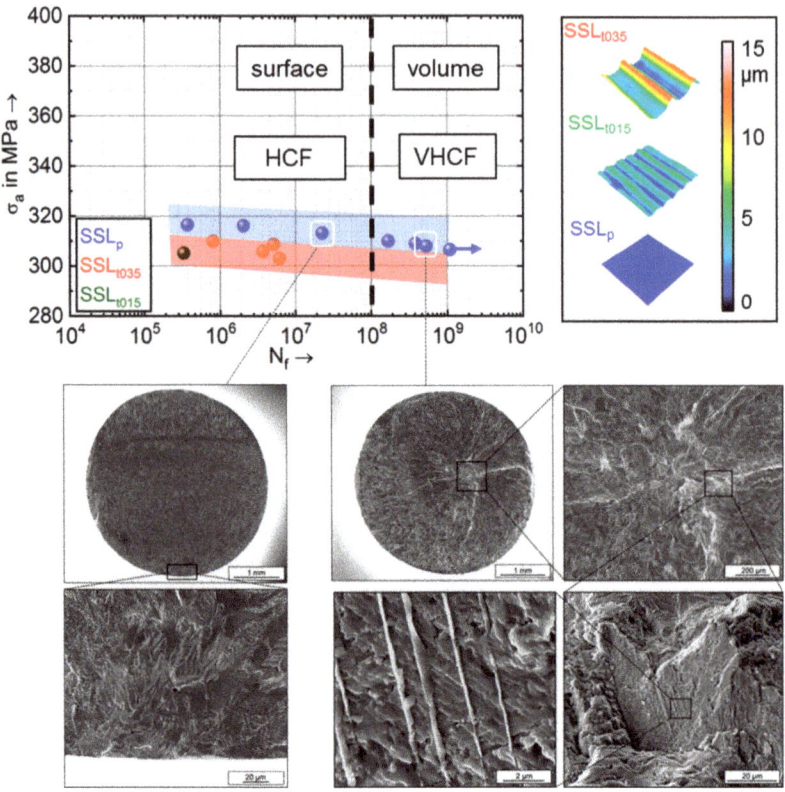

Figure 6. S-N diagram of AISI 904L with different surface morphologies and investigation of the fracture surface, run outs are marked with an arrow.

3.3. Fatigue Behavior of Metastable AISI 347 with Different Surface Morphologies

The influence of metastability on the VHCF behavior is very clearly seen by comparison of the S-N curves of both ASSs (Figures 6 and 7); run outs are marked with an arrow. The specimens from metastable AISI 347 with purely austenitic microstructure in the initial state and with different surface morphology only failed in the HCF regime at $N < 10^7$. In the VHCF regime, no specimen failure occurred, which shows that a true fatigue limit exists for this metastable austenite. Only the pre-deformed specimens failed in the VHCF regime, which is consistent with literature for different metastable austenitic steels [8,29,30]. During cyclic loading, the metastable AISI 347 transforms from paramagnetic γ-austenite to ferromagnetic α'-martensite, which is confirmed by Feritscope™ measurements. Hence, the surface morphology influences the fatigue behavior of metastable AISI 347 completely differently in comparison to the stable AISI 904L. The specimens with morphology MSL_{t035}, which exhibit nanocrystalline and martensitic layers show higher fatigue strength despite of the higher roughness values (R_z = 11 μm) and tensile residual stresses (504 MPa) than the conventionally turned and polished specimens with the ASL_p morphology and resulting small roughness of R_z = 1 μm and compressive residual stresses (−20 MPa). The fatigue strength of specimens increased after cryogenic turning about 50 MPa. The scatter of MSL_{t035} specimens failed in the HCF-regime was higher than that of the variant ASL_p, which is caused by the turning process resulting in irregularities by chip formation and chip breaking. The results show that the martensitic surface layer in combination with the nanocrystalline structure had a very positive effect on the fatigue behavior in VHCF regime, whereas only a nanocrystalline structure did not lead to higher fatigue life as shown for stable AISI 904L (comp. fatigue life SSl_p with SSL_{t035} in Figure 6). This results from the low roughness, the martensitic surface layer, and the compressive residual stresses at the specimens' surface. The fatigue tests with MSL_{dt035} specimens were demanding the maximum displacement amplitude of the ultrasonic using testing system. It was not possible to perform higher stress amplitudes than $\sigma_a = 381$ MPa due to the high internal friction of the austenite in the specimens' volume. Nevertheless, an increase of fatigue strength for the morphology MSL_{dp} was determined. In contrast to the results from specimens without pre-deformation, failure in the VHCF regime occurred in specimens with the morphology MSL_{dt035} and the data points show a drop in the S-N diagram. Such behavior could also be observed in a metastable austenite with a α'-martensite content of 54 vol.% achieved by pre-deformation [30]. Using scanning electron micrographs, the fracture surfaces of specimens with the morphology MSL_{dt035} were examined. Figure 7 shows SEM micrographs of the MSL_{dp} specimen failed at $N = 2.8 \times 10^8$ cycles. The fracture surface contains an AlCaO inclusion, and a niobium carbide a few microns below the surface. The fracture topography around the NbC indicates that crack initiation originated from the carbide. Hence, the deeper located subsurface AlCaO inclusion was fractured by crack growth. In general, the pre-deformation with deformation-induced martensite formation increases the HCF fatigue strength. However, in the VHCF regime, a drop of fatigue strength occurred. Taking into account the fatigue life of the specimen with the morphology MSL_p, it becomes clear that, even compared to the cryogenically turned samples with partially converted volume (morphology MSL_{dt035}), the MSL_p samples have a higher fatigue strength in the whole fatigue regime.

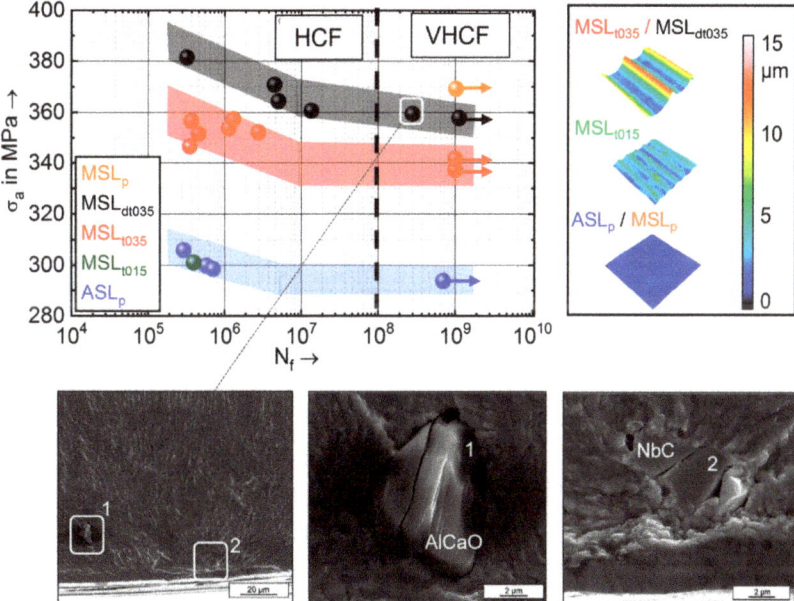

Figure 7. S-N curves of metastable AISI 347 with different surface morphologies and investigation of the surface fracture, run outs are marked with an arrow.

4. Conclusions

The stable austenitic steel AISI 904 and the metastable austenitic steel AISI 347 were investigated with different surface morphologies up to the VHCF regime using an ultrasonic testing system. Microstructural investigations of initial surface morphologies and fracture surface were performed. The study provides the following results:

1. Cryogenic turning of stable AISI 904L (SSL_{t035}) and metastable AISI 347 (MSL_{t035}) results in higher roughness ($R_z \sim 11$ µm) and tensile residual stresses at the specimen's surface. Moreover, a nanocrystalline layer of a few micrometers under the surface was formed in both materials. In the metastable AISI 347, phase transformation from austenite to martensite was observed additionally. Polishing of the cryogenically turned metastable austenitic specimens (MSL_p) leads to very low average roughness ($R_z < 1$ µm) values and compressive residual stresses at the surface.
2. Fatigue tests of conventionally turned and polished specimens with purely austenitic microstructure in the initial state leads to decreasing S-N curve of stable AISI 904L (SSL_p) in the VHCF regime. With increasing number of cycles to failure, a change from surface to subsurface crack initiation was observed. Opposed to AISI 904L, the AISI 347 (ASL_p) specimens only failed in the HCF regime.
3. Fatigue tests at cryogenic turned specimens show shorter lifetime for specimens of stable AISI 904L (SSL_{t015} and SSL_{t035}) and a lifetime extension for specimen of metastable AISI 347 (MSL_{t035}) even in case of significantly higher surface roughness in comparison to conventionally turned and finally polished samples. The increase in fatigue strength of the cryogenic turned AISI 347 is mainly caused by the machining induced martensitic surface layers.
4. In spite of its smaller monotonic strength, the metastable AISI 347 (MSL_{t035}) reaches a higher HCF and VHCF strength after surface modification by cryogenic turning.
5. The increase of material strength due to pre-deformation and martensitic formation in specimen volume of metastable AISI 347 (MSL_{dt035}) was successful; however, in the VHCF regime a drop in S-N curve was observed.
6. The best fatigue properties are achieved with specimens with surfaces polished after cryogenic turning of metastable AISI 347 (MSL_p).

Author Contributions: Conceptualization, M.S., A.B., D.E., and T.B.; methodology, A.B. and M.S.; experiments and data analysis, A.B. and M.S.; writing original draft, M.S.; review and editing, A.B., D.E., and T.B. All authors have read and agreed to the published version of the manuscript.

Funding: This work was supported by the Deutsche Forschungsgemeinschaft (DFG, German Research Foundation)—project number 172116086—SFB 926 "Microscale Morphology of Component Surfaces".

Institutional Review Board Statement: Not applicable.

Informed Consent Statement: Not applicable.

Data Availability Statement: The data presented in this study are available on request from the corresponding author.

Acknowledgments: The authors thank the German Research Foundation (DFG) for the financial support within the CRC 926 "Microscale Morphology of Component Surfaces". The fatigue specimens were turned at the Institute for Manufacturing Technology and Production Systems (FBK), TU Kaiserslautern, Germany. We thank Aurich J.C. and Kirsch for their support.

Conflicts of Interest: The authors declare no conflict of interest.

References

1. Lai, J.K.L.; Shek, C.H.; Lo, K.H. *Stainless Steels: An Introduction and Their Recent Developments*; Bentham Science Publishers, eBook: United Arab Emirates, 2012; ISBN 1608053059.
2. Smaga, M.; Boemke, A.; Daniel, T.; Klein, M.W. Metastability and fatigue behavior of austenitic stainless steels. *MATEC Web Conf.* **2018**, *165*, 4010. [CrossRef]
3. de Bellefon, G.M.; van Duysen, J.C. Tailoring plasticity of austenitic stainless steels for nuclear applications: Review of mechanisms controlling plasticity of austenitic steels below 400 °C. *J. Nucl. Mater.* **2016**, *475*, 168–191. [CrossRef]
4. Carstensen, J.V.; Mayer, H.; Brønsted, P. Very high cycle regime fatigue of thin walled tubes made from austenitic stainless steel. *Fatigue Fract. Eng. Mater. Struct.* **2002**, *25*, 837–844. [CrossRef]
5. Lago, J.; Jambor, M.; Nový, F.; Bokůvka, O.; Trško, L. Giga-cycle Fatigue of AISI 316L After Sensitising of Structure. *Procedia Eng.* **2017**, *192*, 528–532. [CrossRef]

6. Takahashi, K.; Ogawa, T. Evaluation of Giga-cycle Fatigue Properties of Austenitic Stainless Steels Using Ultrasonic Fatigue Test. *J. Solid Mech. Mater. Eng.* **2008**, *2*, 366–373. [CrossRef]
7. Grigorescu, A.; Hilgendorff, P.-M.; Zimmermann, M.; Fritzen, C.-P.; Christ, H.-J. Fatigue behaviour of austenitic stainless steels in the VHCF regime. In *Fatigue of Materials at Very High Numbers of Loading Cycles: Experimental Techniques, Mechanisms, Modeling and Fatigue Life Assessment*; Christ, H.-J., Ed.; Springer Fachmedien Wiesbaden: Wiesbaden, Germany, 2018; pp. 49–71. ISBN 978-3-658-24531-3.
8. Grigorescu, A.C.; Hilgendorff, P.M.; Zimmermann, M.; Fritzen, C.P.; Christ, H.J. Cyclic deformation behavior of austenitic Cr–Ni-steels in the VHCF regime: Part I—Experimental study. *Int. J. Fatigue* **2016**, *93*, 250–260. [CrossRef]
9. Cherif, A.; Pyoun, Y.; Scholtes, B. Effects of Ultrasonic Nanocrystal Surface Modification (UNSM) on Residual Stress State and Fatigue Strength of AISI 304. *J. Mater. Eng. Perform.* **2010**, *19*, 282–286. [CrossRef]
10. Meyer, D. Cryogenic deep rolling—An energy based approach for enhanced cold surface hardening. *CIRP Ann.—Manuf. Technol.* **2012**, *61*, 543–546. [CrossRef]
11. Aurich, J.C.; Mayer, P.; Kirsch, B.; Eifler, D.; Smaga, M.; Skorupski, R. Characterization of deformation induced surface hardening during cryogenic turning of AISI 347. *CIRP Ann.—Manuf. Technol.* **2014**, *63*, 65–68. [CrossRef]
12. Mayer, P.; Kirsch, B.; Müller, C.; Hotz, H.; Müller, R.; Becker, S.; von Harbou, E.; Skorupski, R.; Boemke, A.; Smaga, M.; et al. Deformation induced hardening when cryogenic turning. *CIRP J. Manuf. Sci. Technol.* **2018**, *23*, 6–19. [CrossRef]
13. Smaga, M.; Skorupski, R.; Eifler, D.; Beck, T. Microstructural characterization of cyclic deformation behavior of metastable austenitic stainless steel AISI 347 with different surface morphology. *J. Mater. Res.* **2017**, *32*, 4452–4460. [CrossRef]
14. Smaga, M.; Skorupski, R.; Mayer, P.; Kirsch, B.; Aurich, J.C.; Raid, I.; Seewig, J.; Man, J.; Eifler, D.; Beck, T. Influence of surface morphology on fatigue behavior of metastable austenitic stainless steel AISI 347 at ambient temperature and 300 °C. *Procedia Struct. Integr.* **2017**, *5*, 989–996. [CrossRef]
15. Boemke, A.; Smaga, M.; Beck, T. Influence of surface morphology on the very high cycle fatigue behavior of metastable and stable austenitic Cr-Ni steels. *MATEC Web Conf.* **2018**, *165*, 20008. [CrossRef]
16. Koster, M. Ultraschallermüdung des Radstahls R7 im Very-High-Cycle-Fatigue (VHCF)-Bereich. Ph.D. Thesis, Technische Universität, Kaiserslautern, Germany, 2011.
17. Man, J.; Smaga, M.; Kuběna, I.; Eifler, D.; Polák, J. Effect of metallurgical variables on the austenite stability in fatigued AISI 304 type steels. *Eng. Fract. Mech.* **2017**, *185*, 139–159. [CrossRef]
18. Eichelmann, G.H.; Hull, T.C. The effect of composition on the temperature of spontaneous transformation of austenite to martensite in 18-8 type stainless steel. *Trans. Am. Soc. Met.* **1953**, *45*, 77–104.
19. Angel, T. Formation of Martensite in Austenitic Stainless Steels—Effects of Deformation, Temperature, and Composition. *J. Iron Steel Inst.* **1954**, *177*, 165–174.
20. Martin, S.; Fabrichnaya, O.; Rafaja, D. Prediction of the local deformation mechanisms in metastable austenitic steels from the local concentration of the main alloying elements. *Mater. Lett.* **2015**, *159*, 484–488. [CrossRef]
21. Smaga, M.; Boemke, A.; Daniel, T.; Skorupski, R.; Sorich, A.; Beck, T. Fatigue Behavior of Metastable Austenitic Stainless Steels in LCF, HCF and VHCF Regimes at Ambient and Elevated Temperatures. *Metals* **2019**, *9*, 704. [CrossRef]
22. Daniel, T.; Smaga, M.; Beck, T. Cyclic deformation behavior of metastable austenitic stainless steel AISI 347 in the VHCF regime at ambient temperature and 300 °C. *Int. J. Fatigue* **2022**, *156*, 106632. [CrossRef]
23. Hornbach, D.J.; Prevéy, P.S.; Mason, P.W. X-ray diffraction charcterization of the residual stress and hardness distributions in induction hardened gears. In Proceedings of the First International Conference on Induction Hardened Gears and Critical Components, Indianapolis, IN, USA, 15–17 May 1995; pp. 69–76.
24. Talonen, J.; Aspegren, P.; Hanninen, H. Comparison of different methods for measuring strain induced alpha-martensite content in austenitic steels. *Mater. Sci. Technol.* **2004**, *20*, 1506–1512. [CrossRef]
25. M'Saoubi, R.; Outeiro, J.C.; Changeux, B.; Lebrun, J.L.; Morão Dias, A. Residual stress analysis in orthogonal machining of standard and resulfurized AISI 316L steels. *J. Mater. Process. Technol.* **1999**, *96*, 225–233. [CrossRef]
26. Chai, G.; Zhou, N. Study of crack initiation or damage in very high cycle fatigue using ultrasonic fatigue test and microstructure analysis. *Ultrasonics* **2013**, *53*, 1406–1411. [CrossRef] [PubMed]
27. Schwerdt, D.; Pyttel, B.; Berger, C. Fatigue strength and failure mechanisms of wrought aluminium alloys in the VHCF-region considering material and component relevant influencing factors. *Int. J. Fatigue* **2011**, *33*, 33–41. [CrossRef]
28. Khan, M.K.; Wang, Q.Y. Investigation of crack initiation and propagation behavior of AISI 310 stainless steel up to very high cycle fatigue. *Int. J. Fatigue* **2013**, *54*, 38–46. [CrossRef]
29. Müller-Bollenhagen, C.; Zimmermann, M.; Christ, H.J. Very high cycle fatigue behaviour of austenitic stainless steel and the effect of strain-induced martensite. *Int. J. Fatigue* **2010**, *32*, 936–942. [CrossRef]
30. Müller-Bollenhagen, C.; Zimmermann, M.; Christ, H.J. Adjusting the very high cycle fatigue properties of a metastable austenitic stainless steel by means of the martensite content. *Procedia Eng.* **2010**, *2*, 1663–1672. [CrossRef]

Article

Surface or Internal Fatigue Crack Initiation during VHCF of Tempered Martensitic and Bainitic Steels: Microstructure and Frequency/Strain Rate Dependency

Ulrich Krupp [1] and Alexander Giertler [2,*]

1. IEHK Steel Institute, RWTH Aachen University, Intzestraße 1, 52072 Aachen, Germany
2. Laboratory for Material Design and Structural Integrity, University of Applied Sciences Osnabrück, Albrechtstr. 30, 49076 Osnabrück, Germany
* Correspondence: a.giertler@hs-osnabrueck.de; Tel.: +49-541-969-3215

Abstract: By means of comparing the VHCF response of heat-treated alloy steel, several factors governing the transition from surface (type I) to internal (type II) VHCF failure, and, in the case of internal inclusion and non-inclusion type II VHCF failure, are discussed: differences in strength, differences in grain size and strength gradients. Therefore, the steel grades (i) 50CrMo4 (0.5 wt%C–1.0 wt%Cr–0.2 wt%Mo) in two different tempering conditions (37HRC and 57HRC) but of the same prior austenite grain size, and (ii) 16MnCrV7 7 (0.16 wt%C–1.25 wt%Mn–1.7 wt%Cr) in the bainitic and martensitic thermomechanical treatment state, were studied. It is concluded that steels of moderate strength (37HRC) exhibit a real endurance limit (10^9 cycles), while the fatigue strength of high strength (43–57HRC) or coarse-grained steels (37HRC) decreases with increasing number of load cycles.

Keywords: VHCF; ultrasonic testing; tempered martensitic steel; bainitic steel; frequency effect; self-tempered martensite

1. Introduction

Heat-treated alloy steels, i.e., tempered martensitic or bainitic steels are used in virtually all kinds of engineering applications that involve cyclic deformation under high or very high cycle fatigue conditions (HCF, VHCF), e.g., power trains in transportation, and gearboxes or crankshafts in power generation. In most cases, the VHCF life of heat-treated alloy steels is limited by internal crack initiation manifesting as a characteristic fisheye-like fracture surface. As a typical feature, the crack initiation site is surrounded by a fine-granular area (FGA), also known as an optically dark area (ODA) [1–3], from which the fatigue crack propagates within the fisheye until the residual cross-section fails by final fracture. This is called type II VHCF behavior [4]. Contrary to this, LCF, HCF, and sometimes VHCF conditions lead to surface crack initiation, which can be attributed to (i) plane stress, and (ii) environmentally-induced slip irreversibility, both promoting pronounced stress concentration (type I VHCF behavior [4]). At the very low stress amplitudes in the macroscopically elastic regime prevailing during VHCF, any dislocation plasticity at the surface vanishes due to work hardening (shakedown). In this case, the material may exhibit a real infinite life since any initial fatigue damage is permanently stopped. In the case of internal microstructure inhomogeneities, such as non-metallic inclusions, selected grain-boundary triple points, or favorably oriented slip planes [5,6], strong stress concentrations in the bulk become dominant. Under plane-strain conditions, these high stress levels initiate local plastic slip, which accumulates during millions of cycles, eventually leading to fracture at the inhomogeneity, appearing as a so-called "fisheye", where the fatigue crack propagates within a circular shape around the crack initiation site until rupture of the residual cross-section. Since the local stress concentration is associated with the size of the microstructure inhomogeneity, i.e., expressed by the projected *area* of a non-metallic inclusion normal to the applied load direction, and the strength of the material (that is expressed by means of the Vickers hardness HV), Murakami [7] suggested a simple, but widely-applied equation to estimate the fatigue limit σ_{FL}:

$$\sigma_{FL} = \frac{C(HV + 120)}{(\sqrt{area})^{1/6}} \left(\frac{1-R}{2}\right)^\alpha \quad (1)$$

with R referring to the stress ratio, and α and C being material constants. When having a closer look at the crack initiation site, it has been found many times that the origin is surrounded by an FGA/ODA. Its granular appearance has been attributed to a cyclic rearrangement of dislocations, polygonization, and, eventually, the formation of nano-sized grains [8–10]. VHCF crack propagation at a stress intensity range ΔK lower than the threshold ΔK_{th} may be attributed (i) to the particularly low resistance of the nano-grained FGA structure to fatigue cracking, (ii) accumulation of hydrogen [1,11], or (iii) to the repeated pressing of the crack faces on top of each other leading to crack closure within the FGA [12,13]. Only if the stress intensity range exceeds the threshold value ΔK_{th} for long fatigue cracks, does the FGA turn into a conventional fatigue fracture pattern [9,14], showing, e.g., striations, being characteristic of long fatigue cracks [15].

To date, the correlation of the material microstructure with the VHCF strength remains challenging and often impossible. Data from different research institutes show a pronounced scatter of service life data often in the range of three orders of magnitude. This can be attributed to two major aspects:

- The significance of microstructure inhomogeneity becomes prevailing in VHCF, since the lower the applied stress amplitude, the smaller the number of stress concentration sites that lead to accumulated cyclic plasticity.
- The frequency effect: testing at higher frequencies often leads to an apparent increase in the VHCF strength.

According to Jeddi and Palin-Luc [16], one can distinguish between (i) intrinsic effects, such as strain rate sensitivity of dislocation plasticity, adiabatic local heating and recrystallization, or dynamic strain ageing; and (ii) extrinsic effects, such as oxide-induced crack closure or atmospheric slip irreversibility. In addition, sample size effects have to be considered, i.e., the larger the stressed volume, the lower the fatigue strength. As an example, Figure 1 shows the pronounced frequency effect of plain 0.15% carbon steel [17].

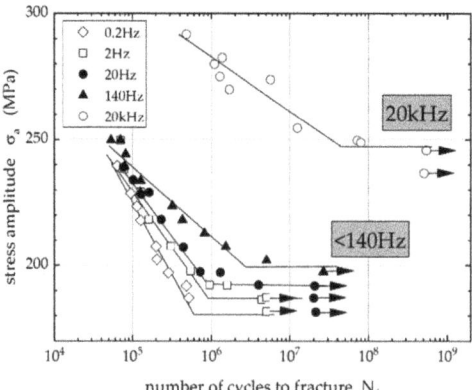

Figure 1. Wöhler S-N Diagram showing the fatigue life for C15 (0.15% carbon) steel at various testing frequencies [17].

In addition, Geilen et al. [18] suggested an influence of the control mode. Although resonance and servohydraulic HCF/VHCF testing is usually stress-controlled, ultrasonic testing is displacement-controlled, i.e., the displacement is adjusted according to linear elastic behavior. However, when cyclic plasticity needs to be taken into account, the adjusted displacement is below the actual value, hence leading to longer fatigue lives and apparently higher VHCF strength values, which can be interpreted as a frequency effect. However, in the case of the steel grades and strength values that are the subject of the present study, the macroscopic stress level should be completely within the elastic regime.

The present paper highlights microstructure effects on the existence or non-existence of a fatigue limit and the occurrence or non-occurrence of an FGA. The frequency effect is attributed to the strain rate dependency of dislocation plasticity that varies with the tempering state and the respective strength of the steel.

2. Materials Processing and VHCF Testing Approach

Two different grades of low-alloy steels 50CrMo4 (1.7228, German designation) and 16MnCrV7 7 (1.8195, German designation), respectively, were used for this study. The chemical composition of the alloys is given in Table 1.

Table 1. Chemical composition of the two steels A and B used in this study (in wt%).

Material	C	Cr	Mo	Mn	Ni	V + Nb + Ti	Fe
50CrMo4	0.48	1.00	0.18	0.71	-	-	bal.
16MnCrV7 7	0.16	1.7	-	1.7	0.16	0.17	bal.

A martensitic microstructure was adjusted for 50CrMo4. For this purpose, round bar sections of 50CrMo4 were austenitized at a temperature of 860 °C and then quenched in oil. By suitably modifying the tempering temperature, two hardness conditions of 37HRC at a tempering temperature of 550 °C and 57HRC at 200 °C were achieved. The tempering duration did not alter the prior austenite grain size of 12 µm and was kept for 90 min each. Figure 2 shows the microstructures of the two 50CrMo4 tempering conditions, 37HRC and 57HRC, respectively.

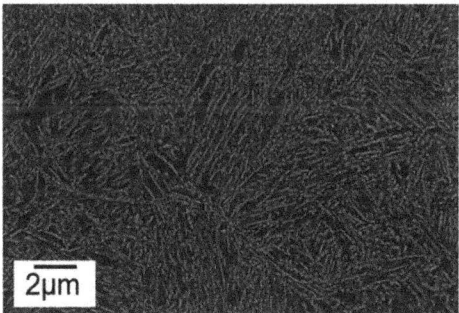
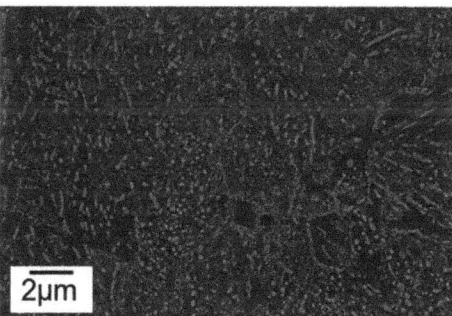

(a)   (b)

Figure 2. SEM micrographs of the tempered microstructure of 50CrMo4 austenitized at 860 °C, quenched in oil, and tempered at (**a**) 200 °C leading to 57HRC, and (**b**) 550 °C leading to 37HRC.

In the case of 16MnCrV7 7, cylindrical samples were forward hot-extruded at 1250 °C and 950 °C, respectively, within a 1000 kN screw press (Weingarten) and subsequently quenched according to defined conditions: (i) pressurized-air quenching to 260 °C followed by slow cooling within the furnace throughout the lower-bainite transformation regime (dashed line in the TTT diagram, Figure 3), and (ii) water-quenching below martensite start (M_s) followed by ambient air cooling (self-tempered martensite, dotted line in the TTT diagram, Figure 3). Due to the different forging temperatures, the prior austenite grain size and the non-equilibrium ferrite lath size are larger in the lower bainite structure as compared to the self-tempered martensite structure, as shown in the EBSD inserts in Figure 3.

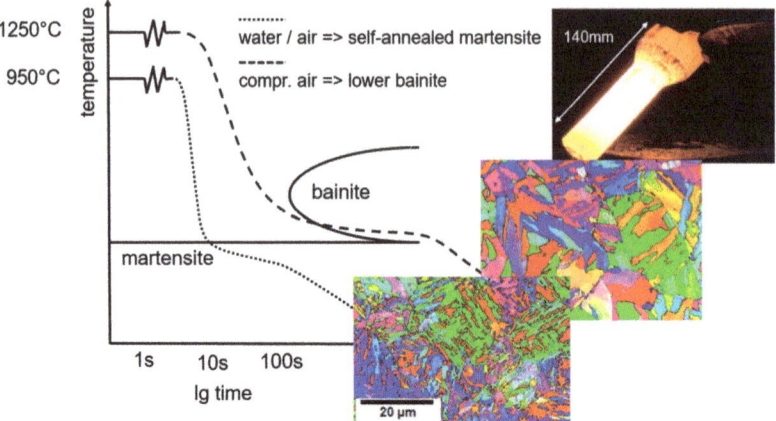

Figure 3. Thermomechanical processing by means of two different cooling routes in a schematic TTT diagram for the steel 16MnCrV7 7 applied to forward hot-extruded cylindrical bars (diameter 50 mm).

The thermomechanical heat treatment outlined above led to a lath-type microstructure with cementite platelets of about 200 nm length, tilted by 60° towards the laths axes in the case of the lower bainitic structure (prior austenite grain size 53 µm, cf. Figure 4a), and with nano-sized cementite precipitates of about 50–100 nm size in the case of the self-tempered martensite (prior austenite grain size 8 µm, cf. Figure 4b). It is worth mentioning that the self-tempered martensite shows not only high static stress values but also a surprisingly high toughness of A_v = 104 J at room temperature (Charpy test data, Table 2).

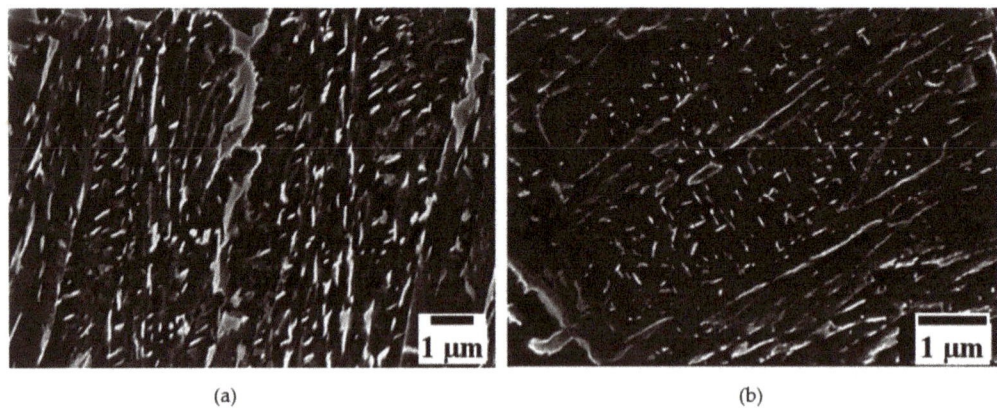

Figure 4. SEM micrographs of thermomechanically-processed steel 16 CrMnV7 7: (**a**) lower bainite with cementite platelets, and (**b**) self-tempered martensitic microstructure with homogeneously distributed small cementite precipitates.

Table 2. Mechanical properties of the steel grades used in the present study.

Material	$R_{p0.2}$ [MPa]	UTS [MPa]	A [%]	A_v [J]
50CrMo4—37HRC	992	1095	-	-
50CrMo4—57HRC	1561	2128	-	-
16MnCrV7 7 lower Bain.—37HRC	885 ± 57	1197 ± 34	57 ± 3.0	17 ± 2
16MnCrV7 7 selftemp.—43HRC	1000 ± 31	1370 ± 34	63 ± 0.7	104 ± 11

Fatigue specimens were machined from the heat-treated sections according to the geometries provided in Figure 5. The gauge lengths of the specimens were electrolytically polished for 3 min using Struers A2 electrolyte at a voltage of 23 V and a temperature of −15 °C. For in situ damage monitoring by light optical microscopy (Hirox long-distance digital microscope, Limonest, France, Figure 5a), high-resolution thermography, and SEM, the specimens were given a shallow notch with a notch factor of 1.1 (details and results are reported in [18,19]). The shallow notch allows the restriction of surface fatigue damage to a limited area without changing the microstructure-related crack initiation process. Fatigue testing was carried out in a resonance testing machine (Rumul Testronic, Neuhausen, Switzerland, 95 Hz, Figure 5a) and an ultrasonic fatigue testing system (Boku Vienna, 20 kHz, Figure 5b) at fully reversed loading conditions ($R = -1$). The temperature increase in the specimens during high-frequency testing was limited to 40 °C by pressurized air-cooling and applying the pulse–pause operation mode of the ultrasonic fatigue testing system, i.e., 50 ms pulse and 950 ms pause in the moderate strength conditions (37HRC) and 200 ms pulse and 800 ms pause in the higher strength conditions (43HRC–57HRC).

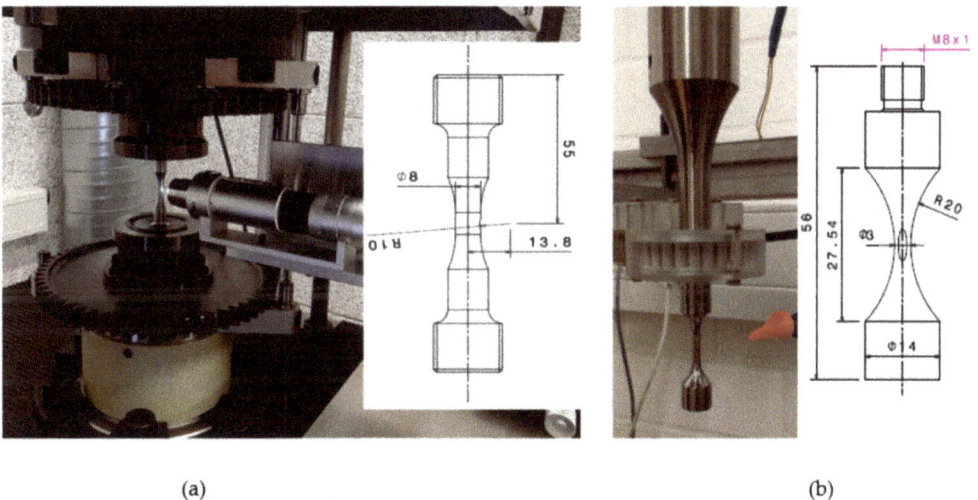

(a) (b)

Figure 5. Specimen geometries (all dimensions in millimeters) for (a) the resonance fatigue testing system RUMUL Testronic (95 Hz), and (b) the ultrasonic fatigue testing equipment of the type BOKU Vienna (20 kHz).

In addition to the quasi-static tensile tests, high strain rate tests were carried out using a 15 kN RoellAmsler servohydraulic high-rate testing system with a specimen geometry, as shown in Figure 6.

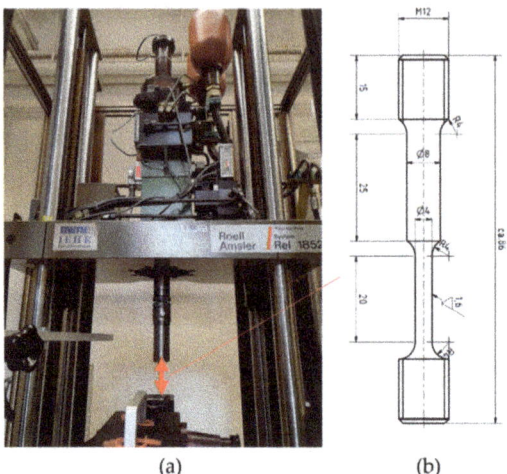

Figure 6. (a) Servohydraulic high-speed testing system (15 m/s) and (b) respective sample geometry (all dimensions in millimeters). The larger diameter is used for optoelectronic strain measurement type Zimmer 200XH.

3. Results

The different microstructure states and strength levels of the steels resulted in a variety of different damage mechanisms, which are briefly introduced in the following. The 50CrMo4 steel in the moderate strength condition (50CrMo4—37HRC) shows type I fatigue behavior, i.e., during cycling, persistent slip markings (PSM) are formed at the surface within Cr-depleted segregation bands, leading to crack initiation. Crack propagation was found only at stress amplitudes higher than 680 MPa. In these cases, failure occurred generally in the HCF regime ($N_f < 10^7$ cycles). At lower stress amplitudes, PSMs and cracks were identified (Figure 7), but the propagation of these was eventually blocked by microstructural barriers, i.e., martensite packet boundaries and prior austenite grain boundaries, respectively. Therefore, it was concluded that in a moderate strength condition, tempered martensitic steels may exhibit a true fatigue limit.

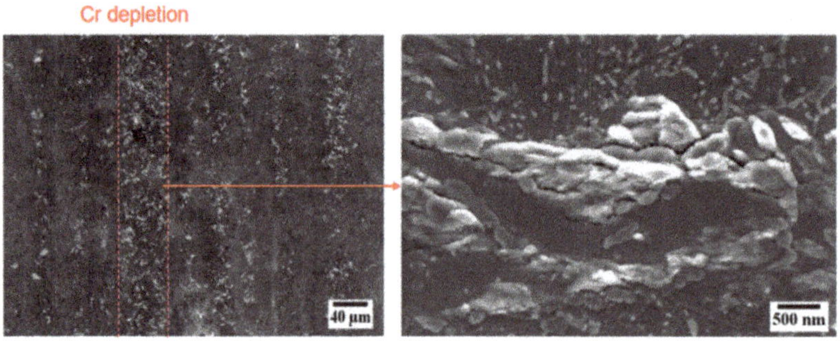

Figure 7. (a) Localized fatigue damage in Cr depleted zones, (b) persistent slip markings (PSM) at the surface of the 50CrMo4—37HRC steel, tested at 95 Hz, σ_a = 480 MPa.

In the high-strength condition steel, 50CrMo4—57HRC, the fatigue damage behavior is completely different. Internal VHCF crack initiation at 10–20 μm-sized nonmetallic $Al_2O_3 \cdot CaO$-type inclusions forming the center of the fisheye (red circle) prevails (Figure 8a).

As outlined in the introduction, the crack initiation site is surrounded by a fine granular area (FGA), as shown in Figure 8b.

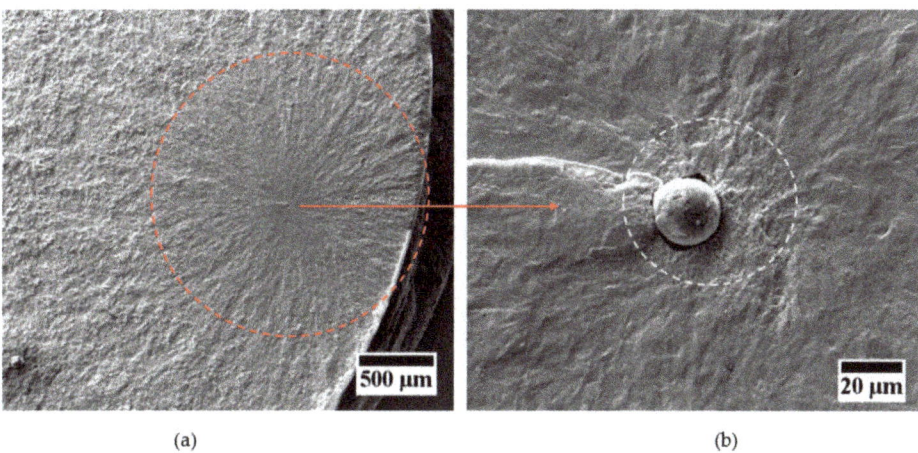

Figure 8. Fracture surface of the 50CrMo4—57HRC steel with a fisheye ((**a**), red circle) and crack initiation at a non-metallic inclusion (type Al_2O_3-CaO) surrounded by FGA ((**b**), white circle), tested at 95 Hz, σ_a = 480 MPa.

The Wöhler S/N data were determined using both 95 Hz and 20 kHz fatigue testing at $R = -1$ for the four different steel grades, as shown in Figures 9 and 10. Two major observations should be highlighted:

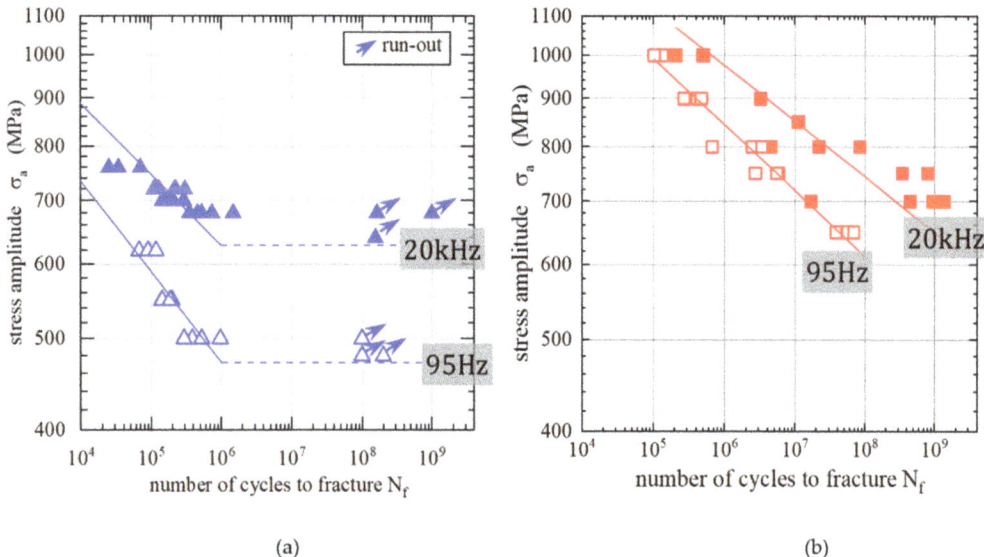

Figure 9. Wöhler S-N Diagram showing the fatigue life for the steel 50CrMo4 at two testing frequencies of f = 95 Hz and f = 20 kHz for (**a**) the 37HRC hardness condition, and (**b**) the 57HRC hardness condition.

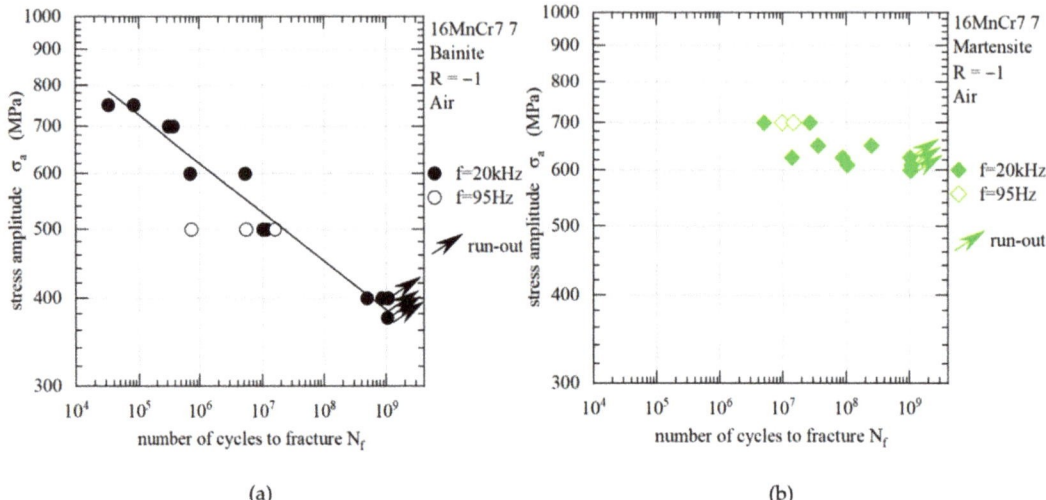

Figure 10. Wöhler S-N Diagram showing the fatigue life for the 16 MnCrV7 7 steel at two testing frequencies of f = 95 Hz and f = 20 kHz for (**a**) the bainitic, and (**b**) the self-tempered martensitic condition.

The moderate-strength 50CrMo4—37HRC grade seems to reveal a true fatigue limit. No fractures were observed beyond 10^7 cycles. SEM studies supported by FIB (focused ion beam sections) through PSMs at the surface proved the occurrence of fatigue cracks (cf. Figure 7b and [20]) as a result of the classical extrusion–intrusion scheme (cf. [15,21]). In the case of the high-strength 50CrMo4—57HRC grade, a continuous decrease in the fatigue strength with increasing numbers of cycles to fracture was observed. Fracture was found even at 10^9 cycles, and in all cases, the fatigue damage mechanism was internal crack initiation along with FGA formation. The size of the FGA was found to increase with decreasing stress amplitude, supporting the hypothesis of a crack propagation mechanism that is driven by accumulated dislocation plasticity and nanograin formation at stress intensity ranges lower than the threshold value ΔK_{th} from linear elastic fracture mechanics (LEFM).

As a second observation, a pronounced frequency shift of the fatigue stress data only in the case of the moderate-strength 50CrMo4—37HRC grade was observed. The high strength 50CrMo4—57HRC grade reveals only slightly higher fatigue strength values for 20 kHz testing as compared with the 95 Hz testing. Although the strong effect is attributed to a strain rate dependency of dislocation plasticity (see discussion), the smaller effect is probably due to a size effect, since the stressed volume of the specimens tested at 20 kHz (V_{20kHz} = 125 mm^3,) is substantially smaller than the ones for the 95 Hz resonance testing (V_{95Hz} = 402 mm^3).

The fatigue strength data for the thermomechanically-processed 16MnCrV7 7 specimens are shown in Figure 10. Analogous to the tensile test data, the fatigue strength of the tempered martensite processing state is higher than that of the lower bainite processing state. It is worth mentioning that the self-tempered martensite outperforms the lower bainite. The fatigue limit is estimated as σ_{FL} = 600 MPa, which is surprisingly high and corresponds to 44% of the quasi-static UTS.

In the lower bainite state, a continuous decrease in the fatigue strength was observed with fractures occurring even at 10^9 cycles at the rather low stress amplitude of σ_a = 400 MPa (34% UTS). A pronounced frequency effect was not observed, but the amount of 95 Hz data is not sufficient to draw a sound conclusion on that aspect. In thermomechanically-processed 16MnCrV7 7 samples, VHCF crack initiation was found to occur exclusively internally at $N_f > 10^7$ (Figure 11). Although in the self-tempered martensite microstructure 10–20 μm-sized non-metallic inclusions act as relevant stress concentrators (Figure 11a), internal crack initiation in the lower bainite microstructure does not come with any inclusions (Figure 11b). This was attributed to the rather large prior austenite grain size of 53 µm, which may cause high stress concentrations at grain boundary triple points, being higher than those at the non-metallic inclusions, which should, instead, be identical to the tempered martensite condition.

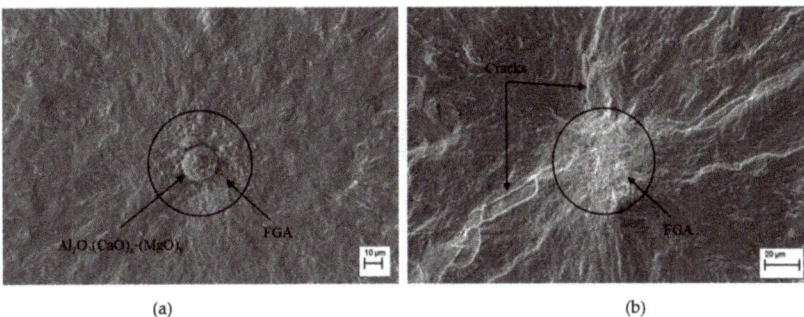

Figure 11. Fatigue crack initiation sites in 16MnCrV7 7: (**a**) internal crack initiation at non-metallic inclusion (tempered martensite, 43HRC, σ_a = 625 MPa, N = 0.9 × 10^8), and (**b**) internal crack initiation at a grain boundary triple point (presumed) (37HRC, σ_a = 400 MPa, N = 8.7 × 10^8).

4. Discussion

The two steel grades 50CrMo4 and 16MnCrV7 7 in two different heat treatment conditions each showed a strong relationship between the microstructure, static strength and the VHCF behavior. The moderate strength condition of 50CrMo4—37HRC revealed only surface fatigue damage manifesting itself in persistent slip markings—typically parallel to the martensite block boundaries. Generally, crack initiation was identified at those slip markings, but at sufficiently low stress amplitudes (see Figure 12) these cracks were considered as non-propagating., i.e., the conditions of a true fatigue limit seemed to be fulfilled.

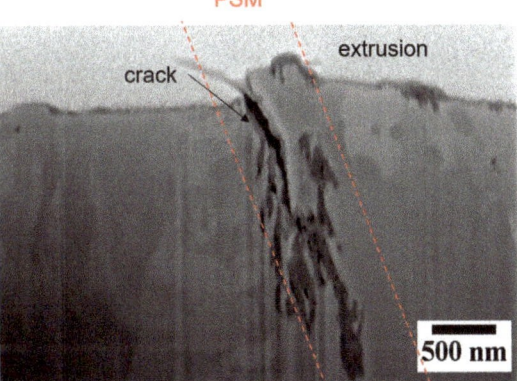

Figure 12. FIB section through a persistent slip marking (PSM) at the surface of the 50CrMo4—37HRC steel, tested at 95 Hz, σ_a = 480 MPa, run out after 2 × 10^8 cycles, showing crack initiation and void formation.

In the case of the high-strength 50CrMo4—57HRC grade and the two 16MnCrV7 7 microstructure variants, internal crack initiation at inhomogeneities was observed. In many cases, initiation was attributed to Al_2O_3-CaO-type non-metallic inclusions of a size of approximately 10–20 µm that cause stress concentrations due to elastic mismatch. The inclusions themselves were found to be surrounded by fine granular areas (FGA). The size of the FGA (measured as the projected area normal to the loading axis) was observed to increase with decreasing applied stress amplitude.

Evaluating the different sizes of the FGAs of many VHCF samples revealed that the stress intensity range associated with the FGA ΔK_{FGA}, as calculated according to Murakami (internal circular crack):

$$\Delta K_{FGA,incl.} = 0.5\Delta\sigma\sqrt{\pi\sqrt{area}}, \qquad (2)$$

is of a constant value, i.e., $\Delta K_{FGA} \approx 8.5$ MPa m$^{0.5}$ which is close to the value measured by a fracture mechanics threshold analysis, where a threshold stress intensity of $\Delta K_{th} = 9.5$ MPa m$^{0.5}$ was obtained (cf. [19,20]). Accordingly, it was concluded that the stress intensity calculated for the inclusion is too small for technical crack initiation, $\Delta K_{incl.} < \Delta K_{th}$. Due to accumulated microplasticity, dislocation patterning and polygonization, a nanograin structure is formed around the inclusion. As the stress intensity threshold for fine-grained microstructures is smaller than for coarse-grained microstructures, a fatigue crack is initiated at the inclusion, since now $\Delta K_{incl.} > \Delta K_{th,nanograin}$. This kind of crack propagation within the nanograin microstructure forms the FGA and prevails until, at the crack tip, $\Delta K = \Delta K_{FGA} > \Delta K_{th}$ is valid. Then, the crack propagation mechanism changes towards the long fatigue crack regime within the fish-eye until at K_{Ic} the transition to unstable crack propagation (final rupture) is reached.

Figure 13 shows FIB sections through FGAs emanating from an inclusion and a microstructure inhomogeneity (probably a grain-boundary triple point). In particular, Figure 13a shows the rim of nanograins that form the FGA at an Al$_2$O$_3$-type inclusion.

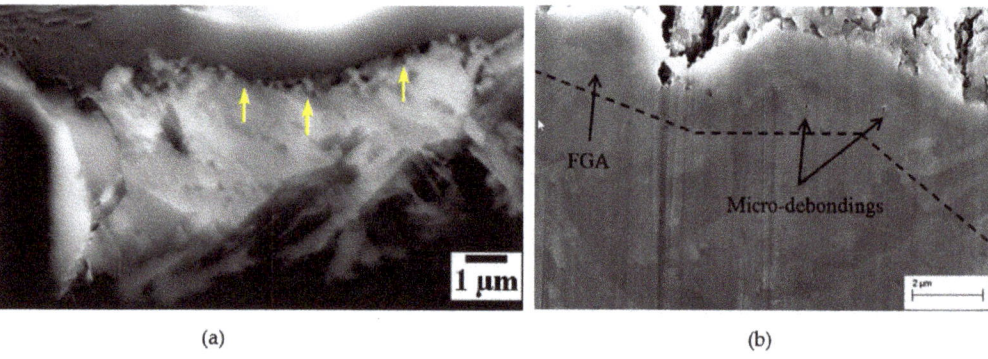

Figure 13. FIB sections through the internal crack initiation sites in (**a**) 50CrMo4—57HRC: nanograins within the FGA (yellow arrows) at an Al$_2$O$_3$-type inclusion (57HRC, $\sigma_a = 750$MPa, N = 3.5 × 10^8), and (**b**) 16MnCrV7 7 lower bainite: an FGA without any non-metallic inclusion ($\sigma_a = 400$ MPa, N = 8.7 × 10^8).

The results revealed that VHCF strength data are not independent of the testing frequency, which has been observed by others (cf. [22,23]). Although it is well-known that effects of the environment, e.g., corrosion fatigue or oxide-induced crack closure, lead to a time-dependent fatigue behavior [24], the use of ultrasonic testing frequency implies substantially high strain rates:

$$\dot{\varepsilon} = 2\pi f \cdot \frac{\sigma_a}{E} \cdot \cos(2\pi f \cdot t). \qquad (3)$$

This, in turn, is necessary to determine the VHCF material properties in a reasonable time. Accordingly, at 20 kHz, a maximum strain rate of $\dot{\varepsilon} = 500$ s^{-1}, and at 95 Hz, a maximum strain rate of $\dot{\varepsilon} = 2$ s^{-1} is reached. For bcc materials such as carbon steel, a temperature/strain rate dependence of dislocation plasticity is known. According to Seeger [25], the critical shear stress of a bcc material can be subdivided into an athermal contribution τ_0 (dislocation shearing/passing) and a thermally-activated contribution τ^* (Peierls stress):

$$\tau_c = \tau_0 + \tau^*(\dot{\varepsilon}, T) \tag{4}$$

Figure 14a,b shows the two contributions. In the case of moderate strength (Figure 14a), τ_0 is rather low and at high strain rates, the increase in τ^* leads to an apparent increase in the yield stress. In the high-strength condition (Figure 14b), the high-strain rate effect vanishes due to the high τ_0 contribution. This hypothesis (cf. [26]) is supported by dynamic tensile testing at strain rates between $\dot{\varepsilon} = 1\ \text{s}^{-1}$ and $\dot{\varepsilon} = 500\ \text{s}^{-1}$. The results (Figure 14c) show for the 50CrMo4—57HRC high-strength condition a negligible strain rate dependence, while for the 50CrMo4—37HRC moderate-strength condition, the strain rate dependence leads to an approximately 30% increase in yield stress.

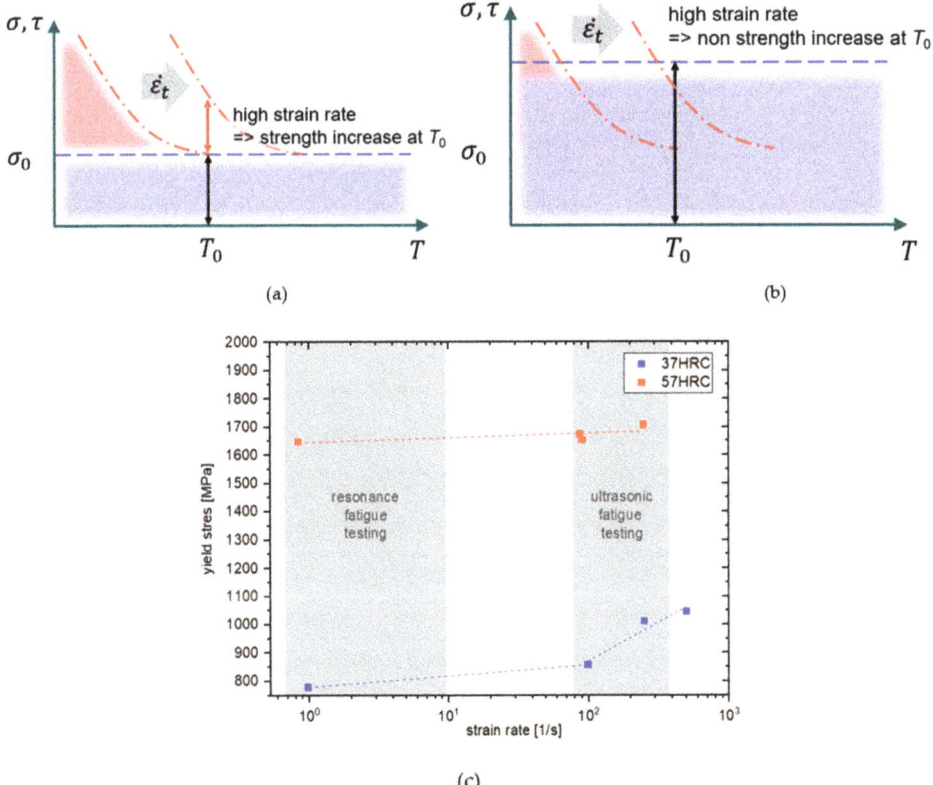

Figure 14. Schematic sketch of the temperature dependence of the critical shear/yield stress: red lines: the thermally-activated contribution (increases with increasing strain rate), blue lines: the athermal contribution in (**a**) moderate-strength alloys, and (**b**) high-strength alloys; and (**c**) yield stress vs. strain rate for the two 50CrMo4 steel grades.

The given fatigue life data and high-strain static stress values show a positive correlation between fatigue test frequency and static strain rate and provide evidence that the slip behavior of the dislocations is generally influenced. The reason for this is assumed to be a restriction of movement of the dislocations. Due to the lattice structure, bcc materials have a very high Peierls potential. This results in a significant expenditure of energy to overcome a maximum of the Peierls potential in order to slide into the next minimum. This leads to an essential property at low temperatures or high strain rates: the mobility of edge dislocations is much higher than the mobility of screw dislocations. Because of this restriction of motion, only segments of long screw dislocations are usually found under high-strain cyclic plastic deformation and no cell or ladder structures, as is otherwise common under fatigue loading [27–29]. A key aspect in identifying the influence of the test frequency is the consideration of the material condition, e.g., heat treatment. A higher dislocation density leads to an increase in τ_0 according to Figure 14b. However, this also includes a consideration of the distribution of the alloying elements, e.g., carbon, which, in turn, has a significant influence on the strain rate effect itself [30]. Therefore, it becomes clear that knowledge about the relationship between the material structure and the movement of dislocations is essential for a clear explanation of the phenomenon of the frequency influence.

5. Conclusions

The VHCF damage of tempered martensitic and bainitic steels depends strongly on the microstructure and the strength level that are adjusted by heat treatment:

- At moderate strength level (martensitic 50CrMo4, 37HRC), irreversible surface plasticity prevails and leads to the formation of persistent slip markings (PSM). Depending on the local combination of stress concentration and microstructural barrier strength, the steel may exhibit a true fatigue limit, below which the fatigue life can be considered as infinite.
- At high strength levels (martensitic 50CrMo4, 57HRC, martensitic 16MnCrV 7 7), the stress concentration at internal non-metallic inclusions leads to an accumulation of dislocation plasticity, patterning, grain refinement, and crack initiation in combination with the formation of a fine granular area (FGA). The lower the remote stress amplitude, the larger the FGA, the size of which correlates with the stress intensity threshold ΔK_{th} for long fatigue cracks.
- In the case of large inhomogeneities, such as the coarse prior austenite grains in the bainitic 16MnCrV 7 7, pronounced stress concentrations at triple points lead to internal crack initiation, even at lower strength levels. Here, FGAs without non-metallic inclusion were observed.

When applying ultrasonic frequencies for VHCF testing, strain rate effects need to be taken into consideration, which can be evaluated by dynamic tensile testing. In the case of low- and moderate-strength steels, high strain rates ($\dot{\varepsilon} = 500$ s^{-1} in ultrasonic testing) cause an apparently higher VHCF strength ($\Delta \sigma = 50$–200MPa) as compared to low strain rates ($\dot{\varepsilon} < 5$ s^{-1} in resonance/servohydraulic testing).

Author Contributions: Conceptualization, U.K. and A.G.; methodology, U.K. and A.G.; software, A.G.; validation, U.K. and A.G.; formal analysis, A.G.; investigation, A.G.; resources, U.K.; data curation, A.G.; writing—original draft preparation, U.K.; writing—review and editing, U.K. and A.G.; visualization, U.K. and A.G.; supervision, U.K.; project administration, U.K.; funding acquisition, U.K. All authors have read and agreed to the published version of the manuscript.

Funding: This research was funded by the Volkswagen foundation in the framework of the research priority program OPTIHEAT grant number VWZN2841 and the German Ministry of Education and Research (BMBF) grant number 03FH011IX4.

Data Availability Statement: The study did not report any data.

Acknowledgments: The German Ministry of Education and Research (BMBF), and the supply of steel specimen material by Georgsmarienhütte Steel and Bosch GmbH are gratefully acknowledged.

Conflicts of Interest: The authors declare no conflict of interest.

References

1. Murakami, Y.; Nomoto, T.; Ueda, T. On the mechanism of fatigue failure in the superlong life regime (N > 107 cycles). Part 1: Influence of hydrogen trapped by inclusions. *Fatigue Fract. Eng. Mater. Struct.* **2000**, *23*, 893–902. [CrossRef]
2. Murakami, Y.; Nomoto, T.; Ueda, T. On the mechanism of fatigue failure in the superlong life regime (N > 107 cycles). Part II: Influence of hydrogen trapped by inclusions. *Fatigue Fract. Eng. Mater. Struct.* **2000**, *23*, 903–910. [CrossRef]
3. Murakami, Y.; Yokoyama, N.N.; Nagata, J. Mechanism of fatigue failure in ultralong life regime. *Fatigue Fract. Eng. Mater. Struct.* **2002**, *25*, 735–746. [CrossRef]
4. Mughrabi, H. Specific Features and Mechanisms of Fatigue in the Ultrahigh-Cycle Regime. *Int. J. Fatigue* **2006**, *28*, 1501. [CrossRef]
5. Chai, G.C. The formation of subsurface non-defect fatigue crack origins. *Int. J. Fatigue* **2006**, *28*, 1533–1539. [CrossRef]
6. Chai, G.C. On Fatigue Crack Initiation in the Matrix in Very High Cycle Fatigue Regime. *Mater. Sci. Forum* **2014**, *783–786*, 2266–2271. [CrossRef]
7. Murakami, Y.; Endo, M. Effects of defects, inclusions and inhomogeneities on fatigue strength. *Int. J. Fatigue* **1994**, *16*, 163–182. [CrossRef]
8. Sakai, T. Review and prospects for current studies on very high cycle fatigue of metallic materials for machine structural use. *JMMP* **2009**, *3*, 425–439. [CrossRef]
9. Grad, P.; Reuscher, B.; Brodyanski, A.; Kopnarski, M.; Kerscher, E. Mechanism of fatigue crack initiation and propagation in the very high cycle fatigue regime of high-strength steels. *Scr. Mater.* **2012**, *67*, 838–841. [CrossRef]
10. Grabulov, A.; Ziese, U.; Zandbergen, H.W. TEM/SEM investigation of microstructural changes within the white etching area under rolling contact fatigue and 3-D crack reconstruction by focused ion beam. *Scr. Mater.* **2007**, *57*, 635–638. [CrossRef]
11. Murakami, Y.; Kanezaki, T.; Sofronis, P. Hydrogen embrittlement of high strength steels: Determination of the threshold stress intensity for small cracks nucleating at nonmetallic inclusions. *Eng. Fract. Mech.* **2013**, *97*, 227–243. [CrossRef]
12. Hong, Y.; Sun, C. The Nature and the Mechanism of Crack Initiation and Early Growth for Very High Cycle Fatigue of Metallic Materials—An Overview. *Theor. Appl. Fract. Mech.* **2017**, *92*, 331. [CrossRef]
13. Hong, Y.; Lei, Z.; Sun, C.; Zhao, A. Propensities of crack interior initiation and early growth for very- high-cycle fatigue of high strength steels. *Int. J. Fatigue* **2014**, *58*, 144–151. [CrossRef]
14. Krupp, U.; Giertler, A.; Koschella, K. Microscopic damage evolution during very-high-cycle fatigue (VHCF) of tempered martensitic steel. *Fatigue Fract. Eng. Mater. Struct.* **2017**, *40*, 1731. [CrossRef]
15. Suresh, S. *Fatigue of Materials*; Cambridge University Press: Cambridge, MA, USA, 1998.
16. Dalenda, J.; Thierry, P.-L. A review about the effects of structural and operational factors on the gigacycle fatigue of steels. *Fatigue Fract. Eng. Mater. Struct.* **2018**, *41*, 969–990.
17. Guennec, B.; Ueno, A.; Sakai, T.; Takanashi, M.; Itabashi, Y. Effect of the loading frequency on fatigue properties of JIS S15C low carbon steel and some discussions based on micro-plasticity behavior. *Int. J. Fatigue* **2014**, *66*, 29–38. [CrossRef]
18. Geilen, M.B.; Schönherr, J.A.; Klein, M.; Leininger, D.S.; Giertler, A.; Krupp, U.; Oechsner, M. On the Influence of Control Type and Strain Rate on the Lifetime of 50CrMo4. *Metals* **2020**, *10*, 1458. [CrossRef]
19. Giertler, A. Mechanismen der Rissentstehung und -ausbreitung im Vergütungsstahl 50CrMo4 bei sehr hohen Lastspielzahlen. Ph.D. Thesis, RWTH Aachen University, Aachen, Germany, 2020.
20. Giertler, A.; Krupp, U. Investigation of Fatigue Damage of Tempered Martensitic Steel during High Cycle Fatigue and Very High Cycle Fatigue Loading Using In Situ Monitoring by Scanning Electron Microscope and High-Resolution Thermography. *Steel Res. Int.* **2021**, *92*. [CrossRef]
21. Man, J.; Petrenec, M.; Obrtlík, K.; Polák, J. AFM and TEM study of cyclic slip localization in fatigued ferritic X10CrAl24 stainless steel. *Acta Mater.* **2004**, *52*, 5551–5561. [CrossRef]
22. Kikukawa, M.; Ohji, K.; Ogura, K. Push-Pull Fatigue Strength of Mild Steel at Very High Frequencies of Stress Up to 100 kc/s. *J. Basic Eng.* **1965**, *87*, 857–864. [CrossRef]
23. Torabian, N.; Favier, V.; Dirrenberger, J.; Adamski, F.; Ziaei-Rad, S.; Ranc, N. Correlation of the high and very high cycle fatigue response of ferrite based steels with strain rate-temperature conditions. *Acta Mater.* **2017**, *134*, 40–52. [CrossRef]
24. Ebara, R. Corrosion fatigue crack initiation in 12% chromium stainless steel. *Mater. Sci. Eng. A* **2007**, *468–470*, 109–113. [CrossRef]
25. Seeger, A. The temperature dependence of the critical shear stress and of work-hardening of metal crystals. *Lond. Edinb. Dublin Philos. Mag. J. Sci.* **1954**, *45*, 771–773. [CrossRef]
26. Bach, J.; Möller, J.J.; Göken, M.; Bitzek, E.; Höppel, H.W. On the transition from plastic deformation to crack initiation in the high- and very high-cycle fatigue regimes in plain carbon steels. *Int. J. Fatigue* **2016**, *93*, 281–291. [CrossRef]
27. Guennec, B.; Ueno, A.; Sakai, T.; Takanashi, M.; Itabashi, Y.; Ota, M. Dislocation-based interpretation on the effect of the loading frequency on the fatigue properties of JIS S15C low carbon steel. *Int. J. Fatigue* **2015**, *70*, 328–341. [CrossRef]
28. Mughrabi, H.; Herz, K.; Stark, X. The effect of strain-rate on the cyclic deformation properties of α-iron single crystals. *Acta Metall.* **1976**, *24*, 659–668. [CrossRef]
29. Stainier, L. A micromechanical model of hardening, rate sensitivity and thermal softening in BCC single crystals. *J. Mech. Phys. Solids* **2002**, *50*, 1511–1545. [CrossRef]
30. Caillard, D. An in situ study of hardening and softening of iron by carbon interstitials. *Acta Mater.* **2011**, *59*, 4974–4989. [CrossRef]

Article

Characterization of the Isothermal and Thermomechanical Fatigue Behavior of a Duplex Steel Considering the Alloy Microstructure

Steven Schellert [1], Julian Müller [2], Arne Ohrndorf [1], Bronslava Gorr [3,*], Benjamin Butz [2] and Hans-Jürgen Christ [1]

1. Institut für Werkstofftechnik, Universität Siegen, Paul-Bonatz-Str. 9-11, 57076 Siegen, Germany; steven.schellert@uni-siegen.de (S.S.); arne.ohrndorf@uni-siegen.de (A.O.); hans-juergen.christ@uni-siegen.de (H.-J.C.)
2. Lehrstuhl für Mikro- und Nanoanalytik und -Tomographie, Universität Siegen, Paul-Bonatz-Str. 9-11, 57076 Siegen, Germany; julian.mueller@uni-siegen.de (J.M.); benjamin.butz@uni-siegen.de (B.B.)
3. Institut für Angewandte Materialien (IAM), Karlsruher Institut für Technologie (KIT), Hermann-von-Helmholtz-Platz 1, 76344 Eggenstein-Leopoldshafen, Germany
* Correspondence: bronislava.gorr@kit.edu; Tel.: +49-721-608-23720

Abstract: Isothermal and thermomechanical fatigue behavior of duplex stainless steel (DSS) X2CrNiMoN22-5-3 was investigated. The aim of this work was to understand the fatigue behavior by correlation of the isothermal and thermomechanical fatigue behavior with microstructural observations. Fatigue tests at plastic-strain-amplitude of 0.2% were carried out at 20, 300 and 600 °C, while in-phase (IP) and out-of-phase (OP) thermomechanical fatigue (TMF) experiments were performed between 300 and 600 °C. During the 20 °C fatigue test, a continuous softening was observed. Transmission electron microscopy examinations reveal pronounced planar slip behavior in austenite. At 300 °C, deformation concentrates in the ferrite, where strong interactions between Cr_xN and dislocations were observed that explain the pronounced cyclic hardening. DSS studied exhibits softening throughout the whole isothermal fatigue test at 600 °C. In ferrite, during the 600 °C fatigue test, the G phase, γ' austenite precipitated, and an unordered dislocation arrangement was observed. The stress responses of the TMF tests can be correlated to those of the isothermal fatigue tests. In IP mode, a positive mean stress resulted in premature failure. No γ' austenite but the formation of subgrains in the ferrite phase was observed after TMF tests. The plastic deformation of the austenite at high temperatures results in an unordered dislocation arrangement.

Keywords: duplex steel; isothermal fatigue; thermomechanical fatigue; ferrite; austenite; G phase; plastic-strain control

1. Introduction

The duplex stainless steel (DSS) X2CrNiMoN22-5-3 (SAF 2205, German designation 1.4662) represents a two-phase material primarily consisting of a ferritic (α) and an austenitic (γ) phase. Generally, the ferrite phase ensures remarkably high mechanical strength, while the austenite phase provides sufficient ductility. DSS possesses high chemical resistance owing to the high chromium content in both phases. Due to its attractive properties, DSS offers a broad range of applications. DSS is widely used in, e.g., shipbuilding, the food industry and the chemical industry. Furthermore, DSS is also used in high temperature applications such as heat exchangers in power plants. If DSS is exposed to high temperatures, the material properties deteriorate dramatically. In particular, it was found that the load-bearing capacity of the ferrite phase is continually reduced with increasing temperature [1]. Other phenomena such as dynamic strain ageing (DSA) have been investigated for both ferritic steels (temperature range: 200–400 °C) [2–4] and austenitic steels (temperature range: 400–600 °C) [4,5] but also for DSS in the temperature range between 250 and 500 °C [4,6]. Further, detrimental microstructural changes take place at elevated temperatures. Especially in the temperature range 350–550 °C, duplex steels suffer a spinodal decomposition of the ferrite into a chromium-rich α' and an iron-rich α phase [7–11]. This spinodal decomposition results in embrittlement, which is most pronounced at 475 °C. The 475 °C embrittlement leads to a deterioration of the mechanical properties, such as an increase in tensile strength, an increase in hardness and a decrease in fracture toughness and fracture strain [12–14]. The effect of the so-called 475 °C embrittlement has already been studied in detail [7–11]. Above this temperature, precipitation of intermetallic phases is reported in the literature [15]. As a consequence of the spinodal decomposition of the ferrite, in the temperature range of 350 to 500 °C, an intermetallic Mo- and Si-rich G phase with a crystal structure $F\,4/m\,\overline{3}\,2/m$ and with a lattice parameter four times larger than that of the ferrite matrix becomes stable [16,17]. Moreover, the Cr- and Mo-rich intermetallic tetragonal sigma phase can occur between 550 and 1050 °C [18,19]. Other precipitation processes, such as the formation of carbides ($M_{23}C_6$), nitrides (Cr_2N) or γ' secondary austenite, are reported to occur at higher temperatures [15,16]. Mateo et al. [16] concluded that spinodal decomposition, formation of Cr_2N as well as dislocations act as further lattice disturbing elements accelerating the nucleation of new phases due to preferential diffusion paths.

When DDS is utilized, e.g., in heat exchangers as pipes or boilers, high temperatures of up to 600 °C can occur in combination with mechanical loads due to start-up and shut-down processes. The so-called thermomechanical fatigue (TMF) can lead to premature component failure. The behavior of DSS 1.4662 under low cycle fatigue (LCF) and TMF has been studied at temperatures between room temperature and 600 °C [4,7]. In the ferrite steel AISI 430F [3], which resembles the ferrite phase of DSS with respect to the chemical composition, and also in other steels [20], DSA was found to be significantly dependent on the strain rate [4]. The effect of the 475 °C embrittlement and the corresponding spinodal decomposition on the isothermal fatigue properties was extensively studied [7,8,21]. Several investigations deal with the thermomechanical fatigue behavior of DSS 1.4662 in the temperature range between 350 and 600 °C with a mechanical strain amplitude of 0.4% or 0.8% and a test frequency of 0.008 Hz [1,22]. It is reported that cyclic heating above 550 °C during the TMF test counteracts the negative effects of spinodal decomposition; hence, no negative influence on the fatigue behavior is observed [1]. In addition, it was demonstrated that the austenite phase possesses sufficient strength even at high temperatures to compensate for the low strength of the ferrite phase [1,22]. However, there is no link between TMF and the corresponding isothermal fatigue tests under the same conditions. Further, the relevant microstructural aspects/evolution and the phase formations such as plasticity (i.e., dislocation formation, evolution, arrangement, interaction) under isothermal and TMF conditions were not discussed.

The aim of the present work is to characterize TMF behavior of a DSS 1.4662 in the temperature range between 300 and 600 °C in plastic-strain control and to link the mechanical behavior to the microstructural properties. In order to understand the damage behavior as a result of TMF, isothermal fatigue tests at 20, 300 and 600 °C were additionally performed and serve as basis for a sound interpretation of TMF particularities.

2. Materials and Methods

The duplex stainless steel X2CrNiMoN22-5-3 (1.4662) was produced by Santo Tomas de las Ollas (León, Spain). The material was delivered in rods with a diameter of 20 mm in hot-rolled condition. The chemical composition of the material in the as-delivered condition was verified by spark spectroscopy analysis (Table 1). The measured chemical composition corresponds well to the SAF 2205® AISI 318 LN specifications.

Table 1. Chemical composition of the DSS X2CrNiMoN22-5-3 determined by spark spectroscopy analyses (wt.%) and the nominal composition according to SAF 2205® AISI 318 LN.

Composition	Chemical Composition (wt.%)									
	C	Si	Mn	P	S	Cr	Ni	Mo	N	Fe
nominal	≤0.03	≤1.0	≤2.0	≤0.035	≤0.015	21–23	4.5–6.5	2.5–3.5	0.1–0.22	Bal.
measured	0.029 ± 0.01	0.403 ± 0.08	1.68 ± 0.03	0.013 ± 0.001	0.0035 ± 0.0002	21.88 ± 0.31	4.65 ± 0.06	3.34 ± 0.02	0.163 ± 0.08	Bal.

For direct comparison with the literature, the material was annealed at 1250 °C in a muffle furnace for 4 h prior to testing as proposed in References [7,8,21,23]. This heat treatment was followed by continuous furnace cooling for 3 h down to 1050 °C and final water quenching. The cylindrical fatigue specimens with a gauge length of 15 mm and a diameter of 5 mm were machined from the rod material. Before fatigue testing, all specimens were mechanically polished to P4000 and finally electrochemically polished (90 vol.% perchloric acid, 10 vol.% acetic acid, at 20 V/res. current 0.1 A). A servo-hydraulic testing machine of type MTS 810 (MTS Systems Corporation, Eden Prairie, MN, USA) (maximum load 100 kN) equipped with hydraulic clamping jaws and a controller of type MTS TestStar IIs (MTS Systems Corporation, Eden Prairie, MN, USA) was used. A high frequency induction heating system Huettinger TIG 5/300 (TRUMPF Hüttinger GmbH + Co. KG, Baden-Württemberg, Freiburg im Breisgau, Germany) with an Eurotherm temperature controller of type 2704 (EURO-THERM GmbH, Worthing, UK) and a self-made induction coil was applied for heating. The strain was measured with a vacuum-resistant high-temperature rod extensometer of type MTS 632.51F-74 (MTS Systems Corporation, Eden Prairie, MN, USA). All experiments were performed at a test frequency of 0.0059 Hz, which results from the attainable cooling rate of the sample from 600 to 300 °C during air cooling. Thermal and mechanical cycles had a triangular wave shape. Prior to isothermal fatigue testing, the specimens were heated up to the testing temperature at zero load under load control and the strain setpoint was adjusted to a value of zero. To establish true plastic-strain control, the elastic strain was subtracted from the mechanical strain signal by adjusting the stiffness $1/EA$ (E means the Young's modulus, and A denotes the cross-sectional area of the sample) during a preliminary load-controlled test in the pure elastic region. All subsequent tests were performed at a plastic-strain amplitude of 0.2%. For the isothermal and thermomechanical fatigue tests, one sample was tested under respective conditions. In TMF tests, the thermal expansion and the temperature dependence of the Young's modulus were additionally considered in the plastic-strain calculation using results of preliminary tests and a 3rd degree polynomial function. Details on the method of controlling thermomechanical fatigue tests in plastic-strain control can be found elsewhere [24]. In-phase (IP) TMF is defined as temperature and mechanical cycling being in phase, where out-of-phase (OP) TMF is characterized by a phase difference of 180°. The fatigue tests were stopped at a maximum number of 2555 cycles (in the saturation regime), which corresponds to a test duration of five days.

A Lext OLS4000 optical microscope (Olympus Corporation, Tokyo, Shinjuku, Japan) (OM) as well as a focused ion beam-scanning electron microscope (FIB-SEM) DualBeam system of type FEI Helios Nanolab 600 (FEI Company, Hillsboro, OR, USA), equipped with backscatter electron (BSE) imaging and electron backscatter diffraction (EBSD), were utilized to investigate the coarse ferrite/austenite microstructure. The average grain sizes were automatically evaluated based on the EBSD measurements using the TEAM™ V4.5.1 software (Version 4.5.1, Ametek, Berwyn, PA, USA). Thin foils for transmission electron microscopy (TEM) were prepared from the fatigue specimens parallel to the stress direction, i.e., the normal foil is perpendicular to the stress direction. The platelets were extracted by diamond wire sawing and mechanically ground to a thickness of about 100 µm using SiC paper. For the final thinning down to electron transparency, the foils were jet-polished with a Struers TenuPol-5 (Struers GmbH, Nordrhein-Westfalen, Willlich, Germany) at 287 K at a voltage of 20 V until perforation. For this purpose, the same electrolyte was applied as used for the electrochemical surface polishing of the fatigue specimens. TEM investigation was conducted with a Thermo Fisher FEI Talos F200X (FEI Company, Hillsboro, OR, USA) operated at 200 kV acceleration voltage. For scanning TEM (STEM) the high-angle annular dark-field (HAADF) detector (FEI Company, Hillsboro, OR, USA) was applied. To enhance chemical contrast, a small camera length of 77 mm was chosen (called HAADF-STEM). In order to clearly depict dislocations, an intermediate camera length of 205 mm was selected (ADF-STEM) to allow Bragg reflection to hit the detector. To identify the crystal structure of the individual phases, selected area diffraction (SAD) patterns were recorded. The chemical composition of the phases was measured using energy dispersive X-ray spectroscopy (STEM-EDX). The quantification was performed with the standard analysis procedure of the Velox software V3.1.0 (Version 3.1.0, FEI Company, Hillsboro, OR, USA).

3. Results and Discussion

3.1. Initial Microstructure

Figure 1 displays the microstructure of the DSS after initial heat treatment. In the optical micrograph in Figure 1a the austenitic grains appear dark gray, whereas the ferrite phase is brighter. The rolling direction of the hot-rolled material is reflected in the vertical chain-like arrangement of the austenitic grains. Figure 1b depicts a EBSD mapping of the DSS. The red color represents the ferrite, while the green color represents the austenite phases. Based on the EBSD analysis, a phase fraction ratio of about 50% austenite and 50% ferrite was determined. The average grain sizes of the ferrite and the austenitic grains yield were 60 and 20 µm, respectively (see Figure A1). Figure 1c provides a BSE-SEM micrograph with the corresponding EDX mapping of the elements Ni, Cr and Mo. It reveals that the ferritic phase is rich in the elements Cr and Mo, while the austenitic phase is enriched in the austenitic stabilizing elements Ni. The EDX analysis of the other elements (Mn, Fe, Si) yielded no visible differences between the phases in the EDX mapping; thus, the visualization was omitted.

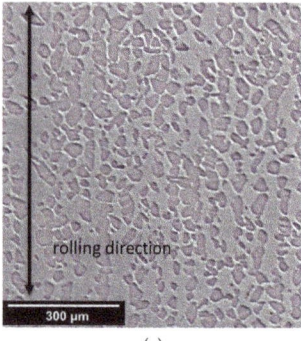
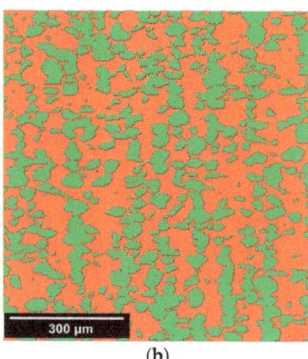
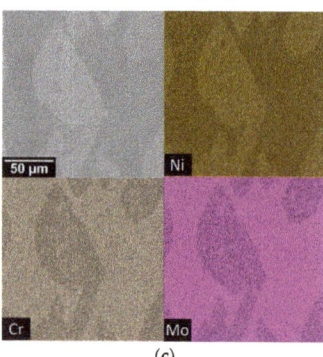

(a) (b) (c)

Figure 1. Microstructure of the hot-rolled DSS X2CrNiMoN22-5-3 after heat treatment. (**a**) Optical microscope image (bright gray: ferrite, dark gray: austenite), (**b**) EBSD phase map (red: ferrite austenite, green: austenite), (**c**) BSE-SEM image and the corresponding EDX mapping of the elements Ni, Cr and Mo. (bright gray: austenite; dark gray: ferrite).

A detailed image of the precipitates within a ferritic grain is depicted in Figure 2a Precipitates of CrN appearing as fine needles in cloud-like clusters and Cr_2N as larger needle-shaped precipitates were identified by chemical composition using STEM-EDX mappings. The results are in good agreement with nitrides reported in References [25–27]. A closer look at the ferritic grains reveals the beginning decomposition into Fe-rich α ferrite and Cr-rich α' ferrite (Figure 2b). By tilting from the zone axis, even dots splitting can be recognized in the diffraction pattern. This is clearly illustrated by the inset in Figure 2c. The decomposed structure in the ferrite phase (Figure 2b) has been proven both experimentally and by phase-field simulation as a key factor of phase separation kinetics in

Fe-Cr-based alloys [28]. Obviously, the spinodal decomposition of the ferrite phase could not be prevented by the applied heat treatment. In contrast to ferrite, the austenitic grains exhibit no peculiarities and are single phase (results not shown here).

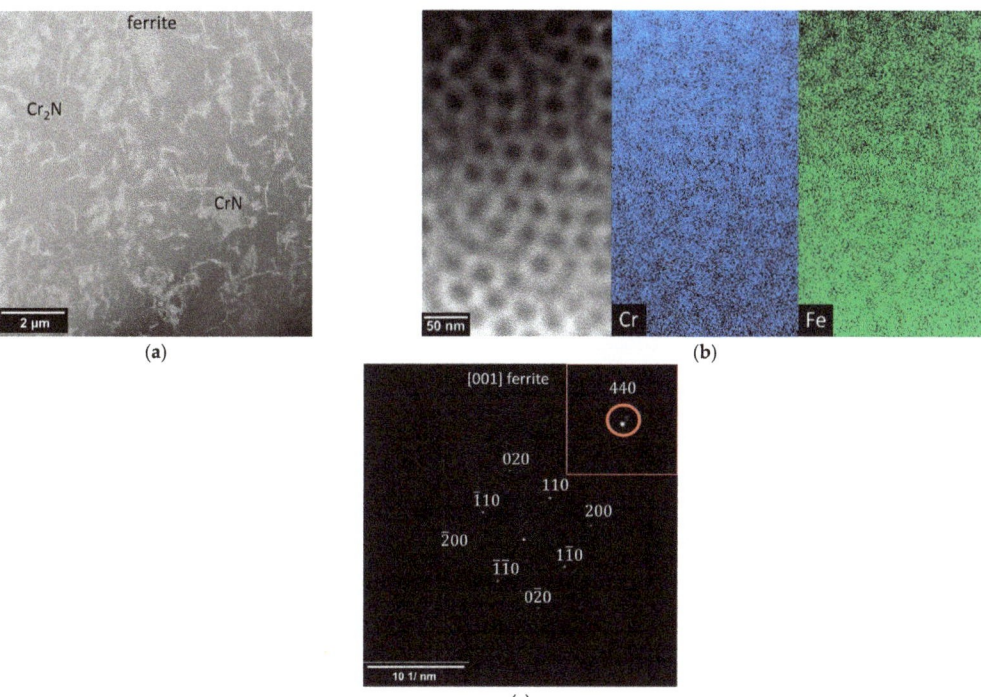

Figure 2. TEM analyses of the ferrite phase of the DSS after heat treatment. (**a**) HAADF-STEM image of a ferritic grain showing Cr_2N and CrN phases that appear as bright contrast. (**b**) HAADF-STEM image of ferrite and the corresponding EDX mapping of the elements Fe and Cr, (**c**) SAD pattern taken close to the [001] zone axis. The inset indicates the splitting of the diffraction spot [440] due to spinodal decomposition.

3.2. Isothermal Fatigue

In Figure 3a, the cyclic stress response of the DSS during the isothermal fatigue tests at 20, 300 and 600 °C at a plastic-strain amplitude of 0.2% is depicted. The ordinate displays the maximum and minimum stress response of a hysteresis. The abscissa represents the respective number of the cycle. The resulting stress response during the room temperature test (black curve) yields a slight initial hardening at the beginning (within the first 10 cycles by approximately ±20 MPa). During the remaining test period, the DSS reveals only a slight softening, which is visible by the decreasing stress amplitude (from approximately ±570 to ±520 MPa). In contrast, the orange curve, which represents the isothermal fatigue behavior at 300 °C, exhibits a pronounced primary hardening up to about 100 cycles (from approximately ±400 to ±600 MPa) followed by a plateau (±600 MPa). Significant deviations in the stress response at 300 °C are explained by serrated flow of the stress signal as depicted in the exemplarily stress–total strain hysteresis loop in Figure 3b. Those rapid stress drops followed by reloading can be attributed to DSA. The occurrence of DSA is additionally indicated by the higher saturation stress amplitude at 300 °C as compared to room temperature. The cyclic deformation curve of the isothermal fatigue test at 600 °C (red curve) exhibits slight softening until failure (from approximately ±290 to ±200 MPa). The results of the fatigue test at 600 °C reveal lower values of the stress amplitude compared to the stress response at the other test temperatures. The drop of the stress response at the end (approximately 2000th cycle) of the test period is most likely caused by crack initiation. In addition, Table A1 gives the yield strength $R_{0.2}$, max. tensile strength R_m and the elongation at fracture A recorded from Kolmorgen during isothermal monotonic tensile tests at 20, 300 and 600 °C for the interested reader [29].

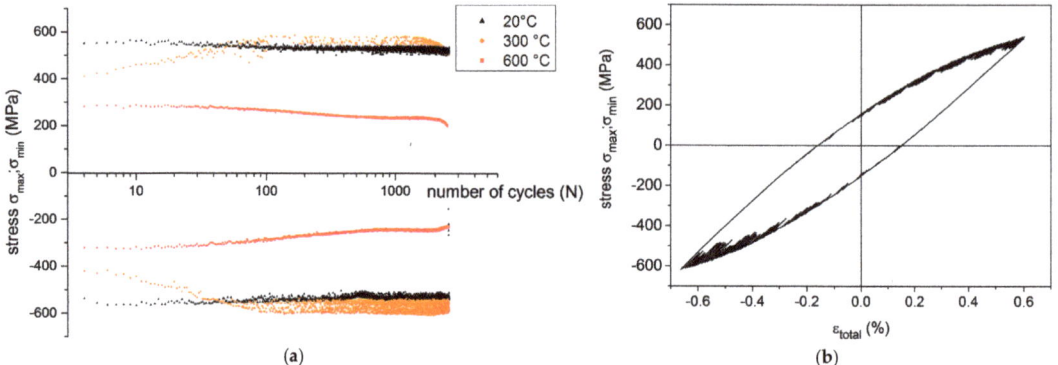

Figure 3. (a) Cyclic stress response curves of isothermal fatigue tests (20, 300 and 600 °C) carried out under plastic-strain control at a plastic-strain amplitude of 0.2% (f = 0.0059 Hz); (b) exemplary single stress–total strain hysteresis loop of the 800th cycle of the isothermal test at 300 °C.

To understand the material behavior and the role of constituting phases, TEM analysis was carried out on the samples after isothermal fatigue testing. After the fatigue test at room temperature, dislocations (white lines, e.g., in Figure 4a) occur in preferred orientations (two directions) in the selected austenitic grain, documenting planar slip in two active slip systems. Furthermore, a higher dislocation density can be observed at the boundaries between neighboring austenite and ferritic grains, indicating the pileup of defects. Obviously, phase boundaries act as barriers to dislocation glide [30,31]. Only individual dislocation loops are visible in the ferritic grain. Detailed micrographs of the dislocation arrangements are given Figure A2. The results of the TEM investigations (Figure 4a) reveal that the plastic deformation is concentrated in the softer austenitic grains. These stress responses are in accordance with the findings of Wackermann et al. [13] who tested the same material at RT with a plastic-strain amplitude of 0.25%. Because of the slightly higher plastic-strain amplitude as compared to the value of 0.20% used in the study presented, the resulting stress response was slightly increased to a stress amplitude of about 610 MPa. The authors also observed softening behavior after initial hardening in the cyclic deformation curves. Initial hardening was attributed to dislocation formation in both phases. Wackermann et al. correlated the extent of softening to the deformation-induced dissolution of spinodal decomposition domains. Moreover, based on the yield strength distribution function, they attributed the global deformation behavior of the DSS at higher plastic strain to the ferrite phase [13]. Whether austenite also contributes to the softening behavior but could not be determined by the yield strength distribution function analysis beyond doubt, as planar sliding in austenitic grain was detected after all stress controlled tests ($\Delta\sigma/2 = 375$ MPa or 400 MPa at 475 °C) independent of the embrittlement state and the stress level [13].

In contrast to the sample tested at room temperature, the microstructure after isothermal fatigue testing at 300 °C reveals a high dislocation density nearby the Cr_xN precipitates in the ferritic grains (Figures 4b and A3a). Detailed micrographs of the microstructure indicate that the cluster-like structure in the ferrite remains unchanged in size (Figure A3b). Apparently, planar gliding is also the predominant glide behavior in austenitic grains during isothermal fatigue at 300 °C (Figure 4b). It can be concluded that in contrast to the room temperature experiment, the plastic response is now governed by the dislocation arrangement in the ferritic grain in accordance with previous studies by Kolmorgens et al. [1,30].

The Cr$_x$N precipitates act as obstacles for the dislocation movement (Figure A3) explaining the massive hardening in the stress response curves (Figure 3a orange curve). The sawtooth-like stress response in the stress–strain hysteresis loops (Figure 3b) is consequently material-related and can be rationalized by the dislocation motion and pinning at Cr$_x$N precipitates within the ferritic grains (Figure A3a). The DSA in ferrite at intermediate temperatures is a well-known phenomenon [6,32].

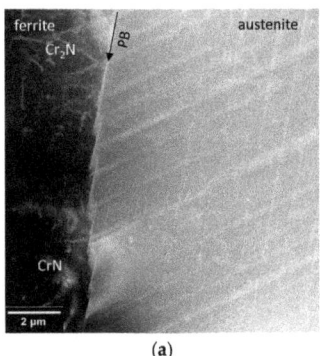 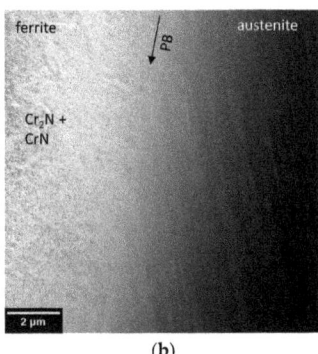 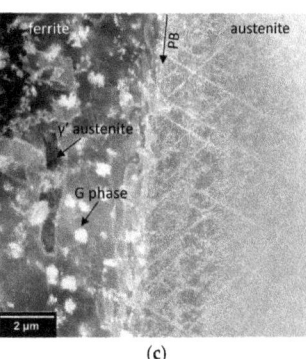

(a) (b) (c)

Figure 4. ADF-STEM images of the DSS after isothermal fatigue at (**a**) 20, (**b**) 300 and (**c**) 600 °C (PB marks the phase boundaries).

Interestingly, new precipitates were identified in ferrite (Figure 4c).

Figure 5a depicts the details of the changed microstructure in the ferrite matrix, after isothermal fatigue testing at 600 °C. Two new phases with bright and dark contrast and with irregular shape preferably form nearby Cr$_2$N precipitates. Furthermore, it can be observed that the spinodal α/α' microstructure of the ferrite dissolves in the vicinity of the new phases, and these areas become iron-rich, which is indicative of the Fe-rich α ferrite. In combination with STEM-EDX (Figure 5a and Table 2) and SAD analysis (Figure 5a,b), the γ' austenite (Ni-type, dark contrast) and the G phase (Ni$_{16}$Si$_7$Mn$_6$-type, bright contrast) were identified. The STEM-EDX measurements of the individual phases reveal that the G phase is rich in the elements Si, Ni and Mo as compared to the ferritic grain. Figure 5b displays the diffraction spots of the G phase as well as the ferrite matrix (red circles) which are visible due to their crystallographic $<111>_\alpha/<111>_G$ relation. This observation is in agreement with results of Mateo et al. who observed that the G phase exhibits a cube-on-cube orientation relationship with the ferrite matrix [16]. Further, the transformation of ferrite to γ' austenite takes place during isothermal fatigue at 600 °C. This is evidenced by the presence of the fcc structure in the diffraction pattern in Figure 5c and the enrichment of Ni and Mn, similar to primary austenite (Table 2). Weidner et al. also observed the formation of the γ' austenite in addition to the G phase even during a TMF experiment in the temperature range of 350 to 600 °C [22]. Only a slight depletion in Cr and Mo in the γ' austenite was identified after aging at 850 °C for few hours, which was also reported by Villanueva et al. in DSS X2CrNiMoN22-5-3 [33].

Clearly, several microstructural transformations occur simultaneously during the isothermal fatigue test at 600 °C. It is stated by Weidner et al. [22] and Meteo et al. [10,16] that Cr$_x$N precipitates as well as dislocations act as nucleation sites for the formation of the G phase, while the precipitation of the γ' austenite (Figure 5a) occurs due to the lattice distortion in the vicinity of Cr$_x$N precipitates. Alternatively, the gradient in the chemical composition, in particular the Cr gradient between the matrix and the Cr$_x$N, acts as the driving force for precipitation [10,16]. In the vicinity of the new phases, no spinodal decomposition of the ferrite phase was observed, most likely because the elements Si, Mo and Ni, which facilitate the process of spinodal decomposition, segregate to the new phases [16,22]. Mateo et al. discussed that the spinodal decomposition of ferrite is often accompanied by Cr_xN precipitation. This assumption is confirmed by the experimental findings in this work (Figure 2) [16]. Tavares et al. also observed the formation of γ' austenite in the decomposed interdomains of the ferrite during ageing treatment of the DSS UNS S31803 at 800 °C [11]. Thus, it can be concluded that the precipitation of Cr_xN and the spinodal decomposition, along with lattice distortion and the establishment of gradients in the chemical composition of corresponding phases, give rise to the formation of secondary phases, namely G phase and γ' austenite.

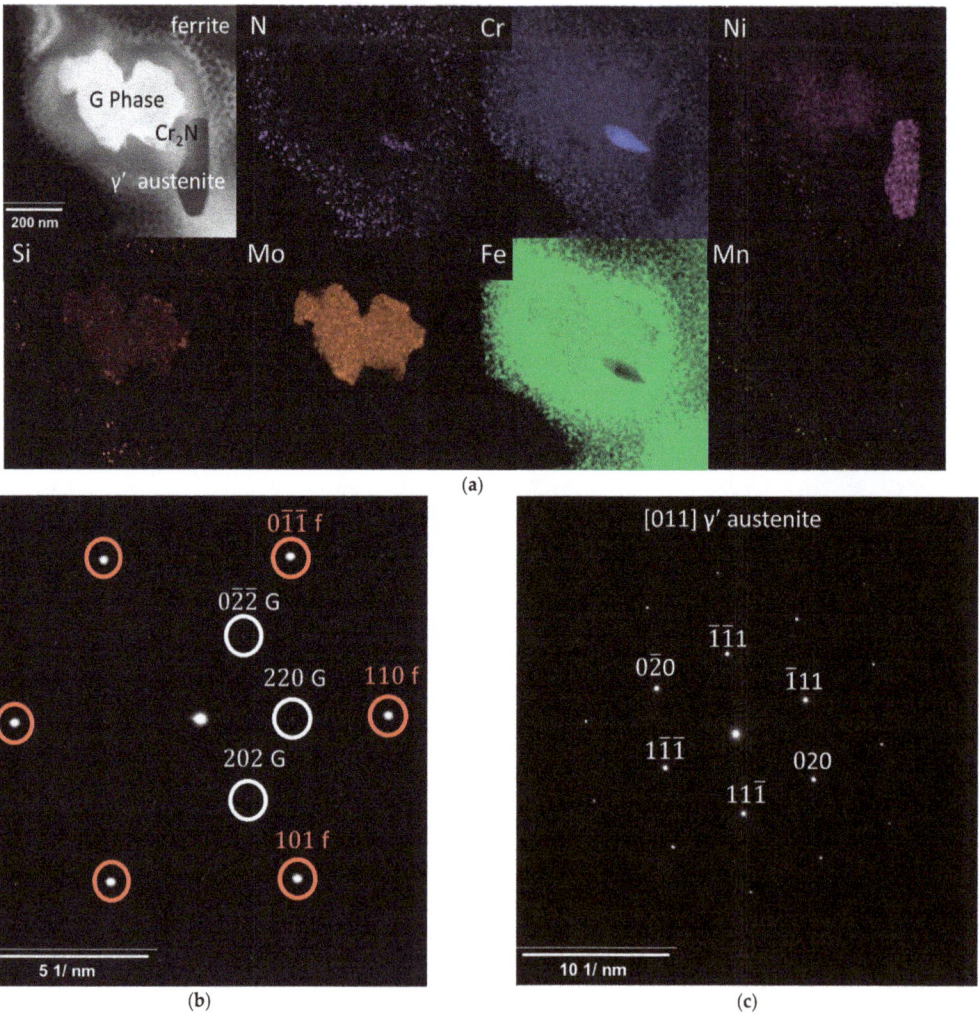

Figure 5. Detailed STEM images of the ferritic grain after isothermal fatigue at 600 °C. (**a**) HAADF-STEM image of the G phase, Cr2N and γ' austenite in a ferritic grain and the corresponding EDX mapping of the elements N, Cr, Ni, Si, Mo, Fe and Mn. Point EDX analysis data of the phases are given in Table 2. (**b**) SAD image of the G phase and the ferrite matrix along $[\bar{1}11]_G/[\bar{1}11]_\alpha$ zone axis. (**c**) SAD image of γ' austenite at [011] zone axis.

Table 2. Chemical composition of the phases after isothermal fatigue at 600 °C determined by STEM-EDX analyses (wt.%).

Element	Si	Cr	Mn	Fe	Ni	Mo
γ austenite	0.22 ± 0.02	20.08 ± 3.41	1.8 ± 0.3	68.4 ± 11.2	6.26 ± 1.03	2.51 ± 0.4
γ' austenite	0.18 ± 0.02	17.94 ± 2.96	2.03 ± 0.17	70.66 ± 11.66	9.11 ± 1.51	1.09 ± 0.18
α/α' ferrite	0.28 ± 0.03	23.79 ± 3.88	1.55 ± 0.25	66.41 ± 10.82	4.05 ± 0.66	3.92 ± 0.62
G phase	1.10 ± 0.08	20.08 ± 3.16	1.66 ± 0.25	40.36 ± 6.13	4.85 ± 0.51	31.24 ± 4.8

Less dislocations were found in the ferrite as compared to the austenite phase after 600 °C fatigue testing (Figure 4c). It is well-known that diffusion determines the dislocation movement at high temperatures [34]. Further, the diffusion processes are generally faster in bcc than in fcc materials [35]. Therefore, the faster dislocation climbing in the ferrite phase and, in turn, the more pronounced dislocation annihilation may explain the low dislocation density within the ferritic grains at high temperature [36]. The hard intermetallic G phase acts as an obstacle to dislocation motion, but due to the high temperature (thermal energy) the dislocations can climb and eventually overcome the G phase. Such a behavior is typical for bcc materials at elevated temperatures [34]. In addition to planar glide behavior in austenitic grains at room temperature and 300 °C testing, further dislocation arrangements with an irregular appearance were found in the austenitic grains after fatigue at 600 °C (Figures 4 and A4a). In the austenitic grain (Figures 4c and A4b) the more irregular dislocation arrangement can be explained by cross slip or climbing, as these processes are favored at high temperatures, making the overall dislocation structure more complex compared to lower temperature fatigue tests (Figure 4a,b). Kolmorgen et al. reported that due to the higher temperature, the ferritic grains take up more of the plastic deformation relative to the austenite. In addition, it was demonstrated that the austenite phase possesses sufficient strength even at high temperatures to compensate for the low strength of the ferrite phase [1]. Both the higher temperature and the high plastic deformation in the ferrite facilitate precipitation of secondary phases. As reported elsewhere, the ferritic phases can only take over part of the load due to the formation of new phases and the consequent strengthening, thus providing a relatively constant level in the stress response [16,22].

3.3. Thermomechanical Fatigue

Figure 6 displays the stress responses during TMF tests under IP and OP conditions in the temperature range of 300 to 600 °C. The purple curve represents the stress response of the DSS during TMF IP fatigue testing, while the dark red curve corresponds to the stress amplitude during TMF OP testing. The stress at the upper load reversal point (ε_{pl} = +0.2%, T = 600 °C) in the tensile stress regime during the IP TMF test loading exhibits a gradual softening. In contrast, the cyclic deformation curve at the lower load reversal point in the compressive stress regime (ε_{pl} = −0.2%, T = 300 °C) reveals initial hardening followed by a plateau. Under OP TMF conditions, the behavior is inverse. Considering the minimum and maximum stress response curves for IP TMF, compressive mean stress evolves. Comparing the results of mean stresses of the two thermomechanical fatigue tests shown in Figure 6, it is visible that the stress response during the OP loading shifts towards a higher tensile stress level. At the end of the TMF OP curve for maximum stress, a rapid drop of the stress amplitude appears. As a result of the high tensile mean stress, the fatigue life during OP testing yields a value of only 1743 cycles, whereas the test in the IP mode was stopped after 2555 cycles without any visible signs of damage.

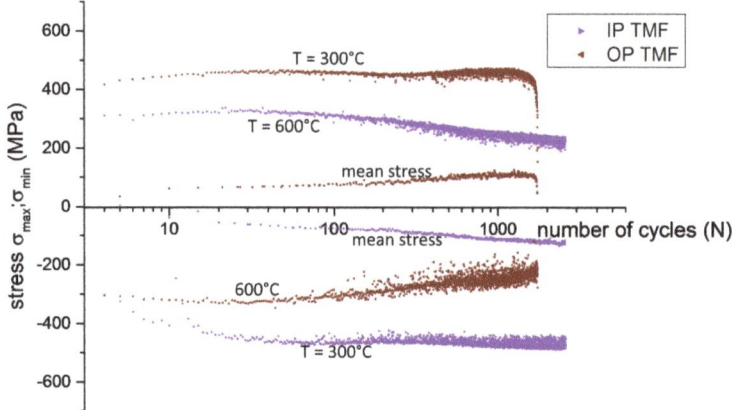

Figure 6. Cyclic stress response curves during thermomechanical fatigue tests (IP TMF and OP TMF) under plastic-strain control at $\Delta\varepsilon_{pl}/2 = 0.2\%$ (f = 0.059 Hz) in the temperature range of 300 to 600 °C.

The results of the TEM analysis after TMF tests in IP and OP mode are presented in Figure 7a,b, respectively. A phase boundary can be seen diagonally in both micrographs. In the upper left half of the figures, ferritic grains in dark contrast can be recognized. The G phase precipitates and the formation of subgrains (visible in Figures A5 and A6) in the ferrite phase were found for both the IP (Figure 7a) and OP (Figure 7b) TMF tests. It should be noted that larger and less frequent G phase precipitates are observable after the TMF test in the IP mode (IP TMF: G phase size 107 ± 24), while in the OP experiment the G phase is slightly smaller but the density is higher (OP TMF: G phase size 67 ± 27 nm). Few dislocations can be identified in the ferritic grain next to the precipitates (Figure A5b). In contrast, a high dislocation density is found in the austenitic grains (Figure 7a,b). In addition to the planar sliding behavior of dislocations, irregular dislocation structures appear in the austenitic grain. Detailed micrographs of the dislocation arrangement and the precipitates are given in Figure A4 (IP condition) and Figure A5 (OP condition).

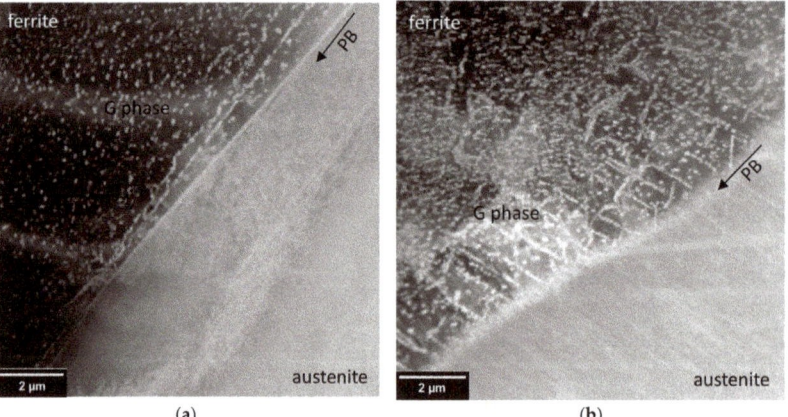

Figure 7. ADF-STEM images of the microstructure and the dislocation arrangement after thermomechanical fatigue in the temperature range from 300 to 600 °C in (**a**) IP condition and (**b**) OP condition. G phase appears in small white particles.

Considering these findings, a correlation between the mechanical behavior during isothermal and thermomechanical fatigue is drawn. In TMF tests, a hardening at 300 °C and a softening at 600 °C is observed. This correlates well to the isothermal fatigue tests at these temperatures leading to cyclic deformation curves that possess a similar shape. This suggests that the stress response of the thermomechanical IP fatigue test at the upper load reversal point (ε_{pl} = +0.2%, T = 600 °C) is governed by the softening typical of the isothermal fatigue at 600 °C, whereas the hardening of the stress response at the lower load reversal point (ε_{pl} = −0.2%, T = 300 °C) corresponds to the 300 °C isothermal fatigue test. The initial strain hardening can be attributed to the interaction between nitrides and dislocations (see isothermal fatigue at 300 °C). For the TMF test in the OP mode, the opposite behavior occurs, as the high temperature regime coincides with the compressive stresses and the low temperature regime with the tensile stresses. The different fatigue life of the IP and OP TMF tests can be explained by the different sign of the mean stress (i.e., IP: negative, and OP: positive) as discussed above. Obviously, the mean compressive stress in the IP mode has a positive influence on the fatigue life of the material studied, while the reduced material strength at higher temperature (compression) results in a tensile mean stress in the OP TMF test and therefore decreases the life due to earlier crack initiation.

Taking into account the results from isothermal 300 and 600 °C fatigue tests, it can be concluded that in the low temperature regime of TMF tests, dislocations are created that pile up at precipitates or phase boundaries (Figure 4b), while at higher temperatures recovery processes take place (dislocation reactions and annihilation) (Figure 4c) resulting in a reduction in the dislocation density in ferrite. Due to the high cyclic plastic deformation of the ferrite phase at alternating temperatures of TMF cycling (300–600 °C), a dislocation rearrangement to subgrain forms (cf. Figures A5 and A6). Kolmorgen et al. reported a similar experimental finding and concluded that the fragmentation of ferrite grains into subgrains during TMF experiments results from the superposition of cyclic mechanical and thermal loading [1]. Furthermore, Mateo et al. found that the subgrain formation during thermal cycles is accompanied by spinodal decomposition, which further promotes the formation of the G phase at the subgrains [16]. The formation of the G phase at the subgrain boundaries was also identified in this study (Figure A6b). Similar to the 600 °C isothermal tests, G phase precipitation was also observed in TMF testing, indicating that the temperature range applied lay sufficiently high to allow G phase formation. These observations are in agreement with those in other works [1,16]. It should be noted that the G phase precipitates found after TMF testing were significantly finer as compared to those formed during the isothermal fatigue test at 600 °C (G phase size 434 nm ± 178 nm) (compare Figures 4c and 7) as the average temperature of TMF is lower. A direct comparison of the G phase size formed during TMF in IP and OP TMF modes demonstrates that the larger G phase precipitates were observed after the IP TMF test (compare Figure 7a,b). The G phase precipitates have a longer time to grow due to the longer test duration of the IP TMF experiment in comparison to the corresponding OP test. The time period at high temperatures during TMF testing seems to be not sufficient to form the γ' austenite that occurs only in the isothermal 600 °C fatigue test (Figure 4c). Finally, the phenomenon of the 475 °C embrittlement was not detected during the TMF tests (in stress responses) in the temperature range between 300 and 600 °C. However, the decomposed structure in the ferrite phase was still present after the TMF experiments (Figure A6). It should be noted that results demonstrated in this work represent post-mortem studies, and the cluster-like ferrite segregation probably occurs during cooling down to room temperature. It should further be stressed that even quenching after the standard heat treatment cannot prevent the spinodal decomposition. These results are consistent with those of Weidner et al. [22] who also did not observe the 475 °C embrittlement during TMF tests of the DSS in the temperature range between 350 and 600 °C as the temporary heating to temperatures above 550 °C counteracts 475 °C embrittlement.

4. Summary

Isothermal (20, 300 and 600 °C) and thermomechanical fatigue tests using the temperature range 300–600 °C were performed on the duplex steel X2CrNiMoN22-5-3 with a plastic-strain amplitude $\Delta\varepsilon_{pl}/2 = 0.2\%$ and a testing frequency f = 0.0059 Hz. The main results can be summarized as follows:

Isothermal Fatigue tests at 20 °C: A continuous softening of the duplex steel was observed during room temperature fatigue test. The austenite phase exhibits a pronounced planar slip behavior.

Isothermal Fatigue tests at 300 °C: A pronounced primary hardening was observed, which is caused by the interaction between dislocations and nitrides. Planer slip prevails in the austenitic grains.

Isothermal Fatigue tests at 600 °C: The cyclic stress amplitude during the isothermal fatigue test is at a low level. The material studied exhibits softening throughout the whole fatigue test. A further precipitation of chromium nitrides as well as the formation of the G phase and secondary γ' austenite leads to a significant change in the microstructure of the ferritic grains. The low dislocation density observed in the ferrite phase is attributed to pronounced recovery processes. In the austenitic grains, a more irregular and wavy dislocation arrangement prevails.

IP TMF: During TMF loading in the IP mode a negative mean stress results, which has a positive effect on fatigue life. The cyclic behavior in the stress response of the TMF test can be correlated with the corresponding stress responses of the isothermal fatigue tests at 300 and 600 °C. Furthermore, initial cyclic hardening (300 °C) can be attributed to interaction between nitrides and dislocations (DSA). While a significant amount of the G phase was found in ferrite, no secondary γ' austenite was observed. Additionally, the formation of subgrains in the ferrite was observed. The plastic deformation of the austenitic grains results in an unordered dislocation arrangement.

OP TMF: TMF in the OP mode results in a tensile mean stress, which leads to premature failure. A comparison of the stress response of OP and IP TMF clearly shows a very similar behavior. Analogously to the corresponding TMF test in IP mode, secondary γ' austenite does also not form under OP conditions. The main microstructural difference caused by the T-ε_{pl}-phasing seem to be the finer morphology of the G phase precipitates.

Author Contributions: Conceptualization, A.O. and H.-J.C.; methodology, S.S., J.M. and A.O.; validation, A.O., H.-J.C. and B.B.; investigation, S.S., J.M. and A.O.; resources, H.-J.C., A.O. and B.G.; data curation, S.S. and A.O.; writing—original draft preparation, S.S.; writing—review and editing, S.S., J.M., A.O., B.G., B.B. and H.-J.C.; visualization, S.S. and J.M.; supervision, H.-J.C., A.O. and B.G.; project administration, H.-J.C. and A.O.; All authors have read and agreed to the published version of the manuscript.

Funding: This research received no external funding. Internal funding was received by Universität Siegen (UoS).

Institutional Review Board Statement: Not applicable.

Informed Consent Statement: Not applicable.

Data Availability Statement: The data that support the finding of this study are available from the corresponding author upon reasonable request.

Acknowledgments: Part of this work was performed at the Micro- and Nanoanalytics Facility (MNaF) of the University of Siegen. We acknowledge support by the KIT-Publication Fund of the Karlsruhe Institute of Technology.

Conflicts of Interest: The funder (UoS) had no role in the design of the study; in the collection, analyses, or interpretation of data; in the writing of the manuscript or in the decision to publish the results. The authors declare no conflict of interest.

Appendix A

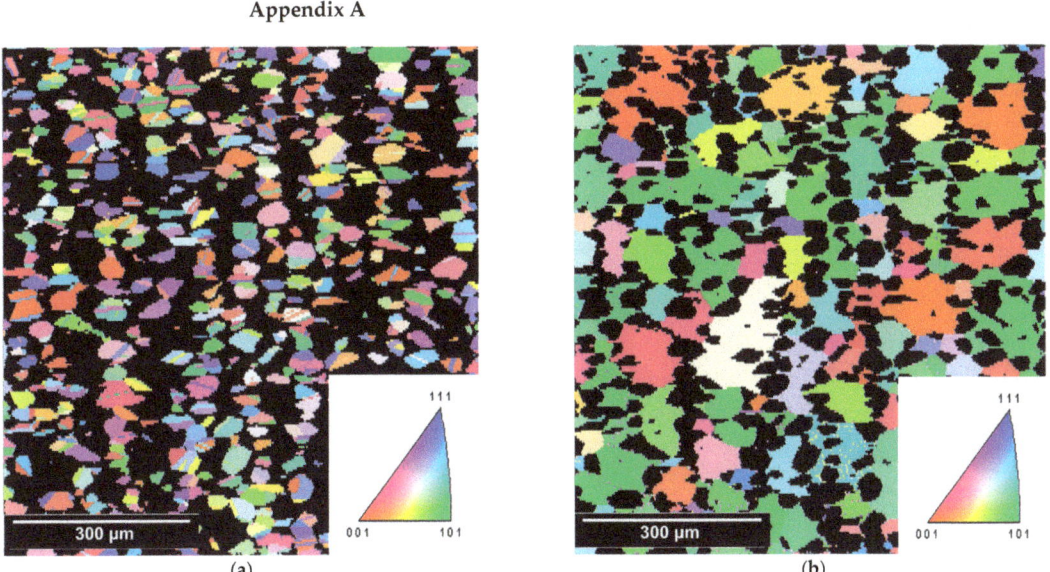

Figure A1. EBSD Crystallographic orientation—inverse pole figure coloring map for austenite (**a**) and ferrite (**b**) of the hot-rolled DSS X2CrNIMoN22-5-3 after heat treatment.

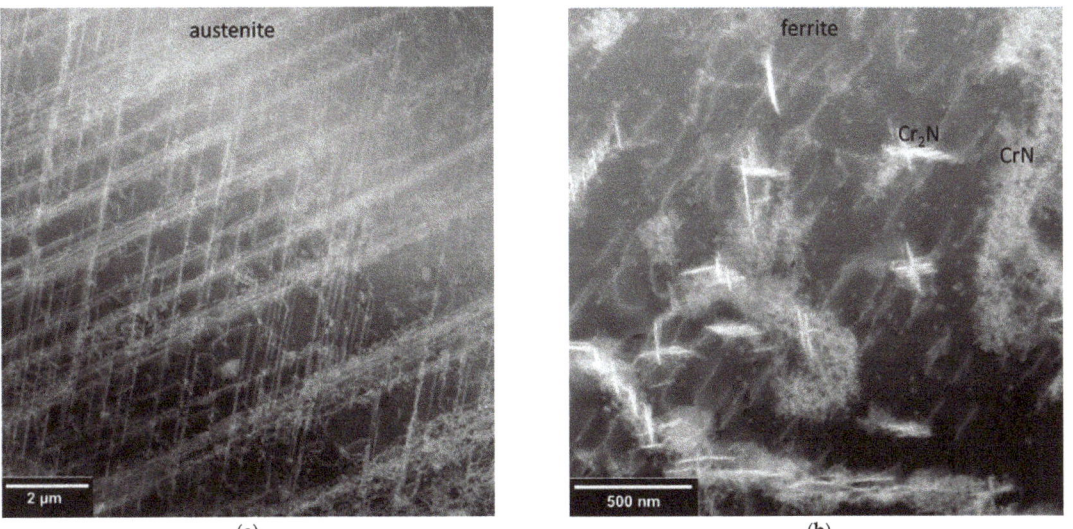

Figure A2. Detailed STEM images of the dislocation arrangement in an austenitic (**a**) and a ferritic grain (**b**) after isothermal fatigue at 20 °C.

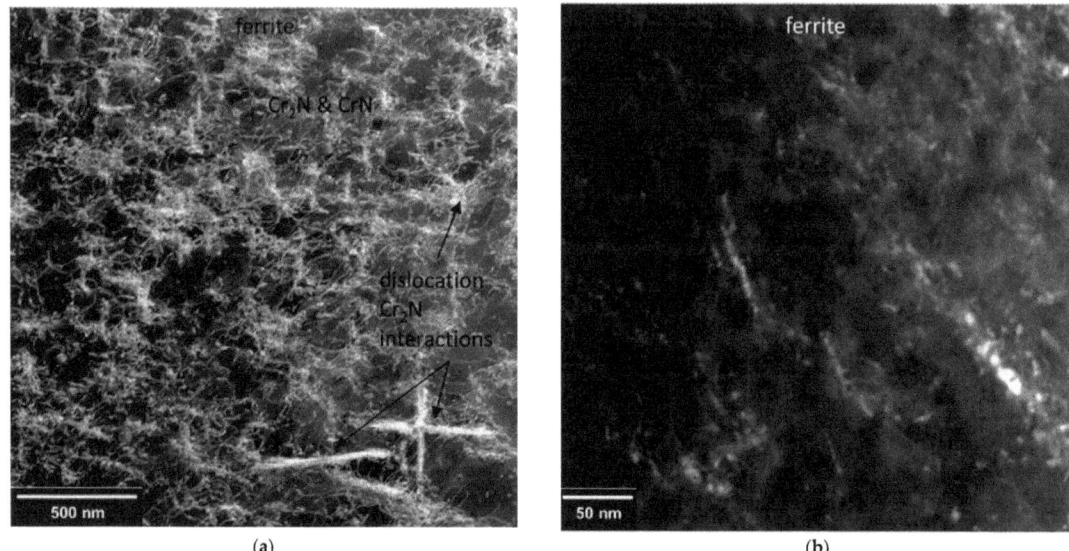

Figure A3. (a,b) detailed STEM images of the dislocation arrangement and precipitates in the ferrite after isothermal fatigue at 300 °C.

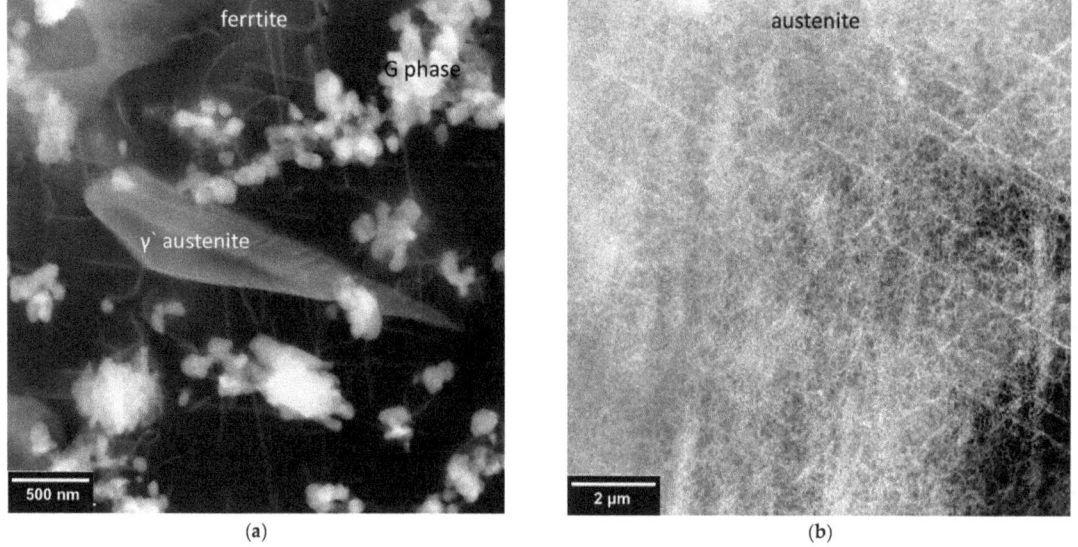

Figure A4. Detailed STEM images of the dislocation arrangement in a ferritic (a) and an austenitic grain (b) after isothermal fatigue at 600 °C.

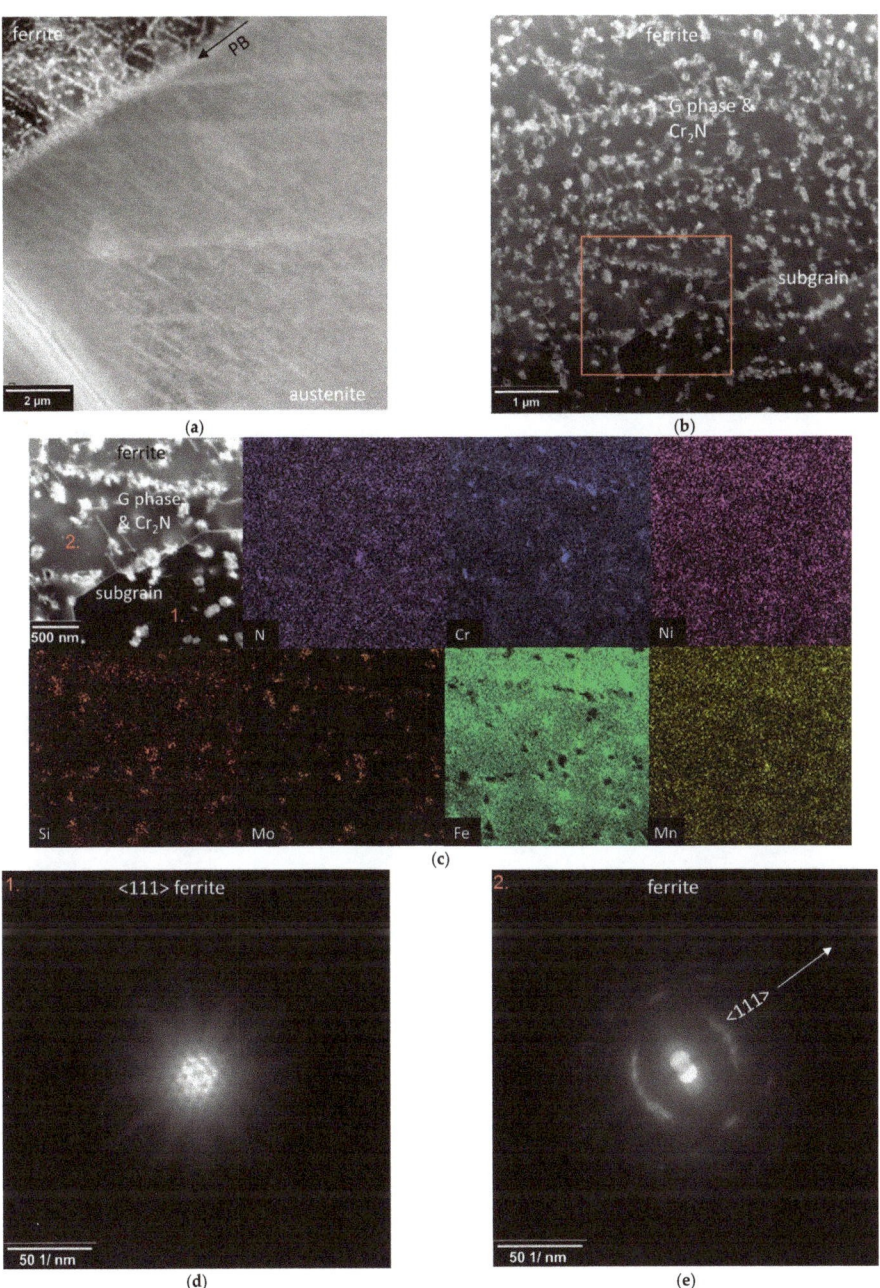

Figure A5. Detailed STEM images of the dislocation arrangement in austenite (**a**) and ferrite (**b**) and STEM-ADF image of precipitates in ferrite and (**c**) the corresponding EDX mapping of the highlighted area in (**b**) of the elements N, Cr, Ni, Si, Mo, Fe and Mn after thermomechanical fatigue in IP condition. (**d**,**e**): Diffraction pattern on the CETA camera in STEM mode of the marked locations in (**c**) (1. corresponds to (**d**), 2. corresponds to (**e**)). Serval degrees of tilt difference between (**d**) in ferrite <111> ZA and (**e**).

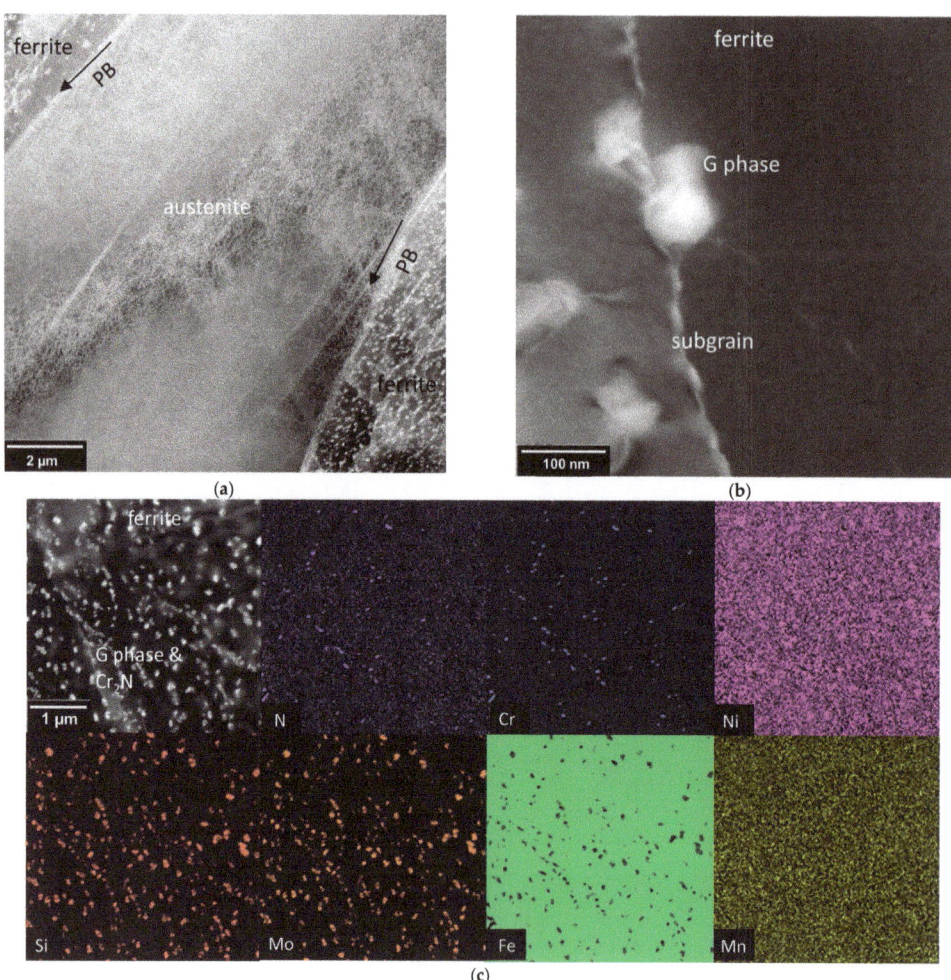

Figure A6. Detailed STEM images of the dislocation arrangement in austenite (**a**), and precipitates in ferritic grain (**b**,**c**) STEM-ADF image of precipitates in ferrite and the corresponding EDX mapping of the elements N, Cr, Ni, Si, Mo, Fe and Mn after thermomechanical fatigue in OP condition.

Table A1. Yield strength $R_{0.2}$, max. tensile strength R_m and the elongation at fracture A of isothermal monotonic tensile tests at a strain rate of 1×10^{-4} s^{-1} of DSS 1.4462 at 20, 300 and 600 °C [29].

Temperature (°C)	20	300	600
$R_{0.2}$ (MPa)	490	330	250
R_m (MPa)	730	680	350
A (%)	42	28	27

References

1. Kolmorgen, R.; Biermann, H. Thermo mechanical fatigue behaviour of a duplex stainless steel in the range of 350–600 °C. *Int. J. Fatigue* **2014**, *65*, 2–8. [CrossRef]
2. Nani Babu, M.; Sasikala, G.; Shashank Dutt, B.; Venugopal, S.; Albert, S.K.; Bhaduri, A.K.; Jayakumar, T. Investigation on influence of dynamic strain ageing on fatigue crack growth behaviour of modified 9Cr–1Mo steel. *Int. J. Fatigue* **2012**, *43*, 242–245. [CrossRef]

3. Avalos, M.; Alvarez-Armas, I.; Armas, A.F. Dynamic strain aging effects on low-cycle fatigue of AISI 430F. *Mater. Sci. Eng. A* **2009**, *513–514*, 1–7. [CrossRef]
4. Kolmorgen, R.; Biermann, H. Thermo-mechanical fatigue behaviour of a duplex stainless steel. *Int. J. Fatigue* **2012**, *37*, 86–91. [CrossRef]
5. Tsuzaki, K.; Hori, T.; Maki, T.; Tamura, I. Dynamic strain aging during fatigue deformation in type 304 austenitic stainless steel. *Mater. Sci. Eng.* **1983**, *61*, 247–260. [CrossRef]
6. Gironès, A.; Llanes, L.; Anglada, M.; Mateo, A. Dynamic strain ageing effects on superduplex stainless steels at intermediate temperatures. *Mater. Sci. Eng. A* **2004**, *367*, 322–328. [CrossRef]
7. Wackermann, K.; Christ, H.-J. Identifying the effect of 475 °C embrittlement on the cyclic stress-strain response of duplex stainless steel by means of the change in the yield stress distribution. *Adv. Mater. Res.* **2014**, *891–892*, 458–463. [CrossRef]
8. Sahu, J.K.; Krupp, U.; Ghosh, R.N.; Christ, H.-J. Effect of 475 °C embrittlement on the mechanical properties of duplex stainless steel. *Mater. Sci. Eng. A* **2009**, *508*, 1–14. [CrossRef]
9. Cortie, M.B.; Pollak, H. Embrittlement and aging at 475 °C in an experimental ferritic stainless steel containing 38 wt.% chromium. *Mater. Sci. Eng. A* **1995**, *199*, 153–163. [CrossRef]
10. Llanes, L.; Mateo, A.; Violan, P.; Méndez, J.; Anglada, M. On the high cycle fatigue behavior of duplex stainless steels: Influence of thermal aging. *Mater. Sci. Eng. A* **1997**, *234–236*, 850–852. [CrossRef]
11. Tavares, S.S.M.; de Noronha, R.F.; da Silva, M.R.; Neto, J.M.; Pairis, S. 475 °C Embrittlement in a duplex stainless steel UNS S31803. *Materials Res.* **2001**, *4*, 237–240. [CrossRef]
12. Weng, K.L.; Chen, H.R.; Yang, J.R. The low-temperature aging embrittlement in a 2205 duplex stainless steel. *Mater. Sci. Eng. A* **2004**, *379*, 119–132. [CrossRef]
13. Wackermann, K. Einfluss einer zyklischen Belastung auf die Versprödungskinetik von Legierungen am Beispiel der 475 °C-Versprödung von Duplexstahl und der Dynamischen Versprödung einer Nickelbasislegierung. Ph.D. Thesis, University of Siegen, Siegener Werkstoffkundliche Berichte, Siegen, Germany, July 2015. Available online: https://nbn-resolving.org/urn:nbn:de:hbz:467-9717 (accessed on 19 May 2022).
14. Sahu, J.K. Effect of 475 °C Embrittlement on the Fatigue Behaviour of a Duplex Stainless Steel. Ph.D. Thesis, University of Siegen, Siegener Werkstoffkundliche Berichte, Siegen, Germany, October 2008. Available online: Https://nbn-resolving.org/urn:nbn:de:hbz:467-3774 (accessed on 19 May 2022).
15. Gunn, R. *Duplex Stainless Steels: Microstructure, Properties and Applications*, 1st ed.; Woodhead Publishing Series in Metals and Surface Engineering: Cambridge, UK, 1997; pp. 94–104.
16. Mateo, A.; Llanes, L.; Anglada, M.; Redjaimia, A.; Metauer, G. Characterization of the intermetallic G-phase in an AISI 329 duplex stainless steel. *J. Mater. Sci.* **1997**, *32*, 4533–4540. [CrossRef]
17. Matsukawa, Y.; Takeuchi, T.; Kakubo, Y.; Suzudo, T.; Watanabe, H.; Abe, H.; Toyama, T.; Nagai, Y. The two-step nucleation of G-phase in ferrite. *Acta Mater.* **2016**, *116*, 104–113. [CrossRef]
18. Tang, X. Sigma Phase Characterization in AISI 316 Stainless Steel. *Microsc. Microanal.* **2005**, *11*, 78–79. [CrossRef]
19. Magnabosco, R. Kinetics of sigma phase formation in a Duplex Stainless Steel. *Mater. Res.* **2009**, *12*, 321–327. [CrossRef]
20. Armas, A.F.; Petersen, C.; Schmitt, R.; Avalos, M.; Alvarez-Armas, I. Mechanical and microstructural behaviour of isothermally and thermally fatigued ferritic/martensitic steels. *J. Nucl. Mater.* **2002**, *307–311*, 509–513. [CrossRef]
21. Krupp, U.; Söker, M.; Giertler, A.; Dönges, B.; Christ, H.-J.; Wackermann, K.; Boll, T.; Thuvander, M.; Marinelli, M.C. The potential of spinodal ferrite decomposition for increasing the very high cycle fatigue strength of duplex stainless steel. *Int. J. Fatigue* **2016**, *93*, 363–371. [CrossRef]
22. Weidner, A.; Kolmorgen, R.; Kubena, I.; Kulawinski, D.; Kruml, T.; Biermann, H. Decomposition and precipitation process during thermo-mechanical fatigue of duplex stainless steel. *Metall. Mater. Trans. A* **2016**, *47*, 2112–2124. [CrossRef]
23. Dönges, B.; Fritzen, C.P.; Christ, H.J. Experimental Investigation and Simulation of the Fatigue Mechanisms of a Duplex Stainless Steel under HCF and VHCF Loading Conditions. *Key Eng. Mater.* **2015**, *664*, 267–274. [CrossRef]
24. Bauer, V. Verhalten Metallischer Konstruktionswerkstoffe unter Thermomechanischer Belastung—Experimentelle Charakterisierung und Modellmäßige Beschreibung. Ph.D. Thesis, University of Siegen, Berichte aus der Werkstofftechnik, Shaker Verlag, Aachen, Germany, July 2007.
25. Chan, K.W.; Tjong, S.C. Effect of secondary phase precipitation on the corrosion behavior of duplex stainless steels. *Materials* **2014**, *7*, 5268–5304. [CrossRef] [PubMed]
26. Sicupira, D.C.; Cardoso Junior, R.; Bracarense, A.Q.; Frankel, G.S.; Lins, V.d.F.C. Cyclic polarization study of thick welded joints of lean duplex stainless steel for application in biodiesel industry. *Mater. Res.* **2017**, *20*, 161–167. [CrossRef]
27. Holländer Pettersson, N.; Lindell, D.; Lindberg, F.; Borgenstam, A. Formation of chromium nitride and intragranular austenite in a super duplex stainless steel. *Metall. Mater. Trans. A* **2019**, *50*, 5594–5601. [CrossRef]
28. Zhou, J.; Odqvist, J.; Höglund, L.; Thuvander, M.; Barkar, T.; Hedström, P. Initial clustering—A key factor for phase separation kinetics in Fe–Cr-based alloys. *Scr. Mater.* **2014**, *75*, 62–65. [CrossRef]
29. Kolmorgen, R. Das Thermomechanische Ermüdungsverhalten eines Ferritisch-Austenitischen Duplexstahls im Temperaturbereich 100 °C bis 600 °C. Ph.D. Thesis, Technische Universität Bergakademie Freiberg Papierpflieger Verlag GmbH, Clausthal-Zellerfeld, Germany, November 2020.

30. Kolmorgen, R.; Weidner, A.; Biermann, H. Deformation and microstructure evolution of a duplex stainless steel under out-of-phase thermo-mechanical fatigue. *Mater. High Temp.* **2013**, *30*, 77–82. [CrossRef]
31. Duber, O.; Kunkler, B.; Krupp, U.; Christ, H.-J.; Fritzen, C. Experimental characterization and two-dimensional simulation of short-crack propagation in an austenitic–ferritic duplex steel. *Int. J. Fatigue* **2006**, *28*, 983–992. [CrossRef]
32. Hereñú, S.; Alvarez-Armas, I.; Armas, A.F. The influence of dynamic strain aging on the low cycle fatigue of duplex stainless steel. *Scr. Mater.* **2001**, *45*, 739–745. [CrossRef]
33. Villanueva, D.M.E.; Junior, F.C.P.; Plaut, R.L.; Padilha, A.F. Comparative study on sigma phase precipitation of three types of stainless steels: Austenitic, superferritic and duplex. *Mater. Sci. Technol.* **2006**, *22*, 1098–1104. [CrossRef]
34. Zhang, J.-S. *High Temperature Deformation and Fracture of Materials*, 1st ed.; Elsevier: Sawston, Cambridge, UK, 2010; pp. 359–365.
35. Heumann, T. *Diffusion in Metallen: Grundlagen, Theorie, Vorgänge in Reinmetallen und Legierungen*, 1st ed.; Springer: Berlin/Heidelberg, Germany, 1992; pp. 128–135.
36. Miller, M.K.; Bentley, J. APFIM and AEM investigation of CF8 and CF8M primary coolant pipe steels. *Mater. Sci. Technol.* **1990**, *6*, 285–292. [CrossRef]

Article

Comparison of the Internal Fatigue Crack Initiation and Propagation Behavior of a Quenched and Tempered Steel with and without a Thermomechanical Treatment

Amin Khayatzadeh *, Stefan Guth and Martin Heilmaier

Institute for Applied Materials (IAM-WK), Department of Mechanical Engineering, Karlsruhe Institute of Technology, Engelbert-Arnold-Str. 4, 76131 Karlsruhe, Germany; stefan.guth@kit.edu (S.G.); martin.heilmaier@kit.edu (M.H.)
* Correspondence: amin.khayatzadeh@kit.edu; Tel.: +49-721-608-44159

Abstract: Previous studies have shown that a thermomechanical treatment (TMT) consisting of cyclic plastic deformation in the temperature range of dynamic strain aging can increase the fatigue limit of quenched and tempered steels by strengthening the microstructure around non-metallic inclusions. This study considers the influence of a TMT on the shape, size and position of crack-initiating inclusions as well as on the internal crack propagation behavior. For this, high cycle fatigue tests on specimens with and without TMT were performed at room temperature at a constant stress amplitude. The TMT increased the average lifetime by about 40%, while there was no effect of the TMT on the form or size of critical inclusions. Surprisingly, no correlation between inclusion size and lifetime could be found for both specimen types. There is also no correlation between inclusion depth and lifetime, which means that the crack propagation stage covers only a small portion of the overall lifetime. The average depth of critical inclusions is considerably higher for TMT specimens indicating that the strengthening effect of the TMT is more pronounced for near-surface inclusions. Fisheye fracture surfaces around the critical inclusions could be found on all tested specimens. With increasing fisheye size, a transition from a smooth to a rather rough and wavy fracture surface could be observed for both specimen types.

Keywords: non-metallic inclusion; thermomechanical treatment (TMT); inclusion area; inclusion shape; inclusion depth; fisheye formation; crack initiation

Citation: Khayatzadeh, A.; Guth, S.; Heilmaier, M. Comparison of the Internal Fatigue Crack Initiation and Propagation Behavior of a Quenched and Tempered Steel with and without a Thermomechanical Treatment. *Metals* **2022**, *12*, 995. https://doi.org/10.3390/met12060995

Academic Editor: Francesco Iacoviello

Received: 12 May 2022
Accepted: 8 June 2022
Published: 10 June 2022

Publisher's Note: MDPI stays neutral with regard to jurisdictional claims in published maps and institutional affiliations.

Copyright: © 2022 by the authors. Licensee MDPI, Basel, Switzerland. This article is an open access article distributed under the terms and conditions of the Creative Commons Attribution (CC BY) license (https://creativecommons.org/licenses/by/4.0/).

1. Introduction

Since fatigue crack initiation and growth is one of the significant causes of structural failure in engineering applications, there is a considerable demand for steels with high fatigue strength in the industry [1,2]. SAE4140 quenched and tempered steel is one of the most favorable steels for applications involving cyclic loading due to its high fatigue strength [3]. It is well-known that the lifetime of quenched and tempered steels in the high cycle fatigue (HCF) and very high cycle fatigue (VHCF) regimes are limited by crack initiation at internal inhomogeneities and non-metallic inclusions [4–7]. Normally, fracture surfaces of HCF and VHCF failures exhibit not only crack initiation at non-metallic inclusions but also fisheye formation around these critical inclusions [7,8]. Fisheye-forming cracks propagate in a vacuum until the free surface of the component or the specimen is reached and then surface crack propagation under the influence of the ambient atmosphere takes place [9]. The fisheye formation may take place in two stages forming a fine granular area (FGA) around the crack initiating inclusion and a neighboring smooth area (SA). Results of Stanzl-Tschegg et al. show that the smooth area of a fisheye may be followed by a rougher fracture surface indicating a more ductile fracture mode [10,11]. The crack propagation stage forming the rougher fracture surface was identified as the Paris regime [10]. However, it is noticeable that FGA formation may not necessarily be detectable for internal crack initiation [12]. There are several studies about steel cleanliness and control of non-metallic inclusions to minimize the negative influence of critical inclusions and improve the fatigue strength [13,14]. In addition, the study of non-metallic inclusions, in particular with respect to their type, area, shape and position, is of high interest in order to predict the fatigue strengths of engineering steels and allow for safe design [1,15–19]. Besides approaches to enhance the purity of steels and minimizing critical inclusion formation during steel production, a thermomechanical treatment (TMT) in the temperature range of maximum dynamic strain aging (DSA) is another approach to increase the fatigue strength of steels by strengthening the microstructure around the inclusions after the steel production [20,21]. Increasing the fatigue limits of SAE4140 in the HCF and VHCF regimes by a TMT was recently shown in [22]. It is assumed that during the TMT, plastic deformation in the temperature range of DSA introduces a more stable dislocation structure around the non-metallic inclusions, which delays or prevents crack initiation thus resulting in higher HCF lifetimes and increased fatigue strengths [20]. However, the detailed mechanisms occurring in the vicinity of inclusions during a TMT are still unknown. Further, it is unclear whether the shape, area and position relative to the surface (e.g., depth) of an inclusion influence the effectiveness of a TMT or whether a TMT may affect the shape, area and position of critical crack-initiating inclusions. The shape, size and inclusion depth of crack-initiating non-metallic inclusions were investigated and analyzed by other researchers. However, these investigations are mostly simulation-based [15,23]. There are experimental investigations reporting no correlation between the areas of critical non-metallic inclusions or inclusion depth and lifetime (N_f) [9,24,25]. However, other studies showed that the area and inclusion depth of critical inclusions decreases with increasing lifetime [18,26]. In order to analyze the influence of non-metallic inclusions on the fatigue behavior, various parameters such as type, shape, area, distribution and applied stress should be taken into account and this makes the fatigue behavior analysis rather complex [17,27]. In this study, we systematically investigate the influence of shape, area and position of critical non-metallic inclusions on the high cycle fatigue lifetime of the steel SAE4140 in a quenched and tempered state. In a second step, we study whether a TMT, which increases the fatigue limit and the fatigue lifetime in the HCF regime [22], affects the shape, area and position of critical inclusions and whether the effectiveness of the TMT is influenced by these parameters. The goal of the study is to gain a better understanding of how the TMT and the inclusion parameters interact. For this, stress-controlled high cycle fatigue tests at room temperature were conducted on round specimens of quenched and tempered specimens and on additionally thermomechanically treated specimens of SAE4140. In order to rule out the influence of load amplitude, all fatigue tests were conducted with a constant stress amplitude. Fracture surfaces, fisheyes and inclusions were investigated using scanning electron microscopy (SEM) and energy dispersive X-ray spectroscopy (EDX).

2. Materials and Experimental Procedure

The investigated material is the steel SAE4140 (according to EN ISO 683-2, German designation: 42CrMo4) in quenched and tempered state. The chemical composition is given in Table 1. The material was delivered in soft-annealed state in the form of round bars from which a near-net-shape geometry of the specimens was machined by turning.

Table 1. Chemical composition of the test material in wt.%.

C	Si	Mn	P	S	Cr	Mo	Fe
0.430	0.259	0.743	0.012	0.039	1.060	0.207	Balance

After the quenching and tempering, the specimen geometries were again machined to avoid dimensional changes after heat treatment. The final specimen geometry for fatigue testing can be seen in Figure 1.

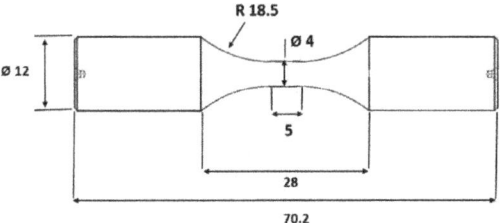

Figure 1. Specimen geometry used for experiments [28].

The initial heat treatment was conducted in a vacuum furnace and included austenitization at 840 °C for about 20 min, quenching in oil to reach room temperature and finally tempering at 180 °C for 2 h. After the heat treatment, a fully martensitic microstructure was obtained. In this state, the material exhibits a 0.2% yield strength of about 1500 MPa and ultimate tensile strength of 1900 MPa [29], which correlates to a hardness of 594 ± 5 HV 0.5. The hardness reduction due to the TMT, which takes place at 265 °C, is not significant [22]. The fatigue tests as well as the TMT were conducted on a servohydraulic push–pull testing machine with a capacity of 100 kN. The force was measured with a 100 kN force transducer. The TMT was conducted at a temperature of 265 °C, which was identified as temperature where the DSA effects are most pronounced [22]. The specimens were heated inductively to this temperature at zero stress and were kept in this state for 15 s soaking time. Then the cyclic mechanical treatment of the TMT was applied with a sinusoidal waveform at a frequency of 1 Hz and a gradually increasing stress amplitude according to the procedure described in [20]. The starting stress amplitude was 600 MPa and the maximum stress amplitude was 1600 MPa with a step of 100 MPa between individual stress amplitudes. At each stress amplitude, 5 cycles were applied. The total time of a specimen at 265 °C during the TMT was 70 s. A more detailed description of the TMT can be found in [22]. Specimens with thermomechanical treatment are designated as TMT specimens, while the specimens after the initial heat treatment served as a reference in the fatigue tests and are designated as heat treated (HT) specimens. Figure 2 shows micrographs of polished and etched longitudinal sections of an HT and a TMT specimen in the initial state. Both specimens exhibit a fully martensitic microstructure. There are no significant differences between HT and TMT specimens, which was expected since the annealing time of 70 s during the TMT at 265 °C is too short to cause changes that can be observed by light microscopy.

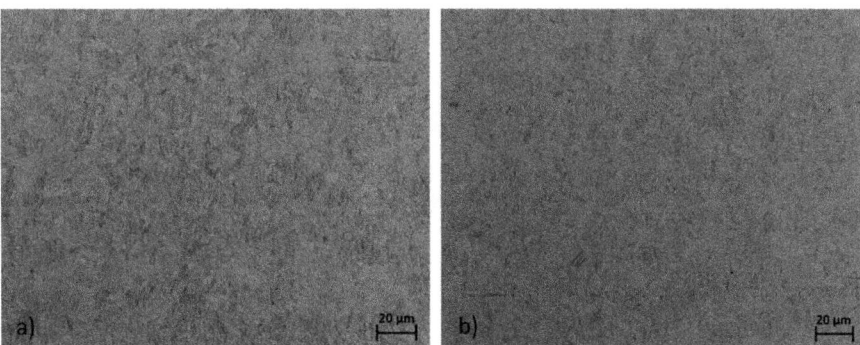

Figure 2. Representative light microscopic images of polished and etched longitudinal sections. The etchant was Nital. (**a**) HT specimen; (**b**) TMT specimen.

Both, TMT specimens and HT specimens were cycled in stress-controlled tests at room temperature until fracture. For cycling, a sinusoidal waveform with a frequency of 50 Hz was applied. The stress amplitude for all tests was 775 MPa with a load ratio of $R = -1$ (fully reversed loading). The stress amplitude of 775 MPa was chosen because it typically produces internal crack initiation at non-metallic inclusions as well as fisheye formation for both HT and TMT specimens [22]. The obtained lifetimes were between 1 and 10 million cycles. An SEM (Carl ZeissAG, Oberkochen, Germany) was used in secondary electron (SE) mode to analyze the fracture surfaces of specimens. The sizes and depths of inclusions and the sizes of fisheyes were measured using SEM images and the software KLONK image measurement (15.1.2.1, Image Measurement Corporation, Cheyenne, WY, USA). The inclusion depth was defined as minimum distance between the inclusion center and the surface. EDX (Thermo Fisher Scientific Inc., Waltham, MA, USA) was applied to determine the chemical composition of the non-metallic inclusions on fracture surfaces.

3. Results and Discussion

3.1. Lifetime Results

Figure 3 presents the fatigue lifetimes of all tested specimens (eight HT specimens and eight TMT specimens) at the constant stress amplitude of 775 MPa. In the considered lifetime regime, the TMT improved the average lifetime by about 40%, which is assumed to be a result of stabilization of the dislocation structure around inclusions and is consistent with the results of previous studies [21,22].

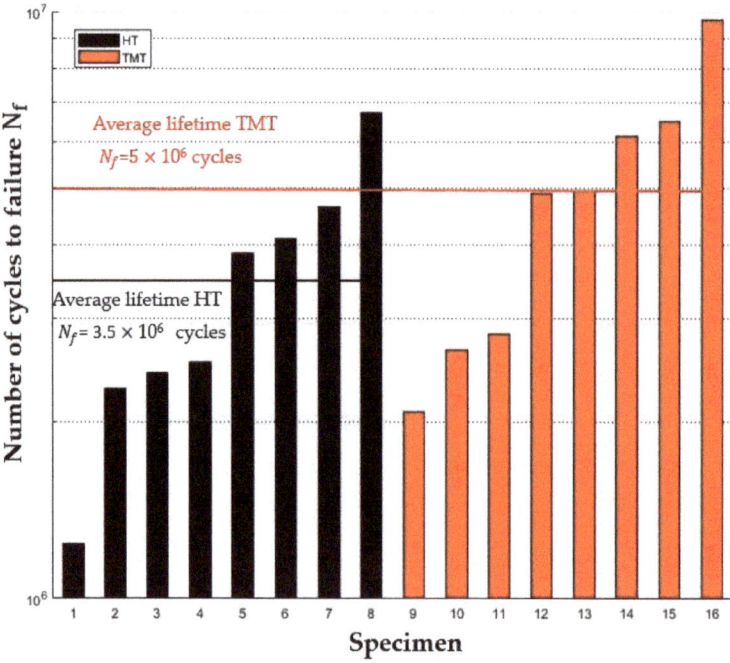

Figure 3. Fatigue lifetime analysis of HT and TMT specimens at given stress amplitude of 775 MPa.

3.2. Type and Form of Critical Non-Metallic Inclusions on the Fracture Surface

All tested specimens fractured due to cracks that initiated at inclusions in the volume. Chemical composition analysis using EDX showed that for both TMT and HT specimens the critical crack initiating inclusions were always oxides of type AlCaO. Figure 4 shows a representative SEM image of a crack-initiating inclusion along with the appropriate EDX mapping.

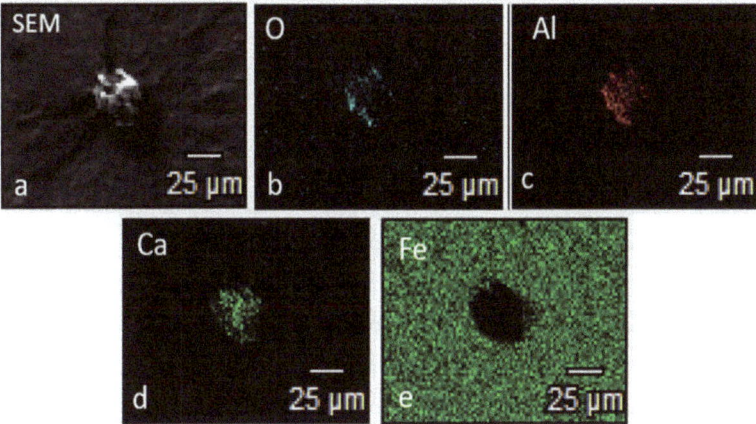

Figure 4. EDX mapping analysis of a crack-initiating inclusion. (**a**) SEM image; (**b**) O content; (**c**) Al content; (**d**) Ca content; (**e**) Fe content.

It was reported elsewhere that AlCaO oxides are typically the most harmful ones for internal fatigue crack initiation in SAE4140 [30]. Figure 5 shows typical critical inclusions on fracture surfaces. There are eye-shaped inclusions including sharp edges on one axis, which are produced during the rolling process [17], and round inclusions without sharp edges. Both forms of critical inclusions were found on the fracture surfaces of both the HT and TMT specimens.

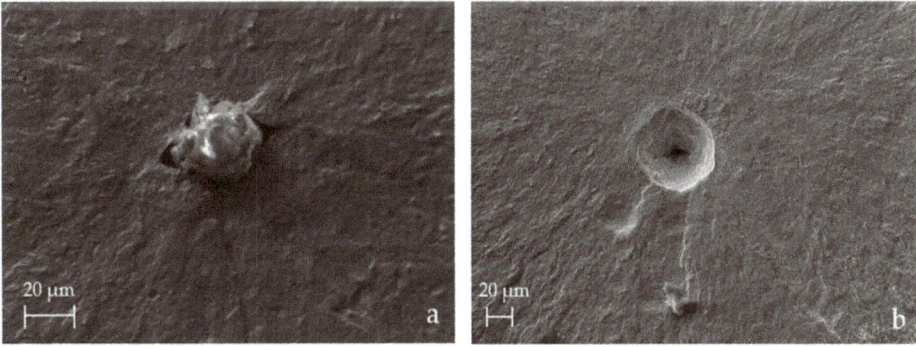

Figure 5. SEM images from two typical non-metallic inclusion shapes. (**a**) Eye-shaped inclusion with two sharp edges—HT specimen; (**b**) Round inclusion—TMT specimen.

Figure 6 shows two critical inclusions of TMT specimen fracture surfaces, which cannot be categorized into round or eye-shaped. Instead, these inclusions exhibit rather angular shapes, but without sharp edges. Nevertheless, also these inclusions were oxides of type AlCaO.

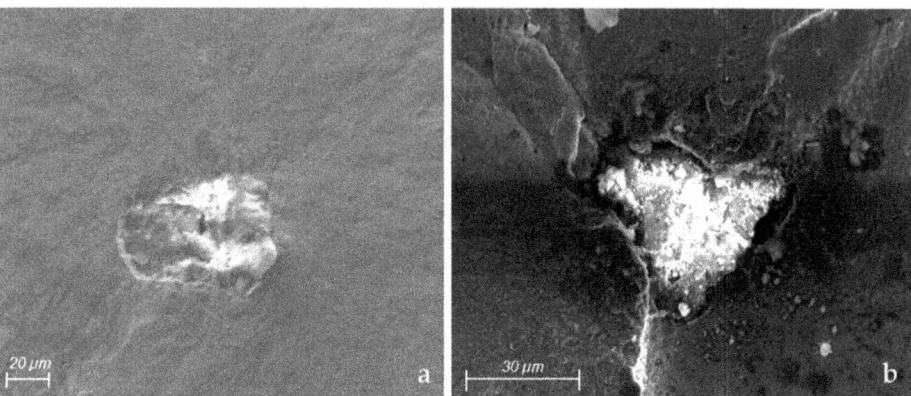

Figure 6. Critical inclusions. (**a**,**b**) with angular shape of TMT specimens.

3.3. Area, Shape and Maximum Stress Intensity Factor ($K_{max,Inc}$) of Non-Metallic Inclusion

Figure 7 shows the area of critical inclusions of TMT and HT specimens versus the fatigue lifetime for the constant stress amplitude of 775 MPa. The markers indicate the form of critical inclusion. For both HT and TMT specimens, the area of critical non-metallic inclusions scatters considerably. However, for both specimen types, no significant influence of the inclusion area on the lifetime can be observed. This is somewhat surprising since one might expect that for the given constant stress amplitude, larger inclusions lead to earlier crack initiation and thus shorter lifetimes. However, other studies have also shown that the HCF lifetime of steels after crack initiation at internal inclusions is independent of the inclusion area [9,24]. The average area of critical inclusions is slightly larger for TMT specimens than for HT specimens. Since the TMT has no influence on the inclusion size distribution, this slight effect is presumably due to the scatter of the inclusion sizes in the specimens. It can be seen that all round and angular-shaped critical inclusions for both HT and TMT specimens exhibit a square root area greater than about 35 μm. All critical inclusions with square root areas smaller than about 35 μm are eye-shaped with sharp edges. Hence, smaller inclusions with sharp edges can be as detrimental as larger ones without sharp edges, which can be explained by the stress concentration near the sharp edges [15,16]. Therefore, the area of inclusion is not the only significant parameter, which determines whether it can become a critical crack initiating inclusion.

With the measured area of the crack-initiating inclusions, the maximum stress intensity factor for subsurface inclusions can be derived with the following Equation (1):

$$K_{max,Inc} = 0.5 \times \sigma_{max} \times \sqrt{\pi \times \sqrt{area_{inc}}} \tag{1}$$

For a square root of the inclusion area of 35 μm, $K_{max,\,Inc}$ is about 4.05 MPa.m$^{1/2}$, which is apparently the required minimum to induce a fatigue crack from inclusions without sharp edges. For eye-shaped inclusions featuring sharp edges, the required value of $K_{max,\,Inc}$ is accordingly lower.

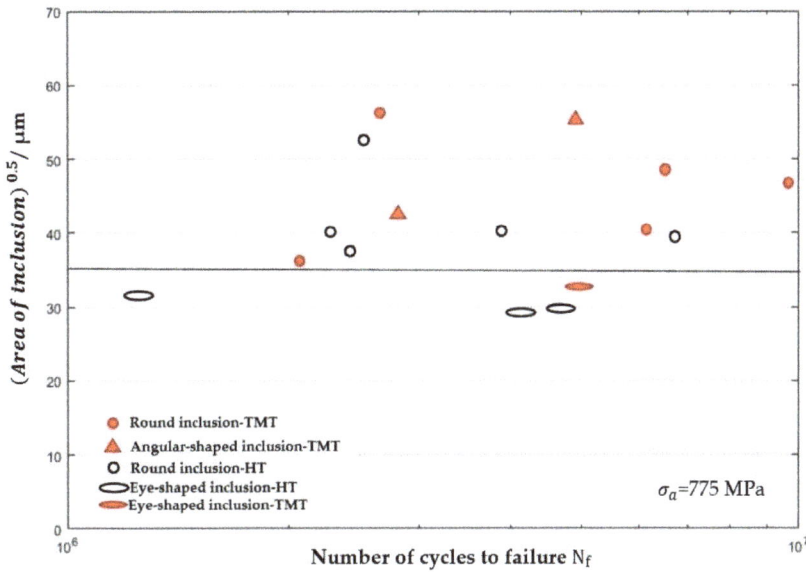

Figure 7. Area of critical inclusions for specimens with TMT and HT versus lifetime at σ_a = 775 MPa.

3.4. Influence of Inclusion Depth on the Fatigue Lifetime

Figure 8 compares the lifetimes of three HT specimens, which initiated cracks at inclusions of a similar size and form, but at various distances from the surface (inclusion depth). Apparently, the inclusion depth does not correlate with the lifetime. It can be assumed that due to the detrimental effect of air moisture, the crack propagation rate is much higher for an internal crack that has reached the surface than for a crack that is still in the fisheye stadium [11]. Consequently, the lifetime period spent in the crack propagation stage should be longer when the critical inclusion is located at a larger distance from the surface. Since the results indicate no obvious influence of inclusion depth which is in good agreement with other investigations in the HCF regime [24,25], we can infer that the lifetime is mostly governed by the period before crack initiation and the crack propagation stage occupies only a small percentage of the total lifetime. We assume that this is true for both HT and TMT specimens. Hence, the observed longer lifetimes for TMT specimens (see Figure 3) result presumably from longer periods until crack initiation, which is in accordance with the assumed TMT strengthening mechanism of a more stable dislocation structure around the inclusions. The parameter, which causes a lifetime scatter of about factor three for critical inclusions with almost the same size and form under constant stress amplitude and the same specimen state (Figure 8), remains unclear.

3.5. Fisheye Formation and Inclusion Depth

Fracture surface analyses of both TMT and HT specimens show the formation of fisheyes around all the critical inclusions. Figure 9 shows typical fracture surfaces of three specimens. As it can be seen from Figure 9a,b, cracks initiate at the inclusion inside the volume and the inclusion is surrounded by the fisheye. Both fisheyes exhibit a smooth area (SA) until reaching the surface and no obvious changes in the fisheye surface structure can be observed. As soon as the fisheye reaches the surface, oxidation-assisted fatigue crack growth begins. In this stage, the cracks grow predominantly away from the touching surface, as can be seen in Figure 9c. Hence, the formation of the nearly round fisheye ends as soon as the internal crack reaches the surface. The fisheye presented in Figure 9c shows a transition from smooth (SA) to rougher (RA) fracture surface with a wavy structure of radially extended peaks and troughs. A similar transition of the fisheye surface characteristics was reported by Stanzl-Tschegg et al. [10,11] who found that the transition from smooth to rough fisheye surface goes along with a significant increase in crack propagation rate [11].

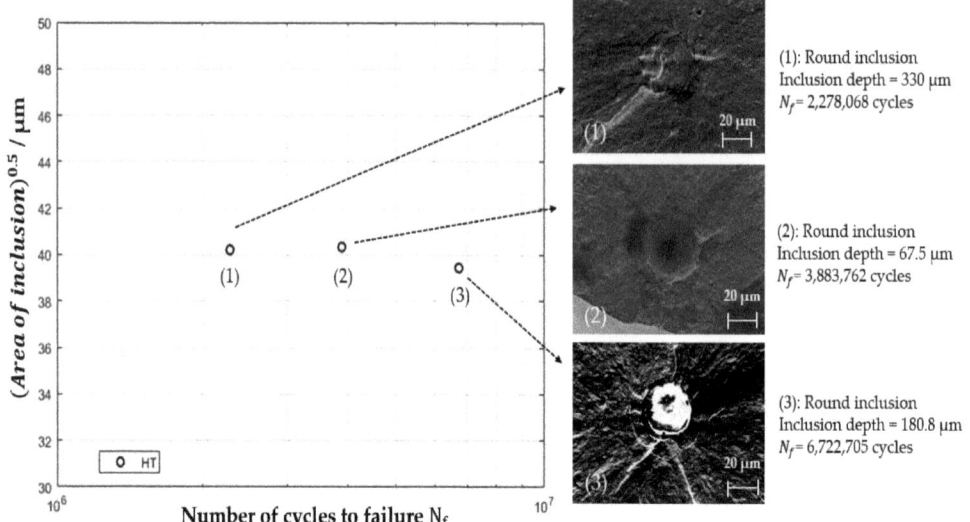

Figure 8. Evaluating the effect of inclusion depth on the lifetime for HT specimens at σ_a = 775 MPa.

Figure 10a,b show the fisheye radius and the inclusion depth over the fatigue lifetime, respectively. The data points in both diagrams correlate strongly, which confirms that fisheyes grow in circular form starting from the critical inclusion until they reach the surface. For both TMT and HT specimens there is no clear relation between inclusion depth or fisheye radius and lifetime. As already discussed in Section 3.4, this means that the lifetime portion in the crack propagation stage is relatively small. The fisheye sizes and the corresponding inclusion depths of TMT specimens are significantly larger than for HT specimens. This indicates that the TMT has a better strengthening effect on inclusions near the surface compared to the inclusions located deeper in the volume, thus shifting the crack initiation site further into the volume. Possibly, the plastic deformation during the TMT is more pronounced in near-surface regions because the respective grains have no neighboring grains in the direction to the surface. It could also be that a radial temperature gradient in the specimen's gauge length occurred during the TMT. The TMT temperature of 265 °C was reached by inductive heating and was measured and controlled at the specimen surface. The soaking time at 265 °C before the mechanical loading starts is only 15 s in order to minimize purely thermal effects. Hence, it is possible that the specimens were not completely heated through when the mechanical loading began. If the temperature in the specimen center would be significantly lower, the strengthening DSA effects might be less effective. If that were the case, an even better increase in the fatigue lifetime and fatigue strength might be possible with a more homogeneous temperature distribution in the treated volume. This may be reached with a longer soaking time at the TMT temperature before the mechanical loading begins. Figure 10 also indicates the fisheye surface structure showing that, for fisheye radii below 300 μm, the fisheye surface has only a smooth structure, while for radii above 300 μm, a smooth and a rougher, wavy structure as in Figure 9c could be observed. This is true for both TMT and HT specimens. With increasing fisheye radius, the roughness and waviness of the fracture surface becomes more pronounced. The transition from smooth to the rough fisheye structure occurred always at a radius of about 300 μm. The corresponding stress intensity factor according to Equation (1) is about 15.8 MPa.m$^{1/2}$, which may be identified as the transition stress intensity factor to the Paris regime [10]. Obviously, if the fisheye reaches the surface before this value is reached, only a smooth fisheye structure is formed. Hence, it depends mainly on the inclusion depth whether a fisheye grows in one or two stages. The results indicate that the TMT has no influence on the fisheye formation and thus on the crack propagation. This was expected since the TMT is supposed to strengthen the microstructure very close to the inclusions and not in the bulk.

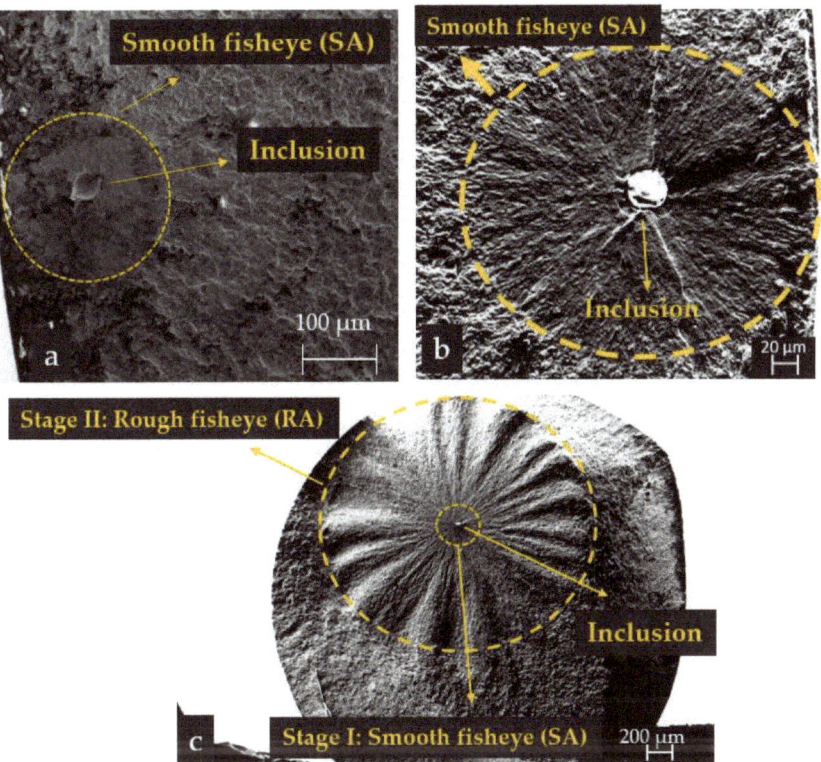

Figure 9. Fisheye formation analysis of fracture surfaces at a stress amplitude of 775 MPa. (**a**) TMT specimen, inclusion is surrounded by a smooth fisheye (fisheye formation in a single stage, N_f = 4,932,070 cycles, inclusion depth = 125 µm); (**b**) HT specimen, inclusion is surrounded by a smooth fisheye (fisheye formation in a single stage, N_f = 6,722,705 cycles, inclusion depth = 180.8 µm); (**c**) TMT specimen, inclusion is surrounded by a small smooth fisheye around the inclusion and a bigger rough fisheye around the smooth one (fisheye formation in two stages, N_f = 9,706,545 cycles, inclusion depth = 1.3 mm).

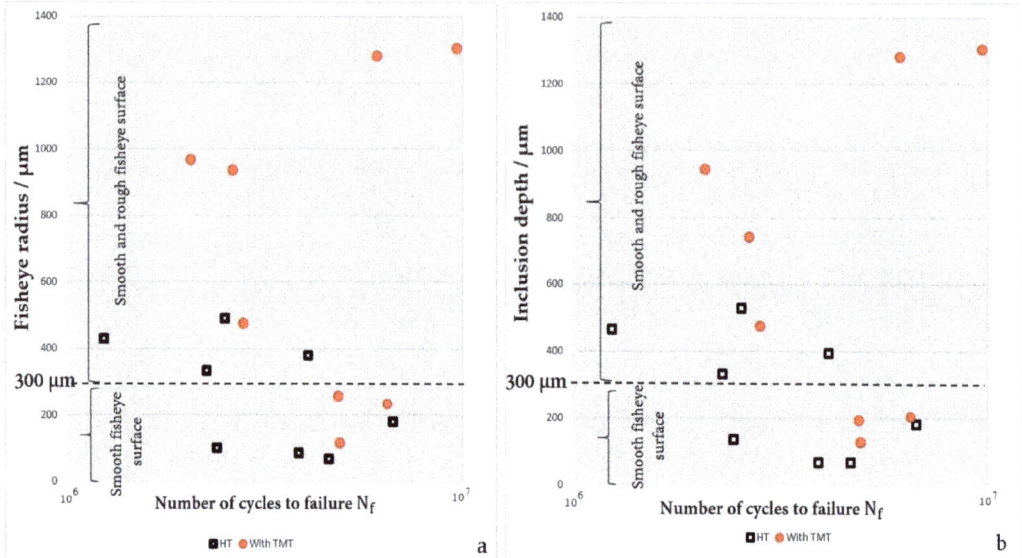

Figure 10. (a) Fisheye radius versus lifetime at σ_a = 775 MPa. (b) Inclusion depth versus lifetime at σ_a = 775 MPa.

4. Conclusions

In this study the influence of thermo-mechanical treatment (TMT) on the lifetime, internal crack initiation and crack propagation behavior under high cycle fatigue loading at a constant stress amplitude of 775 MPa was investigated. The results can be summarized as follows:

(1) For the tested stress amplitude, all specimens fractured from cracks that initiated at oxide inclusions of type AlCaO within the volume. Fisheye fracture surfaces could be observed in all cases. As expected, the TMT increased the average fatigue lifetime by about 40% due to plastic deformation in the temperature regime of dynamic strain aging, which leads to a strengthened dislocation structure around inclusions and delays crack initiation.

(2) The area of the critical inclusion and the inclusion depth has only an insignificant influence on the overall fatigue lifetime for both the TMT and HT specimens. Depending on the shape of critical inclusion, the minimum required inclusion area for the crack initiation and its corresponding stress intensity factor could change.

(3) The inclusion depth correlates strongly with the fisheye radius. This means internal cracks grow in fisheye mode until they reach the surface. When the fisheye reaches a radius of about 300 μm, which corresponds to a stress intensity factor of about 15.8 MPa.m$^{1/2}$, the fracture surface appearance changes from smooth to a rougher wavy form, which is presumably the transition to crack propagation in the Paris regime. TMT had no influence on this behavior.

(4) The TMT increases the average depth of critical inclusions considerably indicating that the strengthening of the microstructure is more effective in the near-surface regions. The reason for this might be that there is a radial temperature gradient in the specimen during the TMT resulting in varying effectiveness of dynamic strain aging effects. With a more homogeneous temperature distribution, an even better effect of the TMT might be possible.

Author Contributions: Conceptualization, A.K. and S.G.; investigation, A.K.; original draft preparation, A.K.; writing—review and editing, S.G. and M.H.; supervision, M.H. All authors have read and agreed to the published version of the manuscript.

Funding: This research project was funded by the Deutsche Forschungsgemeinschaft (DFG, German Research Foundation)—Project number 408139037.

Data Availability Statement: Not applicable.

Acknowledgments: We gratefully acknowledge the financial support by the KIT-Publication Fund of the Karlsruhe Institute of Technology.

Conflicts of Interest: There are no conflicts of interest regarding the presented work.

References

1. Murakami, Y.; Kodamab, S.; Konumac, S. Quantitative Evaluation of Effects of Non-Metallic Inclusions on Fatigue Strength of High Strength Steels. I: Basic Fatigue Mechanism and Evaluation of Correlation between the Fatigue Fracture Stress and the Size and Location of Non-Metallic Inclusions. *Int. J. Fatigue* **1989**, *11*, 291–298. [CrossRef]
2. Kucharski, P.; Lesiuk, G.; Szata, M. *Description of Fatigue Crack Growth in Steel Structural Components Using Energy Approach—Influence of the Microstructure on the FCGR*; AIP Publishing LLC: Fojutowo, Poland, 2016; p. 050003.
3. Starke, P.; Walther, F.; Eifler, D. Fatigue Assessment and Fatigue Life Calculation of Quenched and Tempered SAE 4140 Steel Based on Stress–Strain Hysteresis, Temperature and Electrical Resistance Measurements. *Fatigue Fract. Eng. Mater. Struct.* **2007**, *30*, 1044–1051. [CrossRef]
4. Mughrabi, H. On 'Multi-Stage' Fatigue Life Diagrams and the Relevant Life-Controlling Mechanisms in Ultrahigh-Cycle Fatigue: ON 'MULTI-STAGE' FATIGUE LIFE DIAGRAMS. *Fatigue Fract. Eng. Mater. Struct.* **2002**, *25*, 755–764. [CrossRef]
5. Mughrabi, H. Specific Features and Mechanisms of Fatigue in the Ultrahigh-Cycle Regime. *Int. J. Fatigue* **2006**, *28*, 1501–1508. [CrossRef]
6. Lipiński, T.; Wach, A.; Detyna, E. Influence of Large Non-Metallic Inclusions on Bending Fatigue Strength Hardened and Tempered Steels. *Adv. Mater. Sci.* **2015**, *15*, 33–40. [CrossRef]
7. Li, W.; Deng, H.; Liu, P. Interior Fracture Mechanism Analysis and Fatigue Life Prediction of Surface-Hardened Gear Steel under Axial Loading. *Materials* **2016**, *9*, 843. [CrossRef] [PubMed]
8. Shiozawa, K.; Lu, L.; Ishihara, S. S-N Curve Characteristics and Subsurface Crack Initiation Behaviour in Ultra-Long Life Fatigue of a High Carbon-Chromium Bearing Steel: S-N Curve and Crack Initiation in Ultra-Long Life Fatigue. *Fatigue Fract. Eng. Mater. Struct.* **2001**, *24*, 781–790. [CrossRef]
9. Shiozawa, K.; Murai, M.; Shimatani, Y.; Yoshimoto, T. Transition of Fatigue Failure Mode of Ni–Cr–Mo Low-Alloy Steel in Very High Cycle Regime. *Int. J. Fatigue* **2010**, *32*, 541–550. [CrossRef]
10. Stanzl-Tschegg, S.E. Fracture Mechanical Characterization of the Initiation and Growth of Interior Fatigue Cracks: Fracture Mechanics for Interior Fatigue Cracks. *Fatigue Fract. Eng. Mater. Struct.* **2017**, *40*, 1741–1751. [CrossRef]
11. Stanzl-Tschegg, S.; Schönbauer, B. Near-Threshold Fatigue Crack Propagation and Internal Cracks in Steel. *Procedia Eng.* **2010**, *2*, 1547–1555. [CrossRef]
12. Li, Y.-D.; Zhang, L.-L.; Fei, Y.-H.; Liu, X.-Y.; Li, M.-X. On the Formation Mechanisms of Fine Granular Area (FGA) on the Fracture Surface for High Strength Steels in the VHCF Regime. *Int. J. Fatigue* **2016**, *82*, 402–410. [CrossRef]
13. Guo, J.; Han, S.; Chen, X.; Guo, H.; Yan, Y. Control of Non-Metallic Inclusion Plasticity and Steel Cleanliness for Ultrathin 18 Pct Cr-8 Pct Ni Stainless Steel Strip. *Metall. Mater. Trans. B* **2020**, *51*, 1813–1823. [CrossRef]
14. Yang, Z.G.; Li, S.X.; Zhang, J.M.; Zhang, J.F.; Li, G.Y.; Li, Z.B.; Hui, W.J.; Weng, Y.Q. The Fatigue Behaviors of Zero-Inclusion and Commercial 42CrMo Steels in the Super-Long Fatigue Life Regime. *Acta Mater.* **2004**, *52*, 5235–5241. [CrossRef]
15. Xie, J.P.; Wang, A.Q.; Wang, W.Y.; Li, J.W.; Yang, D.X.; Zhang, K.F.; Ma, D.Q. Stress Field Numerical Simulation of the Inclusions in Large Rudder Arm Steel Casting. *Adv. Mater. Res.* **2011**, *311–313*, 906–909. [CrossRef]
16. Gu, C.; Liu, W.; Lian, J.; Bao, Y. In-Depth Analysis of the Fatigue Mechanism Induced by Inclusions for High-Strength Bearing Steels. *Int. J. Miner. Metall. Mater.* **2021**, *28*, 826–834. [CrossRef]
17. da Costa e Silva, A.L.V. The Effects of Non-Metallic Inclusions on Properties Relevant to the Performance of Steel in Structural and Mechanical Applications. *J. Mater. Res. Technol.* **2019**, *8*, 2408–2422. [CrossRef]
18. Krewerth, D.; Lippmann, T.; Weidner, A.; Biermann, H. Influence of Non-Metallic Inclusions on Fatigue Life in the Very High Cycle Fatigue Regime. *Int. J. Fatigue* **2016**, *84*, 40–52. [CrossRef]
19. Murakami, Y.; Endo, M. Effects of Defects, Inclusions and Inhomogeneities on Fatigue Strength. *Int. J. Fatigue* **1994**, *16*, 163–182. [CrossRef]
20. Kerscher, E.; Lang, K.; Vohringer, O.; Lohe, D. Increasing the Fatigue Limit of a Bearing Steel by Dynamic Strain Ageing. *Int. J. Fatigue* **2008**, *30*, 1838–1842. [CrossRef]
21. Kerscher, E.; Lang, K.-H.; Löhe, D. Increasing the Fatigue Limit of a High-Strength Bearing Steel by Thermomechanical Treatment. *Mater. Sci. Eng. A* **2008**, *483–484*, 415–417. [CrossRef]

22. Khayatzadeh, A.; Sippel, J.; Guth, S.; Lang, K.-H.; Kerscher, E. Influence of a Thermo-Mechanical Treatment on the Fatigue Lifetime and Crack Initiation Behavior of a Quenched and Tempered Steel. *Metals* **2022**, *12*, 204. [CrossRef]
23. Guan, J.; Wang, L.; Zhang, C.; Ma, X. Effects of Non-Metallic Inclusions on the Crack Propagation in Bearing Steel. *Tribol. Int.* **2017**, *106*, 123–131. [CrossRef]
24. Shiozawa, K.; Hasegawa, T.; Kashiwagi, Y.; Lu, L. Very High Cycle Fatigue Properties of Bearing Steel under Axial Loading Condition. *Int. J. Fatigue* **2009**, *31*, 880–888. [CrossRef]
25. Liu, P.; Li, W.; Nehila, A.; Sun, Z.; Deng, H. High Cycle Fatigue Property of Carburized 20Cr Gear Steel under Axial Loading. *Metals* **2016**, *6*, 246. [CrossRef]
26. Lei, Z.; Hong, Y.; Xie, J.; Sun, C.; Zhao, A. Effects of Inclusion Size and Location on Very-High-Cycle Fatigue Behavior for High Strength Steels. *Mater. Sci. Eng. A* **2012**, *558*, 234–241. [CrossRef]
27. Taheri, F.; Trask, D.; Pegg, N. Experimental and Analytical Investigation of Fatigue Characteristics of 350WT Steel under Constant and Variable Amplitude Loadings. *Mar. Struct.* **2003**, *16*, 69–91. [CrossRef]
28. Labisch, S.; Weber, C. *Technisches Zeichnen*; Springer: Wiesbaden, Germany, 2014; ISBN 978-3-8348-0915-5.
29. Kaiser, D. *Experimentelle Untersuchung und Simulation des Kurzzeitanlassens unter Berücksichtigung Thermisch Randschichtgehärteter Zustände am Beispiel von 42CrMo4*; Karlsruher Institut für Technologie: Karlsruhe, Germany, 2019; p. 194.
30. Lang, K.-H.; Korn, M.; Rohm, T. Very High Cycle Fatigue Resistance of the Low Alloyed Steel 42CrMo4 in Medium- and High-Strength Quenched and Tempered Condition. *Procedia Struct. Integr.* **2016**, *2*, 1133–1142. [CrossRef]

Article

Effects of Non-Metallic Inclusions and Mean Stress on Axial and Torsion Very High Cycle Fatigue of SWOSC-V Spring Steel

Ulrike Karr [1], Bernd M. Schönbauer [1], Yusuke Sandaiji [2] and Herwig Mayer [1,*]

[1] Department of Material Sciences and Process Engineering, Institute of Physics and Materials Science, University of Natural Resources and Life Sciences, Vienna (BOKU), Peter-Jordan-Str. 82, 1190 Vienna, Austria; ulrike.karr@boku.ac.at (U.K.); bernd.schoenbauer@boku.ac.at (B.M.S.)

[2] Materials Research Laboratory, KOBE STEEL LTD., 1-5-5 Takatsukadai, Nishi-ku, Kobe 651-2271, Hyogo, Japan; sandaiji.yusuke@kobelco.com

* Correspondence: herwig.mayer@boku.ac.at; Tel.: +43-1-47654-89202

Citation: Karr, U.; Schönbauer, B.M.; Sandaiji, Y.; Mayer, H. Effects of Non-Metallic Inclusions and Mean Stress on Axial and Torsion Very High Cycle Fatigue of SWOSC-V Spring Steel. *Metals* **2022**, *12*, 1113. https://doi.org/10.3390/met12071113

Academic Editors: Martin Heilmaier and Martina Zimmermann

Received: 1 June 2022
Accepted: 23 June 2022
Published: 28 June 2022

Publisher's Note: MDPI stays neutral with regard to jurisdictional claims in published maps and institutional affiliations.

Copyright: © 2022 by the authors. Licensee MDPI, Basel, Switzerland. This article is an open access article distributed under the terms and conditions of the Creative Commons Attribution (CC BY) license (https://creativecommons.org/licenses/by/4.0/).

Abstract: Inclusion-initiated fracture in high-strength spring steel is studied for axial and torsion very high cycle fatigue (VHCF) loading at load ratios of $R = -1$, 0.1 and 0.35. Ultrasonic *S-N* tests are performed with SWOSC-V steel featuring intentionally increased numbers and sizes of non-metallic inclusions. The fatigue limit for axial and torsion loading is considered the threshold for mode I cracks starting at internal inclusions. The influence of inclusion size and Vickers hardness on cyclic strength is well predicted with Murakami and Endo's $\sqrt{area}$ parameter model. In the presence of similarly sized inclusions, stress biaxiality is considered by a ratio of torsion to axial fatigue strength of 0.86. Load ratio sensitivity is accounted for by the factor $((1-R)/2)^\alpha$, with α being 0.41 for axial and 0.55 for torsion loading. VHCF properties under torsion loading cannot appropriately be deduced from axial data. In contrast to axial loading, the defect sensitivity for torsion loading increases significantly with superimposed static mean load, and no inclusion-initiated fracture is found at $R = -1$. Size effects and the stress gradient effective under torsion loading are considered to explain smaller crack initiating inclusions found in torsion ultrasonic fatigue tests.

Keywords: cyclic torsion; non-metallic inclusions; fatigue limit; mean stress sensitivity; defect tolerance; ultrasonic fatigue

1. Introduction

Change of crack initiation mechanism from strain localization at surface grains in the high cycle fatigue (HFC) regime to internal inclusion-initiated fracture in the very high cycle fatigue (VHCF) regime is a primary reason for the absence of a fatigue limit in high strength steels [1–5]. Fatigue cracks initiate at inclusions at stress amplitudes below the conventional fatigue limit and grow in a vacuum-like environment at extremely low initial growth rates [6,7]. Intermitted growth and mean growth rates below 10^{-13} m/cycle can add up to lifetimes of 500 million cycles [8]. With the increasing size of an inclusion, its detrimental influence on cyclic strength becomes stronger [9–11]. For similarly sized inclusions, interface fracture (e.g., at aluminates) is less detrimental than through-particle fracture of inclusions that are tightly bonded to the matrix (e.g., TiN) [12–14]. Cycling at different mean stresses not only shifts the courses of the *S-N* curves but also affects the probability of internal inclusion-initiated fracture [15–20]. However, internal inclusions are not the only crack initiating locations for VHCF failure. Cracks can also start in the metal matrix without an inclusion [12,19,21–24], at grain boundaries [19,22] or at stress raisers (scratches, surface inclusions) at the surface [25–28]. However, for high-strength steels with compressive surface residual stresses introduced, e.g., by grinding, shot peening, carburizing or nitriding, inclusion-initiated fracture is the most important VHCF failure mechanism.

In the literature, the influence of inclusions on fatigue lifetimes is mainly described for cyclic axial loading conditions. However, coil springs or roller bearings are technical components that are subjected to very high numbers of shear load cycles in service with mean shear stress superimposed. Valve springs, for example, are loaded with cyclic shear stresses superimposed on a static torque resulting in load ratios between $R = 0.3$ and 0.5. Crack initiation at internal inclusions was found as an important failure mode in spring steels used for valves in engines that failed above 10^7 cycles [29–31]. Understanding inclusion-initiated fracture under cyclic torsion loading, in particular with superimposed mean shear stresses, is therefore of great scientific as well as technical importance.

With conventional (e.g., servo-hydraulic, rotating bending) testing techniques, fatigue tests in the VHCF regime are very time-consuming, and it is, therefore, favorable to increase the testing frequency. Ultrasonic fatigue testing has been found most appropriate for the rapid collection of VHCF data [32,33]. The high cycling frequency not only shortens testing times but also allows for testing up to extremely high numbers of cycles of 10^{11} [34], a regime that is not accessible with conventional equipment. Ultrasonic fatigue test set-ups for fully reversed torsion loading [35] and torsion loading with superimposed static shear stresses [36] have been invented in the authors' laboratory.

Ultrasonic torsion fatigue tests of SWOSC-V spring steel delivered VHCF failures, however, not starting at inclusions but at the surface [37]. Solely one failure from a TiN inclusion was found in high-frequency torsion tests with VDSiCr spring steel [38]. Failures from MnS inclusions under fully reversed cyclic torsion are reported in the literature, however, not for spring steel but for high carbon chromium-bearing steel [39,40]. Further, soft phases at the surface, such as ferrite grains, may act as crack initiation sites under reversed torsion loading [41]. To better simulate actual loading conditions of coil springs, ultrasonic torsion fatigue tests with VDSiCr spring steel were performed at load ratios $R = 0.1, 0.35$ and 0.5 [36]. These investigations showed that ultrasonic torsion fatigue testing is a powerful method for rapid measurement of the torsion fatigue strength and developed a Haigh diagram for limiting lifetimes of 10^9 cycles. However, due to the high cleanness of commercial spring steels and the relatively small material volume subjected to high-stress amplitudes of ultrasonic torsion fatigue specimens, it is unlikely to encounter inclusions large enough to initiate a fatal crack. Similar conclusions may be drawn from cyclic torsion ultrasonic tests of super clean SWOSC-V spring steel, where excellent cyclic strength was demonstrated even for high load ratios due to the absence of inclusion-initiated failure [42]. Failure from inclusions were found in a commercial spring steel in axial but hardly in torsion ultrasonic fatigue specimens [43].

Testing spring steel SWOSC-V with intentionally increased numbers and sizes of inclusions has been a successful way to study the effect of inclusions on torsion VHCF properties. With this material, the influence of inclusions on cyclic shear strength at load ratios of $R = -1, 0.1$ and 0.35 could be investigated in ultrasonic fatigue tests [44]. Cyclic tension-compression and cyclic tension tests of the same material were used to compare the deleterious effects of inclusions under axial and torsion loading [45]. The $\sqrt{area}$ parameter model proposed by Murakami and Endo [9] was used to compare experimentally determined and predicted torsion fatigue strengths for inclusion-initiated fracture [8].

The present paper aims to evaluate and compare the influence of inclusions on axial and torsion VHCF strengths. Specimens with artificial surface defects are cycled in a vacuum, and near-threshold fatigue crack growth is compared for fully reversed tension-compression and torsion loading, respectively. Inclusions are considered as initial cracks and hence, fracture mechanics principles are applied to interpret the VHCF strengths as thresholds for crack propagation. The $\sqrt{area}$ parameter model [9] is used to quantify the effect of inclusions on cyclic strengths. Loading conditions (axial vs. torsion loading) [46], as well as load ratio effects [47], are considered. Several peculiarities must be taken into account when comparing axial and torsion loading: Types of crack initiating inclusions are different [45]; due to the stress gradient, local stress amplitudes at interior inclusions are lower for torsion but not for axial loading; different specimen geometries (in addition to the aforementioned stress gradient) lead to a size effect which influences the size distribution of crack-initiating inclusions [48–51].

2. Material and Method
2.1. Testing Material

The material used in the present investigation is laboratory-made high-strength spring steel based on the chemical composition of SWOSC-V, as shown in Table 1.

Table 1. Chemical composition of SWOSC-V.

C	Si	Mn	Cr	Al	Fe
0.55	1.50	0.70	0.70	0.003	balance

The material was vacuum-induction furnace melted and cast with a high amount of dissolved oxygen, which was realized by adding iron oxide powder. In this way, the size and density of oxide inclusions, such as aluminate inclusions, were intentionally increased. The material also contains MnS inclusions with an elongated shape aligned approximately in the rolling direction (i.e., specimen's length direction) as a result of forging [44]. Manufacturing of the material included soaking, forging, heating, oil quenching and tempering. The resulting martensite structure with occasionally retained austenite features a mean prior austenite grain size of 18.7 µm.

Specimen shapes used in ultrasonic axial and torsion fatigue tests at different R-ratios can be seen in Figure 1a. The stress distribution along the specimens' length axes as a result of resonance vibrations is shown as well as the stress gradient across the cross-section resulting from torsion loading. The volume subjected to more than 95% of the nominal stress is 174 mm^3 and 1.79 mm^3 under axial and torsion loading, respectively.

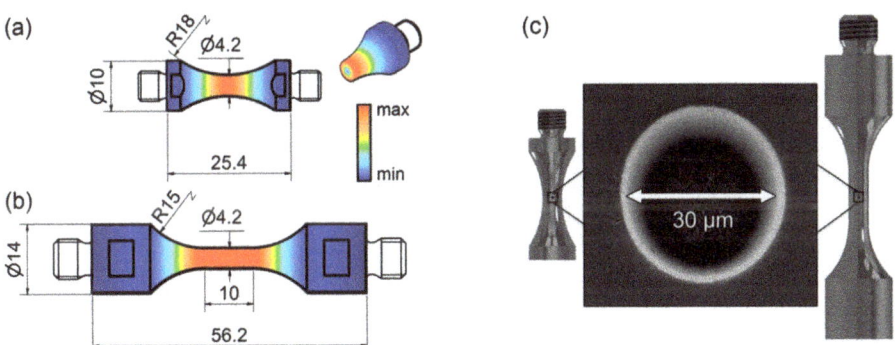

Figure 1. Specimen shapes: for ultrasonic torsion (**a**) and axial (**b**) fatigue tests at different R-ratios (dimensions are in mm); (**c**) for ultrasonic fatigue crack growth rate measurements under fully reversed axial and torsion loading, featuring an artificial defect introduced by Ar$^+$-milling.

For fatigue lifetime measurements in ambient air, the specimens' surface was prepared in order to correspond to in-service conditions: Prior to testing, the specimens were polished to a mirror-like finish and subsequently shot-peened, resulting in compressive residual stresses of 580 ± 130 MPa in the longitudinal and circumferential direction at the surface. Additionally, the specimens were blued. Mechanical properties determined by using fatigue specimens featuring the above-described surface condition are shown in Table 2.

Table 2. Mechanical properties of the investigated material.

Yield Strength (MPa)	Tensile Strength (MPa)	Elongation (%)	Shear Yield Strength (MPa)	Shear Strength (MPa)	Vickers Hardness (HV)
1720	1910	10	1400	1550	530

For fatigue crack growth rate measurements in a high vacuum, the same specimen shapes as in S-N tests were used. The specimens were ground and electro-polished in order to remove residual stresses of the machining process. Small artificial surface defects were then introduced as starter notches by Ar^+-milling (Figure 1b). Prior to the tests, the depth and diameter of each artificial defect were analyzed by SEM.

2.2. S-N Tests under Axial and Torsion Loading

Ultrasonic axial and torsion S-N tests in ambient air were performed at load ratios of $R = -1, 0.1$ and 0.35 up to limiting lifetimes of 5×10^9 load cycles. The hereby employed testing equipment was developed by BOKU and is described extensively in [33,36]. In order to detect nonmetallic inclusions in runout specimens, those surviving 5×10^9 load cycles were further tested as follows: In the case of axial loading, the specimens were reloaded at the same R-ratio at a higher load level. In the case of torsion loading, runouts of all R-ratios were exclusively reloaded at $R = 0.1$, which had proved to be most favorable for internal inclusion-initiated fracture.

In addition to pressurized air cooling, ultrasonic fatigue loading was applied intermittently, i.e., pulses in the range of 100 to 300 ms were followed by pauses of load level-dependent lengths. During the tests, the specimens' temperature was monitored by pyrometry and hereby confirmed to be below 30 °C.

In order to analyze crack initiation sites, fracture surfaces were investigated by scanning electron microscopy (SEM) and energy-dispersive X-ray spectroscopy (EDS). For SEM-imaging, the fracture surfaces were oriented perpendicular to the electron beam, thus corresponding to the projection plane perpendicular to the major principal stress direction. In the case of torsion specimens, this required tilting of the specimens by approximately 45° towards their length axes and takes account of the fact that the major part of crack propagation is under mode I. The SEM images were then used to determine the sizes of crack initiating non-metallic inclusions found at crack initiation sites.

Fracture modes effective during crack initiation under torsion loading were determined. Fracture surfaces generated by an MnS and an aluminate inclusion, respectively, were cut parallel to the specimens' length axes by Ar^+-milling, subsequently polished and analyzed by SEM [8].

2.3. Fatigue Crack Growth Tests Using Specimens with Artificial Defects

Fatigue crack growth rate measurements using ultrasonic fatigue testing equipment were performed in a high vacuum of about 2×10^{-6} mbar in order to eliminate environmental effects and to simulate conditions of very slowly propagating cracks emanating from the interior inclusions. Axial and torsion specimens featuring artificial surface defects were loaded at $R = -1$ at a constant stress amplitude.

In order to choose appropriate stress amplitudes, depths and diameters of the artificial defects were analyzed by SEM prior to testing, and defect sizes represented by $\sqrt{area_{AD}}$ were calculated assuming a semi-elliptical shape according to:

$$\sqrt{area_{AD}} = \sqrt{\left(\frac{2a}{2} \cdot t \cdot \pi\right)/2} \qquad (1)$$

with $2a$ being the diameter at the surface (i.e., $2a = d = 30$ μm) and t being the depth of the artificial defect. $\sqrt{area_{AD}}$ is the square root of the projection area of the artificial defect (AD) perpendicular to the major principal stress direction according to Murakami and Endo [9].

Throughout the tests, the specimens were monitored by means of optical lenses and a charge-coupled device (CCD) camera enabling a maximum magnification of 975-fold. A time interval for saving images was chosen according to crack propagation rates with a maximum of 30 min (corresponding to approximately 5×10^6 load cycles) during crack initiation and early crack propagation. The images were then used to determine surface crack lengths, $2a$, and crack growth, Δa, during the applied number of cycles, ΔN, respectively. Fatigue cracks initiated and propagated in mode I for axial as well as torsion loading, and the stress intensity factor amplitudes, K_a, were calculated as follows [47]:

$$K_a = 0.65 \cdot \sigma_1 \cdot \sqrt{\pi \cdot \sqrt{area}} \quad (2)$$

σ_1 is the amplitude of the major principal stress which is the nominal stress amplitude in the specimen's length direction for cyclic axial loading, $\sigma_1 = \sigma_a$, and the shear stress amplitude for torsion loading, $\sigma_1 = \tau_a$. $\sqrt{area}$ is the square root of the assumed projection area of the crack perpendicular to the major principal stress direction, including the size of the artificial defect, $\sqrt{area_{AD}}$. Based on fractographic investigations in a previous study [28], it is assumed that fatigue cracks initiate at the mouth of the hole (i.e., at the specimen's surface), that the shape of the crack, including the artificial defect, is semi-elliptical and that the crack grows solely in length at the surface, $2a$, (and not in depth) as long as the ratio of half the surface crack length, a, and the depth of the artificial defect, t, is larger than 0.8. If t/a is equal to or less than 0.8, $t = 0.8a$ is used to calculate the crack size $\sqrt{area}$ according to Equation (1).

During fatigue loading in vacuum, a specimen temperature of up to 45 °C was tolerated and estimated by the observed decrease in resonance frequency. Prior to testing, the correlation between specimen temperature and resonance frequency was determined. The specimen temperature was then controlled by adjusting pause times accordingly.

3. Results

3.1. S-N Data

Figure 2 show fatigue data measured at load ratios $R = -1$, 0.1 and 0.35 for cyclic axial loading (2a) and cyclic torsion loading (2b). Fatigue lifetimes between 2×10^5 and 5×10^9 cycles were investigated. Different symbols were used to indicate crack initiation at the surface (without inclusions) or in the interior (at aluminate-, TiN-and MnS inclusions, respectively, or in the metal matrix without an inclusion). Data of specimens that survived 5×10^9 cycles without failure (runout specimens) were marked as open circles with arrows.

Crack initiation at interior inclusions is the most prominent VHCF failure mechanism for axial loading at all load ratios and torsion loading at positive load ratios (Figure 2). At $R = -1$, where crack initiation was at the surface even in the VHCF regime, no inclusion-initiated fracture was found under torsion loading. This shows that the mechanism of crack initiation is not only influenced by the number of cycles to failure but also by the loading condition.

S-N data shown in Figure 2 correlate nominal stress amplitudes with fatigue lifetimes. For axial loading, the stress amplitude was constant over the specimen's cross-section and consequently equal to the nominal stress amplitude in the case of surface as well as interior crack initiation. For cyclic torsion loading, the local stress amplitude, $\tau_{a,loc}$, decreased linearly from the nominal stress amplitude, $\tau_{a,nom}$, at the surface to zero in the center, i.e., the local stress amplitude at the crack initiation site was equal to the nominal stress amplitude in case of surface crack initiation ($\tau_{a,loc} = \tau_{a,nom}$) but lower in the case of interior crack initiation ($\tau_{a,loc} < \tau_{a,nom}$). The location of interior crack initiation was randomly distributed under axial loading. Interior crack initiation in 65% of torsion specimens was between 200 µm and 300 µm below the surface, with a minimum of 160 µm and a maximum of 480 µm corresponding to local stress amplitudes of 92% to 77% of the nominal stress amplitudes. By mean, shear stress amplitudes at crack initiating inclusions, $\tau_{a,loc}$, were $87 \pm 4\%$ of the nominal shear stress amplitudes at the specimens' surfaces, $\tau_{a,nom}$.

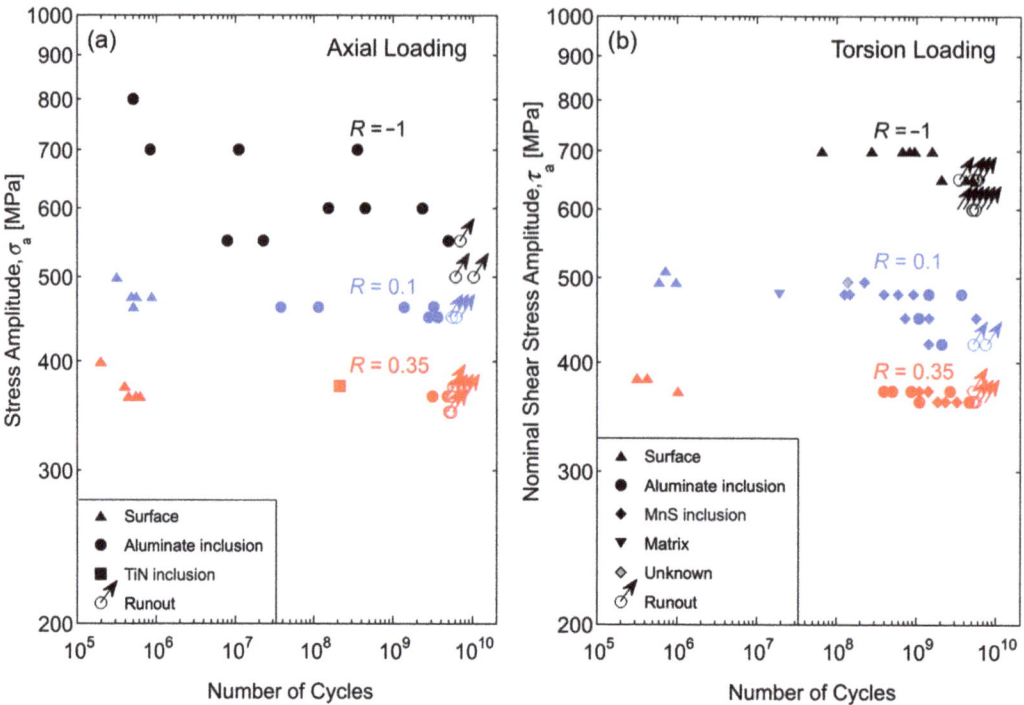

Figure 2. Fatigue data for cyclic axial loading (**a**) and cyclic torsion loading (**b**) at a load ratio of $R = -1$ (black symbols), $R = 0.1$ (blue symbols) and $R = 0.35$ (red symbols); surface crack initiation and internal crack initiation at aluminate, TiN or MnS inclusions or in the metal matrix are distinguished using different symbols; circles with arrows mark runout specimens.

3.2. Crack Initiating Inclusions

Crack initiating inclusions leading to VHCF failure under axial and torsion loading are shown in Figure 3. Under axial loading, aluminate inclusions (Figure 3a) emanating fatigue cracks accounted for all failures but one which was caused by a TiN inclusion (Figure 3b). Under torsion loading, where inclusion-initiated fracture was observed solely at $R = 0.1$ and 0.35, aluminate (Figure 3c) and MnS (Figure 3d) inclusions were found at the crack initiation sites. SEM analyses of crack initiation sites revealed a difference in fracture mode dependent on inclusion type. In the case of aluminate inclusions, the fatigue crack initiates by interface failure between matrix and particle and propagates in mode I, i.e., perpendicular to the major principal stress direction. In contrast, in the case of MnS inclusions, a shear crack initiates within and fractures the particle itself. After initially propagating in mode II/III, the crack branches and continues to grow into the matrix in mode I.

The deleterious effect of an inclusion on the cyclic strength is mainly determined by its size, which is, according to Murakami and Endo's $\sqrt{area}$ parameter model [9], represented by the square root of its projection area perpendicular to the major principal stress direction. Furthermore, in [52,53], it was demonstrated that, similar to axial loading, the threshold condition for mode I crack propagation determines the fatigue limit under cyclic torsion, even if crack initiation occurs in mode II/III. Fracture surfaces formed under both loading conditions were therefore oriented accordingly to correspond to the plane perpendicular to the major principal stress direction in order to evaluate the $\sqrt{area}$ parameter. Thus determined, minimum and maximum values for $\sqrt{area}$ with respect to inclusion type and R-ratio are shown in Table 3. The size ranges shown for aluminate and MnS inclusions refer to specimens that fractured within 5×10^9 load cycles, whereas MnS (runouts) refer to specimens that have been reloaded at $R = 0.1$, after having survived limiting lifetimes, in order to identify the most detrimental inclusion in the respective specimen.

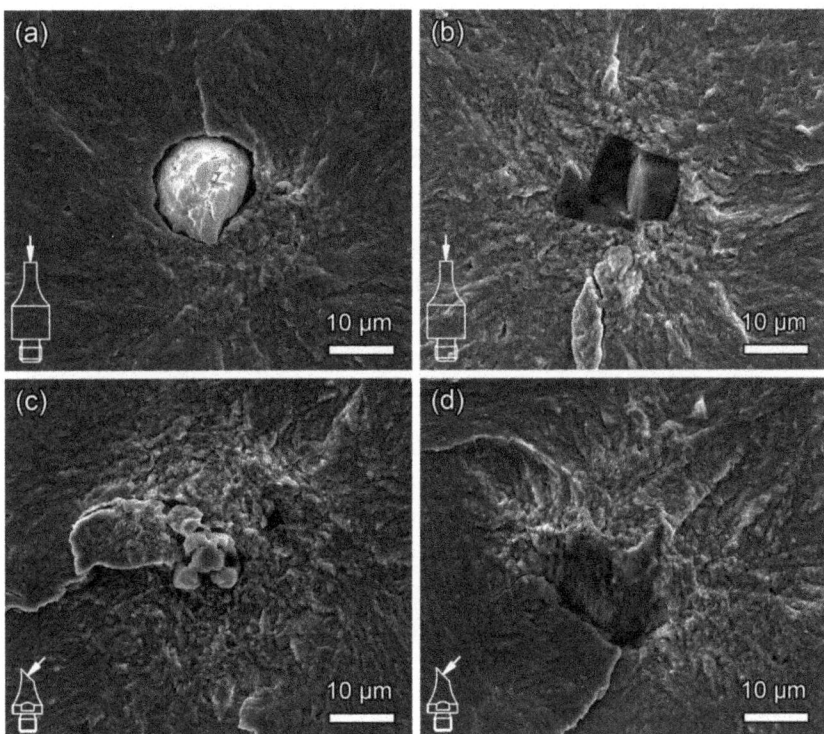

Figure 3. Crack initiation at inclusions under axial VHCF loading: (**a**) aluminate inclusion, $\sigma_a = 450$ MPa, $R = 0.1$, $N_f = 2.8 \times 10^9$ cycles; (**b**) TiN inclusion, $\sigma_a = 375$ MPa, $R = 0.35$, $N_f = 2.2 \times 10^8$ cycles; Crack initiation at inclusions under torsion VHCF loading: (**c**) aluminate inclusion, $\tau_{a,nom} = 479$ MPa, $R = 0.1$, $N_f = 3.7 \times 10^9$ cycles; (**d**) MnS inclusion, $\tau_{a,nom} = 600$ MPa, $R = -1$, $N = 5.1 \times 10^9$ cycles (runout), retested at $\tau_{a,nom} = 450$ MPa, $R = 0.1$, $N_f = 2.3 \times 10^8$ cycles.

Table 3. Minimum and maximum sizes, $\sqrt{area}$, of crack initiating inclusions under axial and torsion loading with respect to inclusion type and R-ratio. All values are given in µm.

	$R = -1$		$R = 0.1$		$R = 0.35$	
	Min	Max	Min	Max	Min	Max
Axial: Aluminate	17.0	61.6	12.4	54.6	15.5	27.6
Torsion: Aluminate			10.2	28.2	13.8	24.0
MnS			9.5	24.5	13.5	21.2
MnS (runouts)	8.8	15.4	7.4	14.4	9.7	20.0

3.3. Near Threshold Fatigue Crack Growth

Results of fatigue crack growth rate measurements in vacuum at $R = -1$ under axial and torsion loading are shown in Figure 4. Artificially initiated holes on the specimens' surfaces featured diameters of approximately 30 µm and depths between 73 and 97 µm resulting in values of $\sqrt{area}$ between 44 µm and 61 µm, according to Equation (1). With these, stress amplitudes near the fatigue limit, according to Murakami and Endo's $\sqrt{area}$ parameter model (see details in Section 4) were estimated prior to the tests. Two specimens were cycled under axial and torsion loading each at nominal stress amplitudes of σ_a = 550 MPa and 575 MPa and τ_a = 480 MPa and 575 MPa, respectively. In three specimens, a single crack was observable that initiated at the artificial defect and grew to fracture propagating in mode I, i.e., perpendicular to the specimen's length axis under axial loading and inclined 45° towards the specimen's length axis under torsion loading, respectively. In the second specimen subjected to cyclic torsion loading, a first crack initiated at the hole but stopped growing after 3.3×10^7 cycles at a length of 11 µm. Subsequently, a second crack originated and grew to fracture.

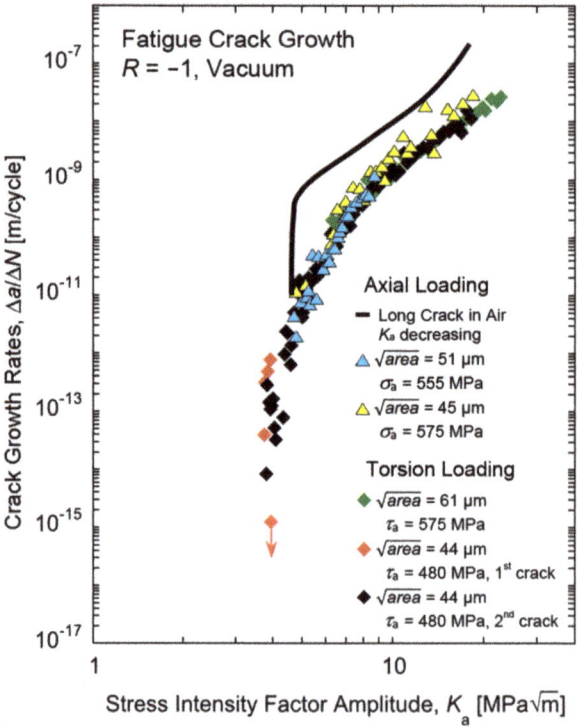

Figure 4. Crack growth rates in vacuum for axial and torsion ultrasonic fatigue loading at $R = -1$; specimens were cycled at constant axial or torsion stress amplitudes until fracture; cracks start at artificial surface defects featuring the indicated size; crack growth rates are presented vs. mode I stress intensity factor amplitudes, K_a; data measured in air [37] are shown for comparison.

Crack propagation rates are plotted vs. stress intensity factor amplitudes, K_a. Data obtained under axial (triangles) and torsion (diamonds) loading fall within the same band of scattering, and no influence of loading condition is visible. The most striking result is that very slow growth rates in the regime between 10^{-14} m/cycle to 10^{-13} m/cycle corresponding to values of K_a between 3.8 MPa$\sqrt{m}$ and 4.0 MPa$\sqrt{m}$ could be experimentally verified. Such slow growth rates correspond to several thousand load cycles that are, by mean, necessary to propagate the crack by one atomic distance. The fracture surface morphology produced by crack growth in vacuum at growth rates below 10^{-12} m/cycle is a so-called optical dark area (ODA) [54] or fine granular area (FGA) [55] and appears similar to those found around interior inclusions as shown in Figure 3. It was demonstrated that the formation of this area consumed 98% of the VHCF lifetime of the torsion specimen [8]. For comparison, the solid line shows crack growth data measured in ambient air at 30 Hz [37].

4. Discussion

Interior inclusions are the main reason for VHCF failures in the investigated spring steel under axial and torsion loading. The testing material is, therefore, most appropriate to compare the deleterious influence of small inherent defects on the cyclic strength under both loading conditions. Inclusions can be considered as initial cracks, and hence the fatigue limit is determined by the stress intensity factor necessary to propagate the crack to fracture. Fracture mechanics principles will be used in the following to interpret the findings described above.

The growth of fatigue cracks in vacuum for axial and torsion loading at $R = -1$ was measured, and similar growth rates were observed if presented versus the mode I stress intensity factor amplitude (Figure 4). Biaxiality under torsion loading did not have a significant influence on crack propagation rates. For cracks starting at artificial surface defects with sizes in $\sqrt{area}$ of approximately 44 µm, the threshold stress intensity factor amplitude, $K_{a,th}$, in vacuum for limiting growth rates of 10^{-13} m/cycle is 3.7 MPa$\sqrt{m}$. Growth rates of 10^{-11} m/cycle were observed for stress intensity factor amplitudes between 4.5 MPa$\sqrt{m}$ and 5.5 MPa$\sqrt{m}$. This is comparable to literature data for SWOSC-V, where a threshold stress intensity factor amplitude of 4.5 MPa$\sqrt{m}$ is found when tested in air at $R = -1$ [37]. The higher growth rates measured in air compared with vacuum can be explained by corrosive influences.

The threshold stress intensity factor amplitude of 3.7 MPa$\sqrt{m}$ determined in a vacuum, however, cannot be used to predict the cyclic strength in the presence of small internal defects. Stress intensity factor amplitudes calculated for five specimens that failed from internal aluminate inclusions after tension-compression loading was lower than this value [45].

The threshold stress intensity factor amplitude decreases with decreasing crack length in the short crack regime. Murakami and Endo [9] suggested the threshold stress intensity factor of short cracks to be proportional to $\left(\sqrt{area}\right)^{1/3}$. VHCF lifetimes and cyclic strengths for inclusion-initiated fracture can be well correlated to the parameter $\sigma_a \cdot \left(\sqrt{area}\right)^{1/6}$ [25] rather than the stress amplitude or the stress intensity factor amplitude, since this parameter considers the crack length-dependent threshold stress intensity [45].

In addition to the crack length-dependent threshold, Murakami and Endo's $\sqrt{area}$ parameter model [9] considers the Vickers hardness of a material, HV, to predict the fatigue limit, σ_w, in the presence of small cracks and defects according to Equation (3).

$$\sigma_w = \frac{b \cdot (HV + 120)}{\left(\sqrt{area}\right)^{1/6}} \tag{3}$$

where b is 1.56 for internal defects, which will be used in the following since all crack initiating inclusions were in the interior in the investigated steel. For surface defects, b would be 1.43. With HV in kgf/mm² and $\sqrt{area}$ in µm, Equation (3) predicts the fatigue limit, σ_w, for tension-compression loading at $R = -1$ in MPa.

The influence of load ratio on the fatigue limit can be considered with the factor $\left(\frac{1-R}{2}\right)^{\alpha_a}$ [47], which leads to Equation (4):

$$\sigma_w = \frac{1.56 \cdot (HV + 120)}{\left(\sqrt{area}\right)^{1/6}} \cdot \left(\frac{1-R}{2}\right)^{\alpha_a}. \tag{4}$$

The exponent α_a accounts for the mean stress sensitivity under axial loading. It is an empirical parameter between 0 and 1.

Murakami and Takahashi [52] extended the $\sqrt{area}$ parameter model to cyclic torsion loading. Endo [46] proposed a criterion for the fatigue limit under biaxial loading according to Equation (5):

$$\sigma_1 + \kappa\sigma_2 = \sigma_w \qquad (5)$$

where σ_w is the uniaxial fatigue limit, σ_1 and σ_2 are the major and minor principal stress amplitudes at the fatigue limit and κ quantifies the effect of compressive stresses acting parallel to the direction of mode I crack propagation. For pure cyclic torsion $\tau_w = \sigma_1 = -\sigma_2$, which leads to Equation (6) for the predicted fatigue limit at $R = -1$:

$$\tau_w = \frac{1}{(1-\kappa)} \frac{b\cdot(HV+120)}{(\sqrt{area})^{1/6}}. \qquad (6)$$

The effect of load ratio on the cyclic torsion strength can be considered similarly as above, using the exponent α_t to account for the mean stress sensitivity under torsion loading. With $b = 1.56$ for interior inclusions, this leads to Equation (7):

$$\tau_w = \frac{1}{(1-\kappa)} \frac{1.56\cdot(HV+120)}{(\sqrt{area})^{1/6}} \cdot \left(\frac{1-R}{2}\right)^{\alpha_t}. \qquad (7)$$

Considering the same location and size of the crack initiating defects, the factor $1/(1-\kappa)$ equals the ratio of torsion to axial cyclic strength, τ_w/σ_w. Studying thresholds for cracks at artificial defects subjected to axial and torsion loading, ratios in the range of 0.83 to 0.88 were found [56]. An average value of $\tau_a/\sigma_a = 0.85$ has been experimentally confirmed on artificial as well as natural surface defects in carbon steels, ductile cast iron, Cr-Mo steel and 17-4PH stainless steel [41,46,57]. In a previous study on the presently investigated spring steel SWOSC-V, a value for $1/(1-\kappa) = 0.86 \pm 0.05$ was determined by comparing the influence of similar size aluminate inclusions on torsion and axial fatigue strength [45]. With this information, Equation (8) predicts the cyclic torsion strength for internal inclusion-initiated fracture:

$$\tau_w = \frac{0.86\cdot 1.56\cdot(HV+120)}{(\sqrt{area})^{1/6}} \cdot \left(\frac{1-R}{2}\right)^{\alpha_t}. \qquad (8)$$

In the following, it will be investigated if Equation (4) for axial loading and Equation (8) for torsion loading with appropriate exponents α_a and α_t, respectively, can be used to accurately predict the fatigue limit in the presence of interior inclusions.

Figure 5 show failures and runouts at different load ratios with respect to relative stress amplitudes. For axial loading at $R = -1$, runouts were found for σ_a/σ_w between 0.86 and 0.92 and failures between 0.95 and 1.27. This means that Equation (3) can predict the endurance limit for fully reversed axial loading accurately within $\pm 10\%$, which is the prediction error reported by Murakami [47].

Furthermore, runouts and failures—independent of inclusion type—serve to evaluate the exponents α_a and α_t that consider the R-ratio influence on the fatigue limit. Data for fitting in Figure 5 were chosen as explained in the following: If failure occurred (i.e., at all R-ratios under axial and at $R = 0.1$ and 0.35 under torsion loading, respectively), the one maximum value of $\sigma_a \cdot \frac{(\sqrt{area})^{1/6}}{1.56\cdot(HV+120)}$ and $\tau_{a,loc} \cdot \frac{(\sqrt{area})^{1/6}}{0.86\cdot 1.56\cdot(HV+120)}$ obtained by a runout was used that was below the lowest respective value obtained by a failed specimen. If no failure occurred, which is the case at $R = -1$ under torsion loading, the maximum value of $\tau_{a,loc} \cdot \frac{(\sqrt{area})^{1/6}}{0.86\cdot 1.56\cdot(HV+120)}$ of a runout was chosen. Through these data points, straight lines were drawn, and the slopes corresponding to the exponents α_a and α_t were determined. The exponent for axial loading α_a was found to be 0.41, and the exponent for torsion loading α_t was 0.55. Values for the exponent α measured for different steels are available in the literature, however, solely for axial and not for torsion loading. The reported exponents are in the range between 0.368 and 0.546 [17,20,28,54]. Hence, the presently determined value fits well with the literature data.

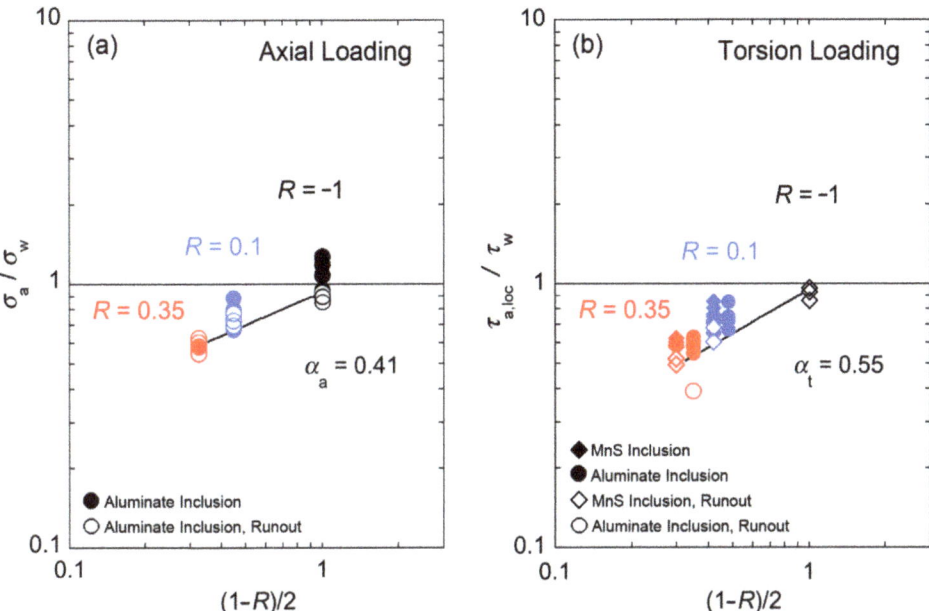

Figure 5. Evaluation of the exponents α_a and α_t that consider the R-ratio dependency of internal inclusion-initiated fracture under axial loading (**a**) and torsion loading (**b**); failures (closed symbols) and runouts (open symbols) are shown vs. the ratio of stress amplitudes, σ_a and $\sigma_w = \frac{1.56 \cdot (HV+120)}{(\sqrt{area})^{1/6}}$ (**a**) and the ratio of shear stress amplitudes at crack initiation, $\tau_{a,loc}$ and $\tau_w = \frac{0.86 \cdot 1.56 \cdot (HV+120)}{(\sqrt{area})^{1/6}}$ (**b**); data points for aluminate and MnS inclusion-initiated fracture under cyclic torsion at $R = 0.1$ and 0.35 have been intentionally set apart for better visibility.

Figure 6 show the ratio of stress amplitudes at crack initiation, σ_a and $\tau_{a,loc}$, and predicted fatigue limits against fatigue lifetimes. The predicted fatigue limit, σ_w for axial loading, is determined by Equation (9):

$$\sigma_w = \frac{1.56 \cdot (HV+120)}{(\sqrt{area})^{1/6}} \cdot \left(\frac{1-R}{2}\right)^{0.41}. \qquad (9)$$

The predicted fatigue limit for torsion loading, τ_w is determined by Equation (10):

$$\tau_w = \frac{0.86 \cdot 1.56 \cdot (HV+120)}{(\sqrt{area})^{1/6}} \cdot \left(\frac{1-R}{2}\right)^{0.55}. \qquad (10)$$

For the sake of conservatism, all displayed data points in Figure 6 representing failures should be larger than 1. Five data points for axial loading and one for torsion loading fall below this line; however, all these data represent failures above 10^9 cycles. Equations (9) and (10) therefore well predict the fatigue limit for limiting lifetimes of 10^9 cycles and only slightly overestimate the fatigue limit at 5×10^9 cycles. Again, the prediction lies within the prediction error of $\pm 10\%$.

Figure 7 show the Kitagawa–Takahashi diagrams for axial and torsion loading. The stress amplitudes are multiplied with $\left(\frac{1-R}{2}\right)^{-\alpha_a}$ and $\left(\frac{1-R}{2}\right)^{-\alpha_t}$, respectively, to eliminate the influence of load ratio.

- Solid lines with a slope of $-1/6$ in Figure 7 predict the fatigue limit in the short crack regime using Equations (9) and (10).
- Solid lines with a slope of $-1/2$ in Figure 7 predict the fatigue limit in the long crack regime using the long crack threshold stress intensity factor amplitude, $K_{a,th,lc}$, according to Equation (11):

$$\sigma_w = \frac{K_{a,th,lc}}{0.5 \cdot \left(\pi \cdot \sqrt{area} \cdot\right)^{1/2}} \quad (11)$$

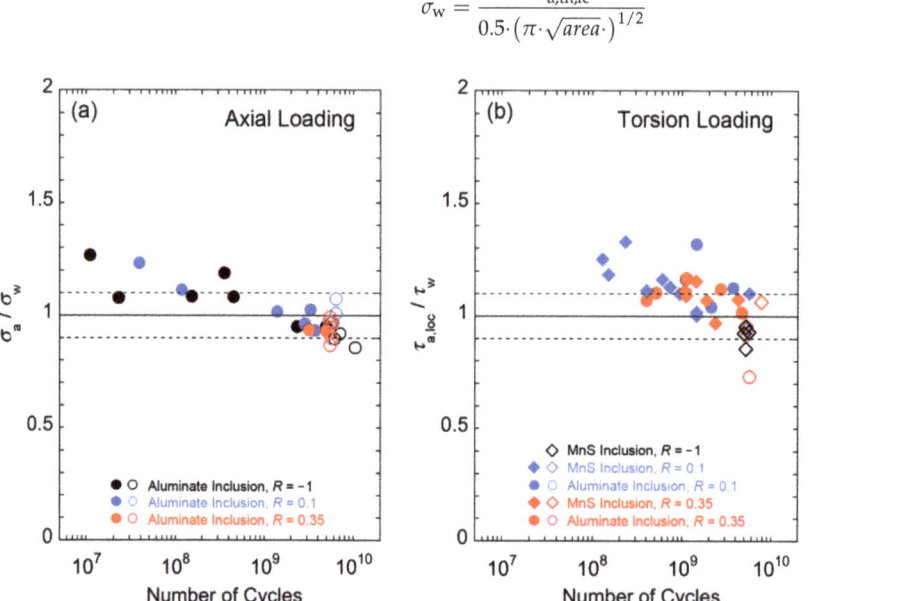

Figure 6. Normalized S-N data for internal inclusion-initiated fracture obtained under cyclic axial and torsion loading; ratios of stress amplitudes, σ_a and predicted cyclic strengths, $\sigma_w = \frac{1.56 \cdot (HV+120)}{(\sqrt{area})^{1/6}} \cdot \left(\frac{1-R}{2}\right)^{0.41}$ (a) and ratios of shear stress amplitudes at crack initiation, $\tau_{a,loc}$, and predicted shear strengths, $\tau_w = \frac{0.86 \cdot 1.56 \cdot (HV+120)}{(\sqrt{area})^{1/6}} \cdot \left(\frac{1-R}{2}\right)^{0.55}$ (b) are plotted against fatigue lifetimes; symbols are distinguished by crack initiation site and load ratio; data of runout specimens are shown with open symbols.

The unit for $\sqrt{area}$ using Equation (11) to predict the cyclic strength in the long crack regime is m. In contrast, $\sqrt{area}$ in μm is used to calculate the cyclic strength for short cracks using Equations (9) and (10). The long crack threshold of $K_{a,th,lc} = 4.5$ MPa$\sqrt{m}$ determined in air at $R = -1$ [37] is used to construct the straight line in the long crack regime in Figure 7. It can be seen that interior inclusions are smaller than $\sqrt{area} = 125$ μm for axial loading and $\sqrt{area} = 196$ μm for torsion loading can be considered as small cracks. Similar transition sizes between small and large defects (short and long cracks) are found for other high-strength steels [28,58].

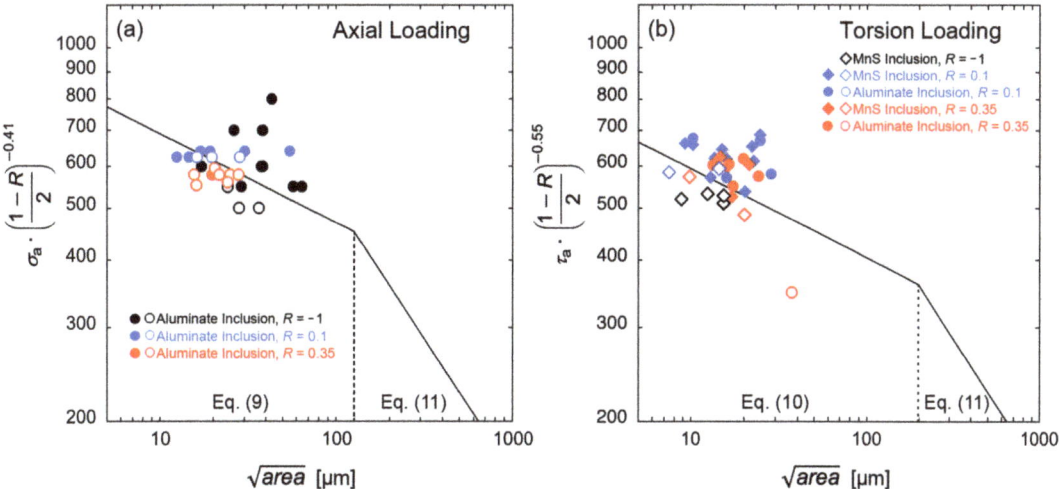

Figure 7. Kitagawa–Takahashi diagrams for axial loading (**a**) and torsion loading (**b**) and limiting lifetimes of 5×10^9 cycles; stress amplitudes are multiplied with $\left(\frac{1-R}{2}\right)^{-0.41}$ and $\left(\frac{1-R}{2}\right)^{-0.55}$ to eliminate the influence of load ratio. Solid and open symbols mark failed and runout specimens, respectively.

Table 3 show that all crack initiating inclusions are in the short crack regime. However, crack initiating inclusions in torsion fatigue specimens are smaller than those found in axially loaded specimens. This is a consequence of the smaller volume subjected to high stresses in torsion compared with axial fatigue specimens. The material volumes where crack initiating inclusions were observed are 108 mm^3 in axial specimens and 9 mm^3 in torsion specimens. Assuming that the origin of failure for each test specimen is the largest inclusion in the highly stressed volume, the size distribution of inclusions can be plotted in an extreme value distribution diagram [47]. This is shown for aluminate inclusions in Figure 8. The median inclusion size for axial specimens is 26 µm which is significantly larger than 18 µm for torsion specimens. This can be explained by the much smaller volume of $V = 9$ mm^3 where crack initiation sites under torsion loading were found. Considering that the highly stressed volume in axial specimens is the reference volume $V_0 = 108$ mm^3, the median size of an inclusion expected in a volume of interest, V, can be calculated with the return period, $T = (V + V_0)/V_0$ [57]. With this, the predicted median size of aluminate inclusions in a volume $V = 9$ mm^3 is 13 µm, which is within the standard deviation of crack initiating inclusions observed in torsion specimens (18 ± 5 µm).

The considerations above suggest that the influence of inclusions can be treated similarly for axial and torsion loading using the $\sqrt{area}$ parameter model. However, this does not mean that the cyclic torsion strength can be deduced from axial loading results since mechanisms causing VHCF failures can be different. The first and most striking difference is that MnS inclusions are solely deleterious for cyclic torsion loading. As schematically shown in Figure 9, this can be explained easily with the slender and elongated form of these inclusions parallel to the specimen's length axis [44]. $\sqrt{area}$ projected in the direction of maximum tensile stress is very small for axial loading but not for torsion loading, and consequently, MnS inclusions are deleterious for torsion but not for axial loading. Axial loading tests therefore underestimate the deleterious influence of inclusions aligned in the length direction.

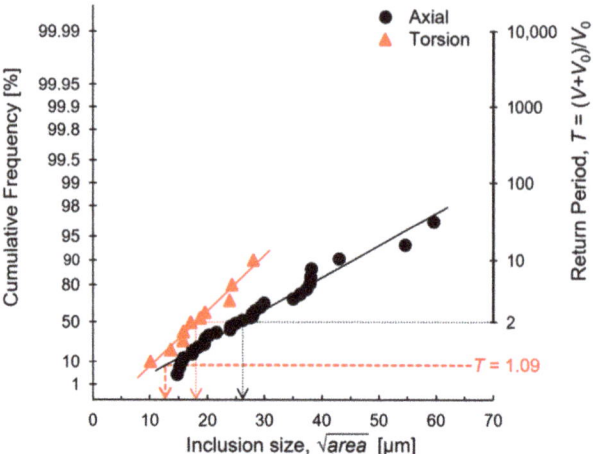

Figure 8. Extreme value distributions of aluminate inclusion size for axial and torsion loading. Dotted lines mark the median inclusion sizes of 26 μm and 18 μm observed in axial (black) and torsion (red) specimens, respectively. The expected median inclusion size for torsion specimens (V = 9 mm^3, T = 1.09) of 13 μm predicted from axial specimens (V_0 = 108 mm^3) is marked with a red dashed line.

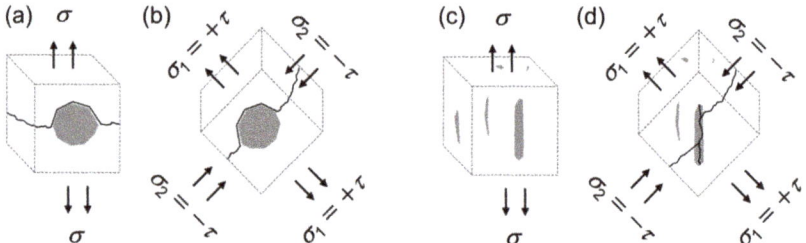

Figure 9. Fracture modes effective under axial ((**a**,**c**)) and torsion ((**b**,**d**)) loading dependent on inclusion type; (**a**,**b**): aluminate inclusion; (**c**,**d**): MnS inclusion.

A second important difference between axial and torsion loading is the insensitivity of cyclic torsion at $R = -1$ against inclusion-initiated fracture. The existence of inclusions in runout specimens could be verified, but they were non-detrimental for fully reversed cyclic torsion. Inclusion-initiated fracture observed at positive load ratios, on the other hand, clearly indicates an increasing defect-sensitivity for cyclic torsion loading with increasing R. In contrast, axial loading tests at $R = -1$ delivered solely inclusion-initiated failures, even in the high cycle fatigue regime, indicating a strong defect sensitivity. This shows that failure mechanisms for axial and torsion loading may be different and influenced differently by mean stresses. Therefore, it is necessary to perform fatigue tests on a material under comparable conditions as in service in order to deduce meaningful data for the cyclic strength.

5. Conclusions

The influence of non-metallic inclusions on very high cycle fatigue (VHCF) strength under axial and torsion loading was investigated with spring steel SWOSC-V featuring an intentionally increased number and size of non-metallic inclusions. S-N tests were performed at ultrasonic cycling frequency up to limiting lifetimes of 5×10^9 load cycles at load ratios $R = -1$, 0.1 and 0.35 in ambient air. Fatigue crack growth was studied in a vacuum with specimens containing small artificial surface defects subjected to cyclic axial and torsion loading. The following conclusions can be drawn:

1. Inclusions can be considered as initial cracks. $\sqrt{area}$ measured on the plane perpendicular to the maximum principal stress determines the deleterious influence of an inclusion. Therefore, aluminates are deleterious for axial and torsion loading, whereas elongated MnS inclusions aligned in a rolling direction solely initiate cracks under torsion loading.
2. Crack initiation and propagation at aluminates are in mode I under axial and torsion loading. Crack initiation at MnS inclusions under torsion loading is in mode II/III, followed by further crack growth in mode I. For both types of inclusions and both loading conditions, the threshold for mode I crack growth determines the fatigue limit.
3. The long crack threshold stress intensity amplitude in vacuum at $R = -1$ for limiting growth rates of 10^{-13} m/cycle is 3.7 MPa $\sqrt{m}$. $\sqrt{area}$ of inclusions is in the short crack regime where the threshold stress intensity decreases with decreasing crack length and failures can occur at lower stress intensities.
4. Murakami and Endo's $\sqrt{area}$ parameter model predicts the axial loading fatigue limit at $R = -1$ within the reported prediction error of $\pm 10\%$. The influence of biaxiality on torsion loading of a material defect is considered with a factor of 0.86. The factor $\left(\frac{1-R}{2}\right)^{\alpha}$ accounts for the mean stress sensitivity. With this, the following equation was successfully used to predict the fatigue limit for cyclic axial loading at different R-ratios:

$$\sigma_w = \frac{1.56 \cdot (HV + 120)}{(\sqrt{area})^{1/6}} \cdot \left(\frac{1-R}{2}\right)^{0.41}$$

The following equation was successfully used to predict the fatigue limit for cyclic torsion loading at different R-ratios:

$$\tau_w = \frac{0.86 \cdot 1.56 \cdot (HV + 120)}{(\sqrt{area})^{1/6}} \cdot \left(\frac{1-R}{2}\right)^{0.55}$$

5. Cyclic torsion properties in the presence of inclusions can hardly be deduced from cyclic axial properties: Shape and orientation of inclusions influence their harmfulness, size effects cause different distributions of inclusion sizes, and R-ratio sensitivity was found to be different for axial and torsion loading. Ultrasonic cyclic torsion tests with different mean loads are most appropriate for rapidly collecting torsion fatigue data of material.

Author Contributions: U.K.: investigation, conceptualization, methodology, formal analysis, validation, visualization, writing—review and editing; B.M.S.: formal analysis, methodology, visualization, writing—review and editing; Y.S.: resources, conceptualization; H.M.: conceptualization, methodology, formal analysis, validation, writing—original draft, supervision, project administration. All authors have read and agreed to the published version of the manuscript.

Funding: This research received no external funding.

Institutional Review Board Statement: Not applicable.

Informed Consent Statement: Not applicable.

Data Availability Statement: The data that support the finding of this study are included in the paper.

Conflicts of Interest: The authors declare no conflict of interest.

Nomenclature

$\sqrt{area}$	square root of the projection area of a small, arbitrary-shaped defect perpendicular to the major principal stress direction
HV	Vickers hardness
K_a	stress intensity factor amplitude
R	load ratio
α_a	exponent accounting for the load ratio sensitivity of the fatigue limit for axial loading
α_t	exponent accounting for the load ratio sensitivity of the fatigue limit for torsion loading
κ	parameter accounting for the influence of a biaxial stress state on the fatigue limit
σ_w	fatigue limit under axial loading
σ_a	normal stress amplitude
σ_1, σ_2	major and minor principal stress
τ_w	fatigue limit under torsion loading
τ_a	shear stress amplitude
$\tau_{a,nom}$	nominal stress amplitude under torsion loading effective at the specimen surface
$\tau_{a,loc}$	local stress amplitude under torsion loading effective at the crack initiation site

References

1. Naito, T.; Ueda, H.; Kikuchi, M. Fatigue behavior of carburized steel with internal oxides and nonmartensitic microstructure near the surface. *Metall. Trans. A* **1984**, *15*, 1431–1436. [CrossRef]
2. Asami, K.; Sugiyama, Y. Fatigue strength of various surface hardened steels. *J. Heat Treat. Technol. Assoc.* **1985**, *25*, 147–150.
3. Murakami, Y.; Takada, M.; Toriyama, T. Super-long life tension–compression fatigue properties of quenched and tempered 0.46% carbon steel. *Fatigue* **1998**, *20*, 661–667. [CrossRef]
4. Nishijima, S.; Kanazawa, K. Stepwise S-N curve and fish-eye failure in gigacycle fatigue. *Fatigue Fract. Engng. Mater. Struct.* **1999**, *22*, 601–607. [CrossRef]
5. Shiozawa, K.; Lu, L.; Ishihara, S. S-N curve characteristics and subsurface crack initiation behaviour in ultra-long life fatigue of a high carbon-chromium bearing steel. *Fatigue Fract. Engng. Mater. Struct.* **2001**, *24*, 781–790. [CrossRef]
6. Furuya, Y. Visualization of internal small fatigue crack growth. *Mater. Lett.* **2013**, *112*, 139–141. [CrossRef]
7. Stäcker, C.; Sander, M. Experimental, analytical and numerical analyses of constant and variable amplitude loadings in the very high cycle fatigue regime. *Theor. Appl. Fract. Mech.* **2017**, *92*, 394–409. [CrossRef]
8. Karr, U.; Schönbauer, B.M.; Sandaiji, Y.; Mayer, H. Influence of load ratio on torsion very high cycle fatigue of high-strength spring steel in the presence of detrimental defects. *Fatigue Fract. Eng. Mater. Struct.* **2021**, *444*, 2356–2371. [CrossRef]
9. Murakami, Y.; Endo, M. Effects of defects, inclusions and inhomogeneities on fatigue strength. *Int. J. Fatigue* **1994**, *16*, 163–182. [CrossRef]
10. Zhang, J.M.; Li, S.X.; Yang, Z.G.; Li, G.Y.; Hui, W.J.; Weng, Y.Q. Influence of inclusion size on fatigue behavior of high strength steels in the gigacycle fatigue regime. *Int. J. Fatigue* **2007**, *29*, 765–771. [CrossRef]
11. Lei, Z.; Hong, Y.; Xie, J.; Sun, C.; Zhao, A. Effects of inclusion size and location on very-high-cycle fatigue behavior for high strength steels. *Mater. Sci. Eng. A* **2012**, *558*, 234–241. [CrossRef]
12. Furuya, Y.; Hirukawa, H.; Kimura, T.; Hayaishi, M. Gigacycle fatigue properties of high-strength steels according to inclusion and ODA sizes. *Metal. Trans. A* **2007**, *38*, 1722–1730. [CrossRef]
13. Spriestersbach, D.; Grad, P.; Kerscher, E. Influence of different non-metallic inclusion types on the crack initiation in high-strength steels in the VHCF regime. *Int. J. Fatigue* **2014**, *64*, 114–120. [CrossRef]
14. Karr, U.; Schuller, R.; Fitzka, M.; Schönbauer, B.; Tran, D.; Pennings, B.; Mayer, H. Influence of inclusion type on the very high cycle fatigue properties of 18Ni maraging steel. *J. Mater. Sci.* **2017**, *52*, 5954–5967. [CrossRef]
15. Sakai, T.; Sato, Y.; Nagano, Y.; Takeda, M.; Oguma, N. Effect of stress ratio on long life fatigue behavior of high carbon chromium bearing steel under axial loading. *Int. J. Fatigue* **2006**, *28*, 1547–1554. [CrossRef]
16. Shiozawa, K.; Hasegawa, T.; Kashiwagi, Y.; Lu, L. Very high cycle fatigue properties of bearing steel under axial loading condition. *Int. J. Fatigue* **2009**, *31*, 880–888. [CrossRef]
17. Kovacs, S.; Beck, T.; Singheiser, L. Influence of mean stresses on fatigue life and damage of a turbine blade steel in the VHCF-regime. *Int. J. Fatigue* **2013**, *49*, 90–99. [CrossRef]
18. Sander, M.; Müller, T.; Lebahn, J. Influence of mean stress and variable amplitude loading on the fatigue behaviour of a high-strength steel in the VHCF regime. *Int. J. Fatigue* **2014**, *62*, 10–20. [CrossRef]
19. Schuller, R.; Karr, U.; Irrasch, D.; Fitzka, M.; Hahn, M.; Bacher-Höchst, M.; Mayer, H. Mean stress sensitivity of spring steel in the very high cycle fatigue regime. *J. Mater.Sci.* **2015**, *50*, 5514–5523. [CrossRef]
20. Schönbauer, B.M.; Yanase, K.; Endo, M. VHCF properties and fatigue limit prediction of precipitation hardened 17-4PH stainless steel. *Int. J. Fatigue* **2016**, *88*, 205–216. [CrossRef]
21. Abe, T.; Furuya, Y.; Matsuoka, S. Gigacycle fatigue properties of 1800 MPa class spring steel. *Fatigue Fract. Engng. Mater. Struct.* **2004**, *27*, 159–167. [CrossRef]

22. Chai, G. The formation of subsurface non-defect fatigue crack origins. *Int. J. Fatigue* **2006**, *28*, 1533–1539. [CrossRef]
23. Yu, Y.; Gu, J.L.; Bai, B.Z.; Liu, Y.B.; Li, S.X. Very high cycle fatigue mechanism of carbide-free bainite/martensite steel micro-alloyed with Nb. *Mater. Sci. Engng A* **2009**, *527*, 212–217. [CrossRef]
24. Li, W.; Sakai, T.; Wakita, M.; Mimura, S. Effect of surface finishing and loading condition on competing failure mode of clean spring steel in very high cycle fatigue regime. *Mater. Sci. Engng A* **2012**, *552*, 301–309. [CrossRef]
25. Mayer, H.; Haydn, W.; Schuller, R.; Issler, S.; Furtner, B.; Bacher-Höchst, M. Very high cycle fatigue properties of bainitic high-carbon-chromium steel. *Int. J. Fatigue* **2009**, *31*, 242–249. [CrossRef]
26. Li, W.; Sakai, T.; Wakita, M.; Mimura, S. Influence of microstructure and surface defect on very high cycle fatigue properties of clean spring steel. *Int. J. Fatigue* **2014**, *60*, 48–56. [CrossRef]
27. Zimmermann, M. Diversity of damage evolution during cyclic loading at very high numbers of cycles. *Int. Mater. Rev.* **2012**, *57*, 73–91. [CrossRef]
28. Schönbauer, B.M.; Mayer, H. Effect of small defects on the fatigue strength of martensitic stainless steels. *Int. J. Fatigue* **2019**, *127*, 362–375. [CrossRef]
29. Kaiser, B.; Berger, C. Fatigue behaviour of technical springs. *Mat. Wiss. U. Werkstofftechn.* **2005**, *36*, 685–696. [CrossRef]
30. Kaiser, B.; Pyttel, B.; Berger, C. VHCF-behavior of helical compression springs made of different materials. *Int. J. Fatigue* **2011**, *33*, 23–32. [CrossRef]
31. Pyttel, B.; Brunner, I.; Kaiser, B.; Berger, C.; Mahendran, M. Fatigue behaviour of helical compression springs at a very high number of cycles—Investigation of various influences. *Int. J. Fatigue* **2014**, *60*, 101–109. [CrossRef]
32. Stanzl-Tschegg, S. Very high cycle fatigue measuring techniques. *Int. J. Fatigue* **2014**, *60*, 2–17. [CrossRef]
33. Mayer, H. Recent developments in ultrasonic fatigue. *Fatigue Fract. Eng. Mater. Struct.* **2016**, *39*, 3–29. [CrossRef]
34. Schönbauer, B.M.; Fitzka, M.; Karr, U.; Mayer, H. Variable amplitude very high cycle fatigue of 17-4PH steel with a stepwise S-N curve. *Int. J. Fatigue* **2021**, *142*, 105963. [CrossRef]
35. Stanzl-Tschegg, S.E.; Mayer, H.R.; Tschegg, E.K. High Frequency Method for Torsion Fatigue Testing. *Ultrasonics* **1993**, *31*, 275–280. [CrossRef]
36. Mayer, H.; Schuller, R.; Karr, U.; Irrasch, D.; Fitzka, M.; Hahn, M.; Bacher-Höchst, M. Cyclic torsion very high cycle fatigue of VDSiCr spring steel at different load ratios. *Int. J. Fatigue* **2015**, *70*, 322–327. [CrossRef]
37. Akiniwa, Y.; Stanzl-Tschegg, S.; Mayer, H.; Wakita, M.; Tanaka, K. Fatigue strength of spring steel under axial and torsional loading in the very high cycle regime. *Int. J. Fatigue* **2008**, *30*, 2057–2063. [CrossRef]
38. Schuller, R.; Mayer, H.; Fayard, A.; Hahn, M.; Bacher-Höchst, M. Very high cycle fatigue of VDSiCr spring steel under torsional and axial loading. *Materialwiss. Werkstofftech.* **2013**, *44*, 282–289. [CrossRef]
39. Xue, H.Q.; Bathias, C. Crack path in torsion loading in very high cycle fatigue regime. *Engng. Fract. Mech.* **2010**, *77*, 1866–1873. [CrossRef]
40. Sandaiji, Y.; Tamura, E.; Tsuchida, T. Influence of inclusion type on internal fatigue fracture under cyclic shear stress. *Proc. Mater. Sci.* **2014**, *3*, 894–899. [CrossRef]
41. Schönbauer, B.M.; Yanase, K.; Chehrehrazi, M.; Endo, M.; Mayer, H. Effect of microstructure and cycling frequency on the torsional fatigue properties of 17-4PH stainless steel. *Mater. Sci. Eng. A* **2021**, *801*, 140481. [CrossRef]
42. Nishimura, Y.; Yanase, K.; Tanaka, H.; Miyamoto, N.; Miyakawa, S.; Endo, M. Effects of mean shear stress on the torsional fatigue strength of a spring steel with small scratches. *Int. J. Damage Mech.* **2020**, *29*, 4–18. [CrossRef]
43. Mayer, H.; Schuller, R.; Karr, U.; Fitzka, M.; Irrasch, D.; Hahn, M.; Bacher-Höchst, M. Mean stress sensitivity and crack initiation mechanisms of spring steel for torsional and axial VHCF loading. *Int. J. Fatigue* **2016**, *93*, 309–317. [CrossRef]
44. Karr, U.; Schönbauer, B.; Fitzka, M.; Tamura, E.; Sandaiji, Y.; Murakami, S.; Mayer, H. Inclusion initiated fracture under cyclic torsion very high cycle fatigue at different load ratios. *Int. J. Fatigue* **2019**, *122*, 199–207. [CrossRef]
45. Karr, U.; Sandaiji, Y.; Tanegashima, R.; Murakami, S.; Schönbauer, B.; Fitzka, M.; Mayer, H. Inclusion initiated fracture in spring steel under axial and torsion very high cycle fatigue loading at different load ratios. *Int. J. Fatigue* **2020**, *134*, 105525. [CrossRef]
46. Endo, M. Effects of small defects on the fatigue strength of steel and ductile iron under combined axial/torsional loading. In *Small Fatigue Cracks: Mechanics, Mechanisms and Applications*; Ravichandran, K.S.R.R.O., Murakami, Y., Eds.; Elsevier Science Ltd.: Amsterdam, The Netherlands, 1999; pp. 375–387.
47. Murakami, Y. *Metal Fatigue, Effects of Small Defects and Nonmetallic Inclusions*; Academic Press: Cambridge, MA, USA, 2019.
48. Furuya, Y. Notable size effects on very high cycle fatigue properties of high-strength steel. *Mater. Sci. Eng. A* **2011**, *528*, 5234–5240. [CrossRef]
49. Fitzka, M.; Pennings, B.; Karr, U.; Schönbauer, B.; Schuller, R.; Tran, M.-D.; Mayer, H. Influence of cycling frequency and testing volume on the VHCF properties of 18Ni maraging steel. *Eng. Fract. Mech.* **2019**, *216*, 106525. [CrossRef]
50. Giertler, A.; Krupp, U. Investigation of Fatigue Damage of Tempered Martensitic Steel during High Cycle Fatigue and Very High Cycle Fatigue Loading Using In Situ Monitoring by Scanning Electron Microscope and High-Resolution Thermography. *Steel Res. Int.* **2021**, *92*, 2100268. [CrossRef]
51. Tridello, A.; Niutta, C.B.; Berto, F.; Paolino, D.S. Size-effect in Very High Cycle Fatigue: A review. *Int. J. Fatigue* **2021**, *153*, 106462. [CrossRef]

52. Murakami, Y.; Takahashi, T. Torsional fatigue of a medium carbon steel containing an initial small surface crack introduced by tension-compression fatigue: Crack branching, non-propagation and fatigue limit. *Fatigue Fract. Engng. Mater. Struct.* **1998**, *21*, 1473–1484. [CrossRef]
53. Endo, M.; Iseda, K. Prediction of the fatigue strength of nodular cast irons under combined loading. *Int. J. Mod. Phys. B* **2006**, *20*, 3817–3823. [CrossRef]
54. Murakami, Y.; Nomotomo, T.; Ueda, T.; Murakami, Y. On the mechanism of fatigue failure in the superlong life regime (>10^7 cycles). Part I: Influence of hydrogen trapped by inclusions. *Fatigue Fract. Engng. Mater. Struct.* **2000**, *23*, 893–902. [CrossRef]
55. Sakai, T.; Sato, Y.; Oguma, N. Characteristic S-N properties of high-carbon-chromium-bearing steel under axial loading in long-life fatigue. *Fatigue Fract. Engng. Mater. Struct.* **2002**, *25*, 765–773. [CrossRef]
56. Beretta, S.; Murakami, Y. SIF and threshold for small cracks at small notches under torsion. *Fatigue Fract. Engng. Mater. Struct.* **2000**, *23*, 97–104. [CrossRef]
57. Makkonen, L.; Rabb, R.; Tikanmäki, M. Size effect in fatigue based on the extreme value distribution of defects. *Mater. Sci. Eng. A* **2014**, *594*, 68–71. [CrossRef]
58. Schönbauer, B.M.; Ghosh, S.; Kömi, J.; Frondelius, T.; Mayer, H. Influence of small defects and nonmetallic inclusions on the high and very high cycle fatigue strength of an ultrahigh-strength steel. *Fatigue Fract. Eng. Mater. Struct.* **2021**, *44*, 2990–3007. [CrossRef]

Article

Influence of Residual Stresses on the Crack Initiation and Short Crack Propagation in a Martensitic Spring Steel

Anna Wildeis *, Hans-Jürgen Christ and Robert Brandt

Department Maschinenbau, Institut für Werkstofftechnik, Universität Siegen, Paul-Bonatz-Str. 9-11, D-57068 Siegen, Germany; hans-juergen.christ@uni-siegen.de (H.-J.C.); robert.brandt@uni-siegen.de (R.B.)
* Correspondence: anna.wildeis@uni-siegen.de; Tel.: +49-271-740-4744

Abstract: The crack initiation and short crack propagation in a martensitic spring steel were investigated by means of in-situ fatigue testing. Shot peened samples as well as untreated samples were exposed to uniaxial alternating stress to analyze the impact of compressive residual stresses. The early fatigue damage started in both sample conditions with the formation of slip bands, which subsequently served as crack initiation sites. Most of the slip bands and, correspondingly, most of the short fatigue cracks initiated at or close to prior austenite grain boundaries. The observed crack density of the emerging network of short cracks increased with the number of cycles and with increasing applied stress amplitudes. Furthermore, the prior austenite grain boundaries acted as obstacles to short crack propagation in both sample conditions. Compressive residual stresses enhanced the fatigue strength, and it is assumed that this beneficial effect was due to a delayed transition from short crack propagation to long crack propagation and a shift of the crack initiation site from the sample surface to the sample interior.

Keywords: martensitic spring steel; fatigue crack initiation; short fatigue crack propagation; residual stress; shot peening; miniaturized testing device

1. Introduction

Fatigue fracture is one of the main causes of failure for many technical components. It is commonly subdivided into five stages, i.e., crack nucleation, crack initiation, short crack propagation, long crack propagation, and eventually failure. Especially, the stages of crack nucleation, crack initiation, and short crack propagation can account for up to 90% of the fatigue life in the high cycle fatigue regime, leading to a particular interest in investigating these stages of fatigue damage evolution [1]. To increase component fatigue lives, various surface strengthening methods, such as shot peening [2,3], deep rolling, and laser shock peening [2], are utilized. Thereby, shot peening is widely used due to its operating comfort, production speed, and good automation capacity. The impact of compressive residual stresses on long crack propagation has been reported extensively in the literature (see, for example, [2,4,5]). However, the impact of compressive residual stresses on short crack propagation requires further research. Thereby, the definition of a short crack in this work is equivalent to the definition of a microstructurally short crack in the work of Miller [6].

Previous studies have shown that crack initiation often occurs at slip bands (see, for example, [1,7–9]) and that short crack propagation is strongly influenced by the local microstructure, leading to an oscillating crack propagation rate (see, for example, [1,7,10,11]). Despite this comprehensive literature base, until now, only a few studies have investigated the fatigue damage evolution in martensitic steel and the corresponding impact of the complex martensitic microstructure [12] on the crack initiation and the short crack propagation, especially, when residual stresses are concerned. The early phase of fatigue damage evolution in martensitic steels seems to be characterized by the formation of slip bands, which, in turn, often serve as crack initiation sites [13–17]. Those slip bands are oriented parallel to martensitic laths and mainly arise at microstructural interfaces such as prior austenite grain boundaries (PAGBs), packet boundaries, or block boundaries [17–19] without crossing them [16,19,20]. Consequently, the crack initiation and early short crack propagation seem to occur often intergranularly at PAGBs [17,18,20,21] or block boundaries [13,17,22]. Generally, intergranular crack initiation can be attributed to grain boundary weakening due to precipitates or segregations and to the mutual superimposing mechanisms of elastic anisotropy and plastic incompatibility. Elastic anisotropy can lead to local stress concentrations at grain boundaries that may exceed the interfacial cohesion. In that case, no pronounced plastic deformation within the adjacent grains occurs. Thereby, the grain boundary itself exhibits a high dislocation density and can be considered as a slip plane [23]. Ohmura et al. showed this experimentally for a martensitic steel by in-situ indentation in a transmission electron microscope [24]. Plastic incompatibility can lead to a dislocation pairing along the active slip band and/or dislocation pileup at the grain boundary, which provokes a slip step and subsequent crack initiation [17,18,21,23]. However, in some studies, transgranular crack initiation and early short crack propagation along martensitic laths were also observed [15,19]. A frequently used explanation for intergranular crack initiation and early short crack propagation at PAGBs is the preferential formation of carbides [25–32] or segregations [25,26,28–31] along microstructural interfaces. Further reasons are the impingement of slip bands at PAGBs [20,33,34] or the anisotropic elastic and plastic properties of the grains [17,18,21,35–37]. The numbers of slip bands and short cracks rise with an increasing number of cycles and with increasing stress amplitude [10,14,15,18,38,39]. The phenomenon of multi-crack initiation is also observed for other loading cases, not only for a uniaxial load [40,41].

Compressive residual stresses seem to shift the initiation site of the fatal crack from the sample surface to the interior of the sample [33,42–48]. Despite existing compressive residual stresses, crack initiation of the fatal crack also happens at the sample surface, particularly at surfaces of unpolished samples with an increased surface roughness due to shot peening [39,49–51]. Regarding the impact of compressive residual stresses on the number of cycles required for crack initiation, contrary observations have been made. De los Rios et al., Bag et al., and Berns and Weber observed a retarded crack initiation due to compressive residual stresses [49,52,53], whereas Misumi and Ohkubo as well as Gao and Wu could not identify an influence of compressive residual stresses on the number of cycles required for crack initiation [54,55]. Mutoh et al. noticed an even accelerated crack initiation due a higher surface roughness because of shot-peening [50].

Furthermore, the martensitic microstructure has a strong impact on the short crack propagation. Short cracks propagate along martensitic laths [14,38,56]. Furthermore, PAGBs [11,14,20,38,57] and block boundaries [11,13,57] act as obstacles leading to an oscillating short crack propagation rate. The magnitude of this barrier effect seems to depend on the amount of the misorientation between the two main slip systems of the adjacent grains [10,11,13,35,58].

Compressive residual stresses do not affect the short crack propagation rate [50,55], but rather, they cause a retarded transition from short crack propagation to long crack propagation [49,52,53,59,60]. However, Hu et al. observed a retardation in both stages, in the short crack propagation and the long crack propagation [61].

The aim of this study was the characterization of the crack initiation and short crack propagation in martensitic spring steel by means of uniaxial in-situ fatigue testing in the high cycles fatigue (HCF) regime, applying a confocal laser microscope (CLM). Furthermore, the impact of compressive residual stresses on the crack initiation and short crack propagation were analyzed by additional testing of shot-peened samples. Thereby, this study particularly focused on the fatigue damage evolution by the formation of short crack networks. To link the local crystallographic orientations of the hierarchical martensitic microstructure with the characteristics of short crack propagation behaviour, complementary electron back-scattered diffraction (EBSD) analyses were conducted.

2. Material and Methods

2.1. Material and Sample Geometry

The test specimens were made of the high-strength martensitic spring steel SAE 9254 (German material number: 1.7102, DIN/EN: 54SiCr6). Its chemical composition, according to the inspection certificate of the supplier, and mechanical properties are shown in Table 1. The mechanical properties were evaluated by means of a tensile test, according to DIN EN ISO 6892-1.

Table 1. Chemical composition and mechanical properties of SAE 9254 (mass fraction in percent).

C	Si	Mn	Cr	P	S	UTS [MPa]	YS [MPa]	E [GPa]
0.53	1.43	0.66	0.63	0.008	0.007	1750	1550	210

Austenization of wire rods with a diameter of d = 12 mm was done under a vacuum condition at 1080 °C for 100 min, with subsequent gas quenching using compressed nitrogen followed by tempering at 400 °C for 1 h in an inert argon atmosphere. The fatigue samples (Figure 1) were manufactured from these wire rods by means of electric discharge machining.

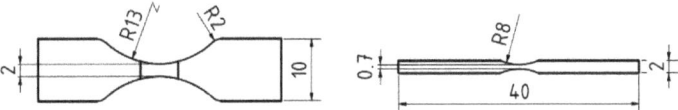

Figure 1. Sample geometry in mm.

2.2. Shot Peening and Residual Stress Measurement

The shot peening of fatigue samples was done by means of a pneumatic machine with two opposing jet nozzles at Sentenso GmbH in Datteln, Germany. Steel wire pieces of a hardness of 700 HV with a diameter of 0.4 mm (G3 according to VDFI 8001) were shot with 1.5 bar jet pressure onto the surface. The resulting Almen intensity was 0.16 A at a surface coverage of 100%.

The residual stresses were determined by means of the energy dispersive measurement method. The measurements were conducted by the Institute for Material Science of the University of Kassel. The used x-ray diffractometer was of type Huber 4 equipped with a W-anode as an x-ray source using an energy of 60 kV and a beam size of 0.5 mm. Thereby, a 2Θ-angle of 25° was adjusted. The analysis of the measured data was performed by means of the multi-wavelength method. Further information regarding the residual stress determination by means of the energy dispersive measurement method and the used equipment is reported in [62–65].

2.3. Testing Strategy

For the sample surface preparation, successive grinding with a stepwise finer SiC paper up to grit 4000 and a final polishing step with colloidal silicon suspension at a grain size of 0.25 μm was performed. The fatigue tests were carried out in the HCF regime with a piezo-driven miniaturized fatigue testing device, shown in Figure 2a. Each sample was clamped between the load frame and the sample slide. A piezo-actuator (PSt 1000/35/125 vs. 45 Thermostable, Co. Piezomechanik GmbH, Munich, Germany) was used for the cyclic loading and was installed in a traverse, which again was connected to the load frame. Additionally, installed disk springs interacted with the piezo-actuator, leading to an uniaxial alternating loading with a maximum force amplitude of F = 1280 N and a maximum test frequency of 30 Hz. A quartz crystal force sensor (type 9134B29, Co. Kistler, Winterthur, Switzerland) was used to determine the force value with an accuracy of better than 0.01 N. This force signal was used as an input signal for the peak power control designed by LabVIEW. For further information regarding the design and the functional principles of the miniaturized testing device, see [66].

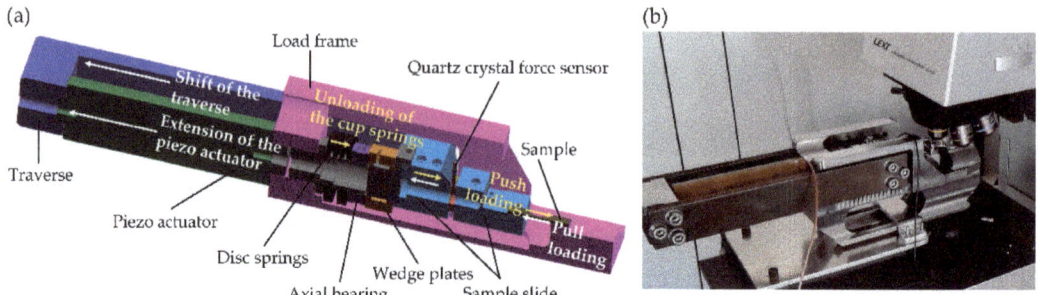

Figure 2. (a) Design and functional principle of the miniaturized testing device; (b) Installation of the miniaturized testing device into the confocal laser microscope.

The in-situ fatigue tests were conducted by installing the miniaturized fatigue testing device into a confocal laser microscope (CLM) of type Olympus LEXT OLS4000 (see Figure 2b). The miniaturized fatigue testing device operated in a stress-controlled manner at a stress ratio of R = −1 with a sinusoidal command signal at a test frequency of 10 Hz.

2.4. Short Crack Growth Monitoring and Microstructural Characterization

To monitor and assess the fatigue damage evolution and to reconstruct the averaged short crack propagation, micrographs of the sample surface were taken after intervals of continuous cyclic loading. The number of cycles in such an interval depended on the applied stress amplitude and the already-endured number of cycles. To measure the crack lengths, a measurement tool provided by the software of the Olympus LEXT OLS4000 microscope was used. The crystallographic orientation analyses were carried out by means of EBSD. The data obtained were used to link the short crack propagation behavior with the local microstructure. Furthermore, to analyze the effect of the PAGBs as well, the software ARPGE [67] was applied to the EBSD data.

3. Results

3.1. Microstrucuture and Residual Stress Profile

The martensitic microstructure of the investigated material is shown in Figure 3a. Its average prior austenite grain size is 125 µm and was determined by means of the linear intercept method according to ISO 643:2012. This was done with a picric acid etching, but the resulting microstructure is not shown here. To identify the orientation relationships (ORs) between the prior austenite grains and the martensitic microstructure, a comparison of the ideal {111} pole figure of the martensite variants inside an austenite grain assuming a Kurdjumov–Sachs (K–S) OR and a Nishiyama–Wassermann (N–W) OR with the measured {111} pole figure was conducted [68]. The results confirmed an austenite to martensite transformation in the material studied according to the K–S OR (Figure 3b–e), which can be described as follows:

$$(111)_\gamma || (011)_{\acute{\alpha}} \quad [10\overline{1}]_\gamma || [11\overline{1}]_{\acute{\alpha}}. \tag{1}$$

The residual stress measurements were made at the surfaces of two samples, and the average value was taken as the value of the residual stress for each measured depth (see Figure 4). The near surface compressive residual stress was 911 MPa, and it stayed close to 900 MPa within a depth of 40 µm.

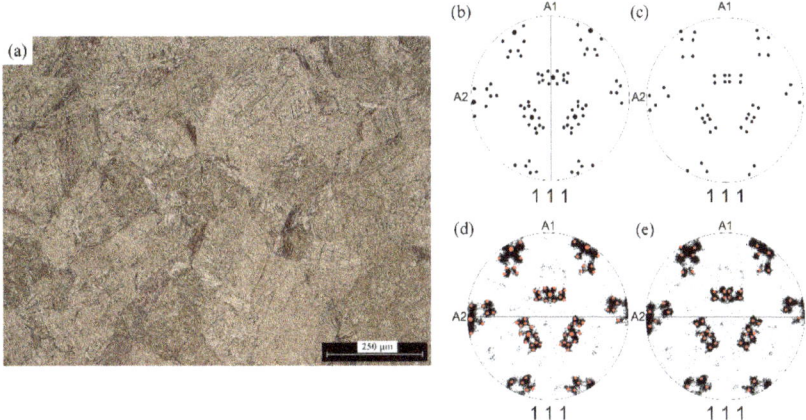

Figure 3. (**a**) Martensite microstructure after heat treatment; ideal {111} pole figure determined according to (**b**) Kurdjumov–Sachs (K–S) orientation relationship and (**c**) Nishiyama–Wassermann (N–W) orientation relationship; measured {111} pole figure superimposed with the (**d**) ideal K–S orientation relationship and (**e**) the ideal N–W orientation relationship.

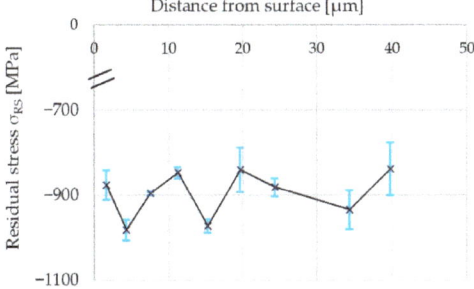

Figure 4. Residual stress profile with corresponding error indicators.

3.2. Crack Initiation

The in-situ observations of the shot-peened (SP) samples and the non-shot-peened (NSP) samples clearly documented that fatigue damage evolution in both conditions started with an early formation of numerous slip bands locally oriented in parallel on the sample surface (see Figure 5). Slip bands indicate a localization of cyclic plastic deformation and, therefore, can give rise to crack initiation. Since the early fatigue damage evolution is characterized by the formation of numerous slip bands, not just one crack was initiated on the sample surface, but a network of short cracks formed. To assess the influence of residual stresses and the applied stress amplitude on the fatigue damage evolution, the crack density ρ_{CD} was determined as a function of the loading cycles. ρ_{CD} was calculated as the total accumulated crack length at the surface divided by the considered surface area. In this context, it is important to note that from the surface observation, it was difficult to distinguish between a slip band and a short crack, as both were depicted as black lines in the CLM images. The same issue was also reported in other studies [14,16,21]. Figure 6a shows the crack density ρ_{CD} of the NSP samples and the SP samples as a function of the number of cycles for different stress amplitudes. The crack densities ρ_{CD} increased with the increasing number of cycles, indicating an accelerating fatigue damage evolution. Thereby, the final data point of each curve represents the number of cycles right before failure. A comparison of those data points near failure between NSP samples and SP samples loaded with similar stress amplitudes revealed that SP samples had a higher fatigue life. The crack density ρ_{CD} in both sample conditions increased with increasing applied stress amplitudes (Figure 6a). The CLM image of the NSP sample loaded with a stress amplitude of 630 MPa revealed a higher extent of fatigue damage evolution as compared to the NSP sample loaded with a lower stress amplitude of 550 MPa (see Figure 6b). The comparison of the crack densities ρ_{CD} of the NSP samples and the SP samples loaded with a similar value of the stress amplitude indicated a lower crack density ρ_{CD} in the case of the SP-samples, especially when a higher number of cycles was concerned. In Figure 6c, this relation is shown based on two CLM images of an NSP sample and an SP sample loaded with a similar stress amplitude.

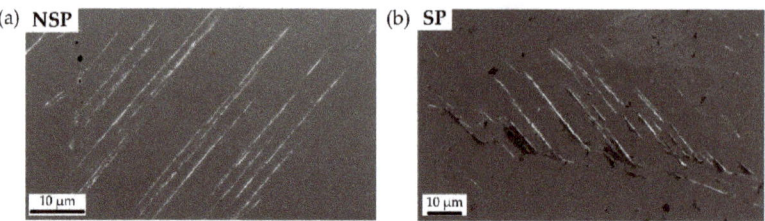

Figure 5. SEM images of slip bands locally oriented in parallel in an (**a**) NSP sample, loaded at $\Delta\sigma/2 = 680$ MPa, recorded after 90,000 loading cycles and (**b**) an SP sample, loaded at $\Delta\sigma/2 = 700$ MPa, recorded after 642,030 loading cycles.

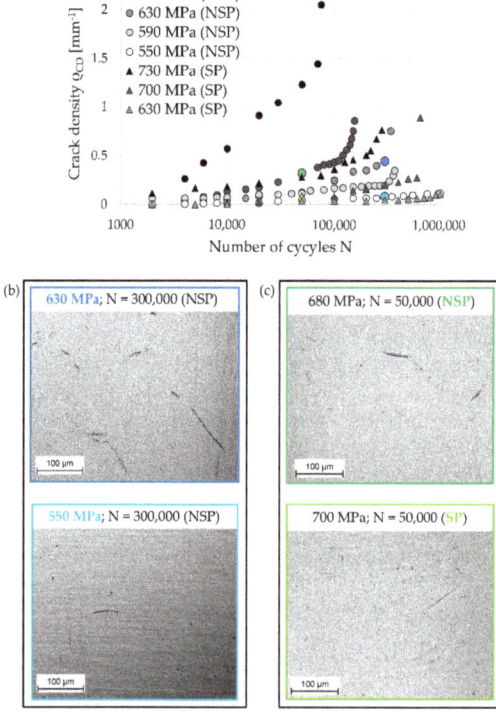

Figure 6. (**a**) Crack density ρ_{CD} of the NSP samples and the SP samples as a function of the number of cycles for different stress amplitudes; (**b**) CLM images of two NSP samples of different stress amplitudes; (**c**) CLM images of an NSP sample and an SP sample of similar stress amplitude.

The majority of the detected slip bands initiated near PAGBs for both types of samples. All short cracks monitored initiated intergranularly at PAGBs with or without compressive residual stresses. Thereby, two types of intergranular crack initiation and early short crack propagation were observed. The first type was characterized by the impingement of slip bands upon a PAGB, leading to an intergranular crack initiation along the PAGB (see Figure 7). The second type exhibited just one isolated and pronounced slip line at the PAGB. An example for this crack initiation type in both sample conditions is shown in Figure 8.

Fracture area analysis revealed the origin of the fracture at the sample surface for NSP samples and at a depth of 130–160 µm for SP samples. Fracture surfaces of an NSP sample and a SP sample, respectively, are shown in Figure 9.

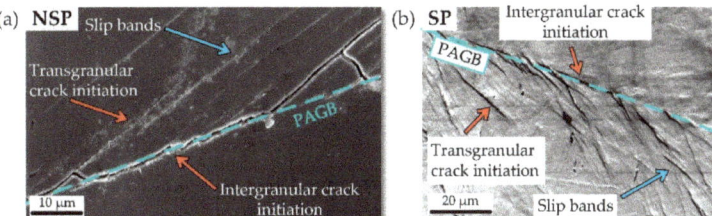

Figure 7. SEM images of intergranular crack initiation caused by the impingement of slip bands on a PAGB in an (**a**) NSP sample, loaded at $\Delta\sigma/2 = 600$ MPa after 90,000 loading cycles, and an (**b**) SP sample, loaded at $\Delta\sigma/2 = 630$ MPa after 300,000 loading cycles.

Figure 8. SEM images of intergranular crack initiation with just one isolated and pronounced slip line at a PAGB in an (**a**) NSP sample, loaded at $\Delta\sigma/2 = 680$ MPa after 155,000 loading cycles, and an (**b**) SP sample, loaded at $\Delta\sigma/2 = 730$ MPa after 286,120 loading cycles.

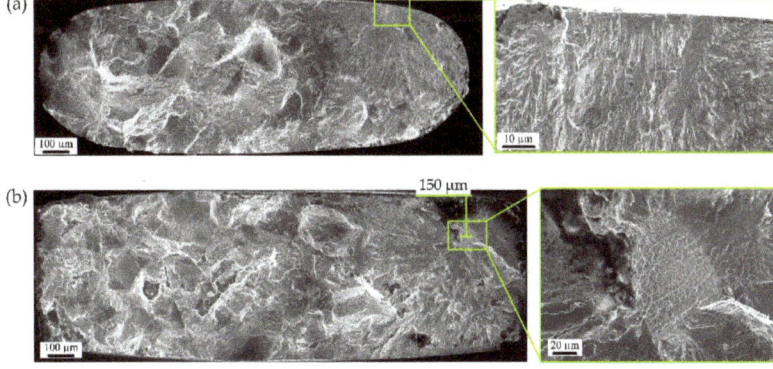

Figure 9. SEM images of the fracture surfaces of an (**a**) NSP sample, loaded at $\Delta\sigma/2 = 590$ MPa for 271,000 loading cycles, and an (**b**) SP sample, loaded at $\Delta\sigma/2 = 730$ MPa for 286,120 loading cycles.

3.3. Short Crack Propagation

To analyze the short crack propagation mechanisms in the investigated martensitic spring steel, short crack propagation rates were related to features of the local microstructure. Figure 10a,b show EBSD orientation mappings of two NSP samples loaded at Δσ/2 = 680 MPa and Δσ/2 = 550 MPa, respectively. Furthermore, reconstructions of the PAGB by means of ARPGE are depicted in the subframes. Accordingly, an intergranular crack path along a PAGB is marked by a white line and a transgranular crack path by a black line. Figure 10c shows the corresponding short crack propagation rates, da/dN, as a function of the stress intensity factor range, ΔK, of these two NSP samples. The oscillations can be attributed to the interaction of the cracks with microstructural obstacles. After intergranular crack initiation along the PAGB, the initial high crack propagation rates decreased as the cracks approached the PAGB of the adjacent grains (① and ①), leading to the assumption that PAGBs act as barriers to short crack propagation. After overcoming these obstacles, the crack propagation rates increased again. The propagating crack of the NSP sample cyclically loaded at Δσ/2 = 680 MPa became partially transgranular, whereas that of the NSP-sample at Δσ/2 = 550 MPa stayed intergranular until the cracks approached another PAGB (② and ②).

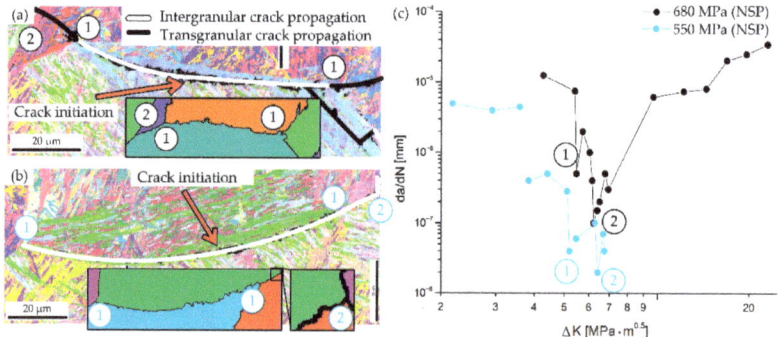

Figure 10. EBSD analysis and reconstructed prior austenite grain boundaries of (**a**) an NSP sample, after 155,000 cycles at Δσ/2 = 680 MPa, and (**b**) an NSP sample, loaded after 1,000,000 cycles at Δσ/2 = 550 MPa; (**c**) crack propagation rate as a function of the stress intensity factor. The numbers in the EBSD analysis and the crack propagation curve mark the breakpoints of the cracks.

For the analysis of the transgranular crack paths, circular inserts indicating glide plane traces were added to the orientation mappings (Figure 11) of the NSP sample fatigued at Δσ/2 = 680 MPa. The active {110} glide planes were highlighted in a red color. Crack propagation to the left (②) happened along a {110} glide plane and stopped again at a PAGB. At the right side (③), crack propagation occurred transgranularly along a {110} glide plane. Thereby, the transgranular crack propagation could be facilitated by previously formed slip bands, which were also parallel to a {110} glide plane (s. SEM image). The almost 20-μm-long transgranular crack then became an intergranular crack that propagated along a PAGB.

Although the origin of fracture was in the sample interior for the SP samples, still, many cracks initiated at the surface. Figure 12 shows the orientation map of an SP sample which has been loaded at Δσ/2 = 630 MPa for 1,000,000 cycles. Furthermore, reconstructions of the PAGB are depicted in the subframes. The main crack (①) propagated along a PAGB and arrested at the PAGB of the adjacent grain. A crack branch (②) propagated transgranularly along a {110} glide plane. The investigation of this short crack suggests that compressive residual stresses do not change the short crack propagation mechanisms fundamentally. Just like the NSP samples, the SP sample showed an intergranular crack initiation and a subsequent short crack propagation along the PAGBs, whereby the PAGBs of the adjacent grains also acted as obstacles to short crack propagation. A transgranular short crack propagated along a {110} glide plane.

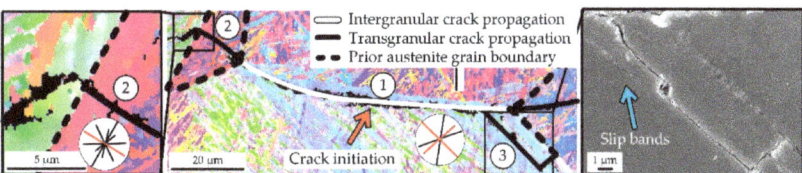

Figure 11. EBSD analysis with calculated {110} glide plane traces (active glide traces marked in red) and SEM image of an NSP sample after 155,000 cycles at $\Delta\sigma/2 = 680$ MPa. The numbers in the EBSD analysis mark the breakpoints of the crack.

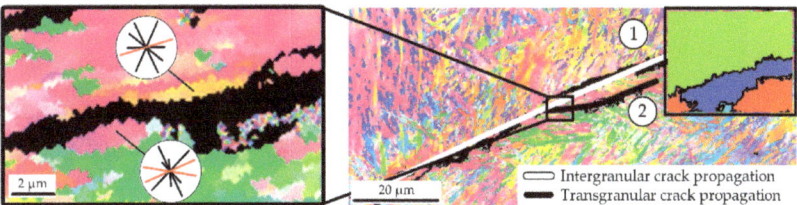

Figure 12. EBSD analysis with calculated {110} glide plane traces (active glide traces marked in red) and reconstructed prior austenite grain boundaries of an SP sample loaded at $\Delta\sigma/2 = 630$ MPa for 1,000,000 cycles. The numbers in the EBSD analysis mark the breakpoints of the crack.

The total length of the fatal crack is plotted for different stress amplitudes as a function of the number of cycles in Figure 13. Crack branches were not considered. Since the origin of fracture of the SP samples was always found to lie in the sample interior, SP samples could not be considered here. During the fatigue tests, the length of the fatal cracks remained under a crack length of 200 μm in all monitored NSP samples during the major fraction of fatigue life, and the crack growth happened by means of the previously discussed short crack propagation mechanisms. After this short crack propagation period, which accounted almost for the entire fatigue life in all examined NSP-samples, a transition to long crack propagation occurred. This transition could be identified by an appreciable increase of crack length taking place within a small number of cycles. This transition happened sooner, the higher the applied stress amplitudes were. CLM images of an NSP sample cyclically loaded at $\Delta\sigma/2 = 680$ MPa are shown in Figure 13 and illustrate the tremendous increase in crack propagation of the fatal crack at around 110,000 cycles. In the case of the NSP sample cycled at $\Delta\sigma/2 = 550$ MPa, no transition could be observed within 10^6 cycles, which represents the ultimate number of loading cycles according to the experimental restrictions.

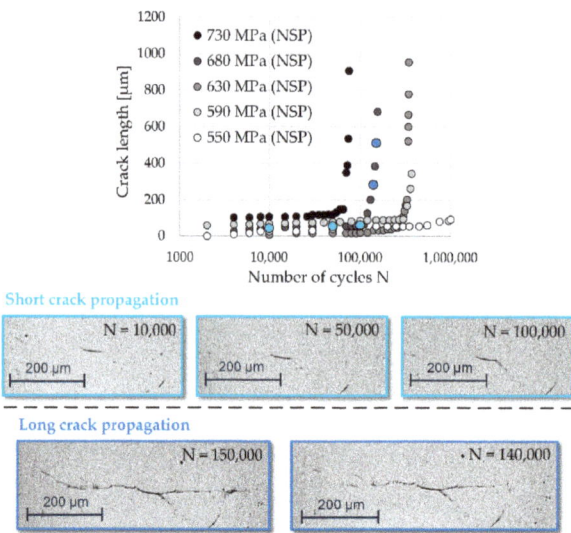

Figure 13. Crack length of NSP samples as a function of the number of cycles for different stress amplitudes and CLM images of the fatal crack development (marked in blue) divided into short crack propagation and long crack propagation, at $\Delta\sigma/2 = 680$ MPa, for selected cycle numbers.

4. Discussion

In previous studies, the early fatigue damage in martensitic steels was characterized by the formation of a crack network of short cracks. Thereby, the resulting crack density seems to strongly depend on the applied stress amplitude [9,14,15,19,38,39]. In this study, the early fatigue damage was also characterized by the formation of a network of short cracks in both sample conditions. The resulting crack density also increased with an increasing number of cycles and increasing applied stress amplitudes. However, a direct comparison of the crack densities of the NSP samples and the SP samples showed a comparatively lower crack density in the case of SP samples, especially when a higher number of cycles was concerned. This leads to the assumption that compressive residual stresses seem to impede the crack initiation on the sample surface. Some earlier studies have reported the same observation [49,52,53]. Misumi and Ohkubo, as well as Gao and Wu, on the contrary, could not identify an impact of compressive residual stresses on the number of cycles required for crack initiation [54,55]. However, in these previous studies, no reference to an emerging fatigue damage evolution in the form of a crack network was made. The findings from our investigation also showed a similar number of cycles required for initiation of the first emerging cracks on the sample surface in both sample conditions. Only the number of initiated surface cracks increased slower in the case of SP samples, resulting in a smaller increase of the crack density. Conversely, Mutho et al. observed an even shorter number of cycles required for crack initiation in the case of peened samples [50]. However, it must be noted that their samples were not polished after shot peening, resulting in a much rougher sample surface compared to the samples investigated in this study. A rough sample surface can lead to stress concentration, which, again, can facilitate crack initiation.

In the material considered here, the crack initiation and early short crack propagation on the sample surface occurred intergranularly along the PAGBs in both sample conditions. In this process, two types of intergranular crack initiation and early short crack propagation along the PAGBs were observed. The first type was characterized by the impingement of slip bands on a PAGB, causing intergranular crack formation, as also described in Batista et al. and Krupp et al. [17,20]. In the second type, the intergranular crack initiation site was characterized as an isolated and pronounced slip step at the PAGB, which is in good agreement with previous studies [17,18,21,69].

A further often-cited explanation for intergranular crack initiation and early short crack propagation, especially in martensitic spring steels, is the preferential alignment of carbides and segregations along microstructural interfaces, as reported in previous studies [25–32]. In this process, the temperature of the tempering process seems to be crucial for the extent of intergranular crack initiation along PABGs. A tempering temperature in the range of 350–450 °C can lead to an enhanced formation of carbides and segregations along the PAGBs, leading to a facilitated intergranular crack initiation due to weakened PAGBs [27–29]. Apart from the carbides along the PAGBs, there was also an additional formation of carbides along the martensitic laths, but the extent seemed to be much smaller and, therefore, negligible [27,69]. Carbides can block the dislocation movement and lead to dislocation pile-ups at the already-weakened PAGBs due to segregations, whereat in martensitic spring steels phosphorous segregations are often considered as particularly detrimental [25,29,31]. This effect again provokes an intergranular crack initiation along the PAGBs. The investigated material was tempered at 400 °C, suggesting that the above-mentioned effects of PAGB weakening are also applicable in our case. In this process, the bigger the average prior austenite grain size is, the more pronounced the detrimental effect of carbides and segregations along PAGBs can be [28,31]. In contrast to Koschella et al., Batista et al., and Ueki et al., an intergranular crack initiation at block boundaries was not observed [13,17,22] in the present study.

Transgranular crack initiation also occurred along martensitic laths in the investigated material, but, unlike in some other studies [13,15,19], this was rarely the case. A possible key factor for those different observations is the influence of the average prior austenite grain size. Morrison and Moosbrugger [34] studied the fatigue behaviour of fine-grained and coarse-grained nickel-270, and, in the case of coarse-grained nickel, cracks initiated almost exclusively at grain boundaries due to slip band impingement. They assumed that there is a critical grain size above which intergranular cracks are formed. In the studied material, the average prior austenite grain size was 125 µm. By comparison, in the work of Koschella et al. [13], who also investigated the fatigue behaviour of a martensitic steel in the HCF regime, the average prior austenite grain size was only 12 µm. In contrast to our results, they observed a frequent transgranular crack initiation within the martensitic blocks. In the work of Bertsch et al. [15] and Seidametova et al. [19] also, a material with a smaller average prior austenite grain size of 75 µm as well of 10–60 µm was investigated.

Many studies have revealed a strong influence of the local microstructure on the short crack propagation (see, for example, [23]), whereat, in the case, of martensitic microstructures, the PAGBs, as well as the boundaries of the different martensitic substructures, can act as barriers to short crack propagation. In this study, the PAGBs were identified to act as obstacles to short crack propagation due to a strong change in the crystallographic orientation. This result is in good agreement with the results in previous studies [11,14,20,38,57]. In some studies, however, block boundaries were considered as the most pronounced barriers for short crack propagation [11,13,57]. According to Morris [70], the decoration of block boundaries with carbides can lead to a barrier effect of those microstructural interfaces. Furthermore, his work also stated that, in the case of tempered martensite, clean block boundaries are often observed, leading to a minor effect of those microstructural interfaces. The material investigated here was a tempered martensite, and no barrier effect of block boundaries could be observed. Rather, the crack growth rate decelerated as soon as a short crack approached the neighboring PAGBs, indicating a pronounced barrier effect of the PAGBs instead. After overcoming this obstacle, the further short crack propagation occured primary intergranularly along the PAGBs and, in some cases, transgranularly. As regards the latter, the short crack propagation seemed to be oriented along a {110} glide plane trace, which is in good agreement with previous studies [13,18,20]. The primary intergranular short crack propagation along the PAGBs could again be a consequence of a relatively high prior austenite grain size and the weaking effect of segregations, leading to a facilitated short crack propagation along the PAGBs as compared to a transgranular short crack propagation. In studies with a primary transgranular short crack propagation, often smaller average prior austenite grain sizes were existent, and higher tempering temperatures had been applied [13,18].

Regarding the influence of residual stresses on the short crack propagation, no difference between the NSP samples and SP samples could be detected. The investigation of the SP samples also led to the assumption that the short crack propagation occurs primarily intergranularly along the PAGBs, with occasional appearing transgranular cracks or crack branches. Furthermore, like in the case of the NSP-samples, PAGBs seem to act as obstacles to short crack propagation. However, in the SP samples, none of the short cracks at the sample surface became a long crack, and the fatal crack always initiated in the sample interior in a depth where no compressive residual stresses were assumed to exist anymore. This leads to the hypothesis that compressive residual stresses impede the transition from short crack propagation to long crack propagation, which is in good agreement with previous studies [49,50,52,53,55,59,60]. In the work of Hu et al., a later transition from short crack propagation to long crack propagation occurred due to an impeded short crack propagation [61]. For the material studied in the presented investigation, no impediment of short crack propagation because of compressive residual stresses could be overserved. An explanation for the different findings could be the primary intergranular crack initiation in the investigated material. It seems that the complete prior austenite grain boundary ruptured after crack initiation, and the short crack propagation stopped at the adjacent neighboring grains. In the work of Hu et al., however, no information was given whether the cracks propagated intergranularly or transgranularly [61].

The SP samples exhibited a higher fatigue life compared to the NSP samples. It is assumed that this beneficial effect of compressive residual stresses can be described by two essential mechanisms. On the one hand, the inhibited transition from short crack propagation to long crack propagation results in a higher fatigue life. On the other hand, compressive residual stresses lead to a shift of the crack initiation site into the interior of the material, as also observed in many other studies [33,42–48]. This is presumably promoted by tensional residual stresses, which occur in sample interior due to the equilibrium of forces [2]. Since crack propagation in the sample interior occurs without environmental effects (i.e., in vacuum), the crack propagation rate is small, resulting in a higher fatigue life [23]. Furthermore, many studies observed a retardation of the long crack propagation due to residual stresses (see for example [2,4,5]).

The transition from short crack propagation to long crack propagation, identifiable by an appreciable increase in the crack length, starts the earlier and the higher the stress amplitude is. Consequently, an incubation phase for long crack propagation could be observed. In Figure 14, the number of cycles of the final data point for different stress amplitudes of NSP samples is plotted, and additionally, the observed values are subdivided into the corresponding number of cycles for crack initiation and for short crack propagation, as well as the number of cycles for long crack propagation. As can be seen in Figure 14, the most significant fatigue life fraction was spent in the stages of crack initiation and short crack propagation, especially at low stress amplitudes. Since for the analyzed SP samples, the crack initiation always occurred in the sample interior and on the sample surface, no transition from short crack propagation to long crack propagation was observed, an identification of the fatigue life fraction of short crack propagation for the SP-samples could not be conducted.

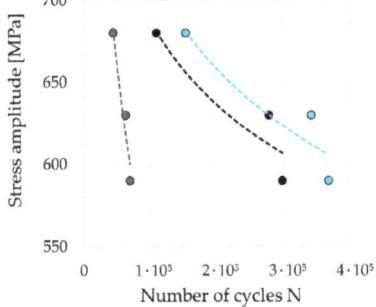

Figure 14. Number of cycles of the final measuring point as a function of the stress amplitude and additional Subdivision into the stages of crack initiation and short crack propagation, as well as long crack propagation for NSP-samples.

5. Conclusions

The crack initiation and short crack propagation was investigated experimentally in a martensitic spring steel with additional consideration of compressive residual stresses. For this purpose, in situ fatigue tests in a CLM and supplementary EBSD analyses have been conducted. The investigated martensitic spring steel SAE 9254 was austenized in a vacuum at 1080 °C for 100 min and gas quenched using compressed nitrogen. Subsequently, a tempering process at 400 °C for 1 h in an argon atmosphere was conducted. Furthermore, some samples were shot peened to assess the influence of compressive residual stresses.

The main results of the study presented are as follows:

- In the first stage of the fatigue tests, which were carried out in the HCF regime, the early fatigue damage in both sample conditions was characterized by the formation of numerous slip bands on the sample surface. These slip plane traces followed the local crystallographic orientations of the martensitic laths. In the slip bands, the cyclic plastic deformation occurred localized, leading to crack initiation. Hence, a network of short cracks formed all over the sample surface.
- The crack density was applied as a parameter for fatigue damage. It was shown that the value of this parameter increased in both sample conditions continuously with increasing number of cycles and that this increase was more pronounced the higher the stress amplitude applied. Regarding the effect of residual stresses on the evolution of crack density, a comparatively lower crack density in the case of the SP samples was observed, especially when a higher number of cycles was concerned.
- In both sample conditions, most of the slip bands and, correspondingly, most of the short fatigue cracks initiated at or close to prior austenite grain boundaries. The subsequent early short crack propagation occurred primarily in an intergranular manner along the prior austenite grain boundaries.
- A detailed analysis of the local short crack propagation rate in correlation with the local microstructure revealed an oscillating crack propagation rate, resulting from a strong interaction of the short fatigue cracks with microstructural features. In this process, the PAGBs were identified to act as obstacles to short crack propagation due to a change in the crystallographic orientation. This relation arose irrespectively of whether the sample surface contained compressive residual stresses or not.
- The SP samples exhibited a higher fatigue life as compared to the NSP samples. It is assumed that this beneficial effect of compressive residual stresses is due to the impediment of the transition from short crack propagation into long crack propagation

on the sample surface and a shift of the fatal crack initiation from the sample surface to the interior of the material.

Author Contributions: A.W. performed the fatigue tests, conducted the electron-microscopic examinations, and analyzed the data regarding the fatigue damage evolution, and H.-J.C. and R.B. contributed by scientific discussions and by reviewing the paper. All authors have read and agreed to the published version of the manuscript.

Funding: This research was funded by Deutsche Forschungsgemeinschaft (Science Foundation, DFG); Funding numbers: Ch 92/57-1 and BR 3276/2-1.

Data Availability Statement: Contact for further information: anna.wildeis@uni-siegen.de.

Acknowledgments: The authors wish to thank the coworkers Alexander Liehr, Artjom Bolender, Sebastian Degener, and Thomas Niendorf from the University of Kassel for the energy dispersive residual stress measurement of the shot-peened samples. We also wish to thank the University of Siegen for the financial support provided through its Open Access Publication Fund. Part of this work was performed at the Micro- and Nanoanalytics Facility of the University of Siegen.

Conflicts of Interest: The authors declare no conflict of interest.

References

1. Christ, H.-J. (Ed.) *Ermüdungsverhalten Metallischer Werkstoffe*, 2nd ed.; Wiley-VCH: Weinheim, Germany, 2009.
2. Totten, G.; Howes, M.; Inoue, T. (Eds.) *Handbook of Residual Stress and Deformation of Steel*; ASM International: Materials Park, OH, USA, 2002.
3. Eleiche, A.M.; Megahed, M.M.; Abd-Allah, N.M. The shot-peening effect on the HCF behavior of high-strength martensitic steels. *J. Mater. Process. Technol.* **2001**, *113*, 502–508. [CrossRef]
4. Almer, J.D.; Cohen, J.B.; Winholtz, R.A. The effects of residual macrostresses and microstresses on fatigue crack propagation. *Metall. Mater. Trans. A* **1998**, *29A*, 2127–2136. [CrossRef]
5. Stacey, A.; Webster, G.A. Fatigue crack growth in autofrettaged thick-walled high pressure tube material. *MRS Online Proc. Libr.* **1984**, *22*, 215–219. [CrossRef]
6. Miller, K.J. The Behaviour of short fatigue cracks and their initiation, Part I—A review of the two recent books. *Fatigue Fract. Eng. Mater. Struct.* **1987**, *10*, 75–91. [CrossRef]
7. McEvily, A.J. The growth of short fatigue cracks: A review. *Trans. Eng. Sci.* **1996**, *13*, 93–107.
8. Efthymiadis, P.; Pinna, C.; Yates, J.R. Fatigue crack initiation in AA2024: A coupled micromechanical testing and crystal plasticity study. *Fatigue Fract. Eng. Mater. Struct.* **2019**, *42*, 321–338. [CrossRef]
9. Hong, Y.; Gu, Z.; Fang, B.; Bai, Y. Collective evolution characteristics and computer simulation of short fatigue cracks. *Philos. Mag. A* **1997**, *75*, 1517–1531. [CrossRef]
10. Miller, K.J. The short crack problem. *Fatigue Fract. Eng. Mater. Struct.* **1982**, *5*, 223–232. [CrossRef]
11. Yang, M.; Zhong, Y.; Liang, Y.-I. Competition mechanisms of fatigue crack growth behavior in lath martensitic steel. *Fatigue Fract. Eng. Mater. Struct.* **2018**, *41*, 2502–2513. [CrossRef]
12. Kitahara, H.; Ueji, R.; Tsuji, N.; Minamino, Y. Crystallographic features of lath martensite in low-carbon steel. *Acta Mater.* **2006**, *54*, 1279–1288. [CrossRef]
13. Koschella, K.; Krupp, U. Investigations of fatigue damage in tempered martensitic steel in the HCF regime. *Int. J. Fatigue* **2019**, *124*, 113–122. [CrossRef]
14. Brückner-Foit, A.; Huang, X. On the determination of material parameters in crack initiation laws. *Fatigue Fract. Mater. Struct.* **2008**, *31*, 980–988. [CrossRef]
15. Bertsch, J.; Möslang, A.; Riesch-Oppermann, H. Fatigue crack initiation in a ferritic-martensitic steel under irradiated and unirradiated conditions. In *ECF 12*; Engineering Materials Advisory Services Ltd.: Sheffield, UK, 1998.
16. Nishikawa, H.-A.; Furuya, Y.; Igi, S.; Goto, S.; Briffod, F.; Shiraiwa, T.; Enoki, M.; Kasuya, T. Effect of microstructure of simulated heat-affected zone on low- to high-cycle fatigue properties of low-carbon steels. *Fatigue Fract. Eng. Mater. Struct.* **2020**, *43*, 1239–1249. [CrossRef]
17. Batista, M.N.; Marinelli, M.C.; Alvarez-Armas, I. Effect of initial microstructure on surface relief and fatigue crack initiation in AISI 410 ferritic-martensitic steel. *Fatigue Fract. Eng. Mater. Struct.* **2019**, *42*, 61–68. [CrossRef]
18. Giertler, A.; Söker, M.; Dönges, B.; Istomin, K.; Ludwig, W.; Pietsch, U.; Fritzen, C.-P.; Christ, H.-J.; Krupp, U. The significance of local plasticity for the crack initiation process during very high cycle fatigue of high strength steels. *Procedia Mater. Sci.* **2014**, *3*, 1353–1358. [CrossRef]
19. Seidametova, G.; Vogt, J.-B.; Serre, I.P. The early stage of fatigue crack initiation in a 12%Cr martensitic steel. *Int. J. Fatigue* **2017**, *106*, 38–48. [CrossRef]

20. Krupp, U.; Giertler, A.; Koschella, K. Microscopic damage evolution during very-high-cycle fatigue (VHCF) of tempered martensitic steel. *Fatigue Fract. Eng. Mater Struct.* **2017**, *40*, 1731–1740. [CrossRef]
21. Motoyashiki, Y.; Brückner-Foit, A.; Sugeta, A. Investigation of small crack behaviour under cyclic loading in a dual phase steel with a FIB tomography technique. *Fatigue Fract. Eng. Mater Struct.* **2007**, *30*, 556–564. [CrossRef]
22. Ueki, S.; Mine, Y.; Takashima, K. Microstructure-sensitive fatigue crack growth in lath martensite of low carbon steel. *Mater. Sci. Eng. A* **2020**, *773*, 138830. [CrossRef]
23. Krupp, U. *Fatigue Crack Propagation in Metals and Alloys: Microstructural Aspects and Modelling Concepts*; Wiley-VCH: Weinheim, Germany, 2007.
24. Ohmura, T.; Minor, A.M.; Stach, E.A.; Morris, J.W. Dislocation–Grain boundary interactions in martensitic steel observed through in situ nanoindentation in a transmission electron microscope. *J. Mater. Res.* **2004**, *19*, 3626–3632. [CrossRef]
25. Choi, S. Optimization of Microstructure and Properties of High Strength Spring Steel. Ph.D. Thesis, Delft University of Technology, Delft, The Netherlands, 2011.
26. Krauss, G. Martensite in steel: Strength and structure. *Mater. Sci. Eng. A* **1999**, *273*, 40–57. [CrossRef]
27. Hayakawa, M.; Matsuoka, S.; Tsuzaki, K. Microstructural analyses of grain boundary carbides of tempered martensite in medium-carbon steel by atomic force microscopy. *Mater. Trans.* **2002**, *43*, 1758–1766. [CrossRef]
28. Horn, R.M.; Ritchie, R.O. Mechanisms of tempered martensite embrittlement in low alloy steels. *Metall. Mater. Trans. A* **1978**, *9A*, 1039–1053. [CrossRef]
29. Briant, C.L.; Banerji, S.K. The fracture behavior of quenched and tempered manganese steels. *Metall. Mater. Trans. A* **1982**, *13*, 827–836. [CrossRef]
30. Bandyopadhyay, N.; McMahon, C.J., Jr. The micro-mechanisms of tempered martensite embrittlement in 4340-type steels. *Metall. Mater. Trans. A* **1983**, *14*, 1313–1325. [CrossRef]
31. Ohtani, H.; McMahon, C.J., Jr. Modes of fracture in temper embrittled steels. *Acta Metall.* **1975**, *23*, 377–386. [CrossRef]
32. Hoseiny, H.; Caballero, F.G.; San Martin, D.; Capdevilla, C. The influence of austenization temperature on the mechanical properties of a prehardened mould steel. *Mater. Sci. Forum.* **2012**, *706*, 2140–2145. [CrossRef]
33. Przybyla, C.; Prasannavenkatesan, R.; Salajegheh, N.; McDowell, D.L. Microstructure-sensitive modeling of high cycle fatigue. *Int. J. Fatigue* **2010**, *32*, 512–525. [CrossRef]
34. Morrison, D.J.; Moosbrugger, J.C. Effects of grain size on cyclic plasticity and fatigue crack initiation in nickel. *Int. J. Fatigue* **1998**, *19*, 51–59. [CrossRef]
35. Christ, H.-J.; Düber, O.; Fritzen, C.-P.; Knobbe, H.; Köster, P.; Krupp, U. Propagation behaviour of microstructural short fatigue cracks in the high-cycle fatigue regime. *Comput. Mater. Sci.* **2009**, *46*, 561–565. [CrossRef]
36. Kolyshkin, A.; Zimmermann, M.; Kaufmann, E.; Christ, H.-J. Untersuchung der Rissinitiierung und -ausbreitung mittels Fernfeldmikroskop im VHCF-Bereich. In *Werkstoffprüfung*; Christ, H.-J., Ed.; Stahleisen GmbH: Düsseldorf; Germany, 2013.
37. Manonukul, A.; Dunne, F.P.E. High- and low-cycle fatigue crack initiation using polycrystal plasticity. *Proc. R. Soc. Lond. Ser. A* **2004**, *460*, 1881–1903. [CrossRef]
38. Meyer, S.; Brückner-Foit, A.; Möslang, A. A stochastic simulation model for microcrack initiation in a martensitic steel. *Comput. Mater. Sci.* **2003**, *26*, 102–110. [CrossRef]
39. Leguinagoicoa, N.; Albizuri, J.; Larranaga, A. Fatigue improvement and residual stress relaxation of shot-peened alloy steel DIN 34CrNiMo6 under axial loading. *Int. J. Fatigue* **2022**, *162*, 1007006. [CrossRef]
40. Branco, R.; Costa, J.D.; Berto, F.; Kotousov, A.; Antunes, F.V. Fatigue crack initiation behaviour of notched 34CrNiMo6 steel bars under proportional bending-torsion loading. *Int. J. Fatigue* **2020**, *130*, 105268. [CrossRef]
41. Doquet, V. Crack initiation mechanisms in torsional fatigue. *Fatigue Fract. Engng Mater. Struct.* **1997**, *20*, 227–235. [CrossRef]
42. Korn, M.; Rohm, T.; Lang, K.H. Influence of near-surface stress gradients and strength effect on the very high cycle fatigue behavior of 42CrMo4 Steel. In *Fatigue of Materials at Very High Numbers of Loading Cycles*; Christ, H.-J., Ed.; Springer Spektrum: Wiesbaden, Germany, 2018; pp. 233–252.
43. Liu, W.; Dong, J.; Zhang, P.; Zhai, C.; Ding, W. Effect of shot peening on surface characteristics and fatigue properties of T5-treated ZK60 Alloy. *Mater. Trans.* **2009**, *50*, 791–798. [CrossRef]
44. Ludian, T.; Wagner, L. Mechanical surface treatments for improving fatigue behavior in titanium alloys. *Adv. Mater. Sci. Eng.* **2008**, *8*, 44–52. [CrossRef]
45. Starker, P.; Wohlfahrt, H.; Macherauch, E. Subsurface crack initiation during fatigue as a result of residual stresses. *Fatigue Fract. Eng. Mater. Struct.* **1979**, *1*, 319–327. [CrossRef]
46. Tange, A.; Takahashi, F. Fatigue strength and shot-peening. In Proceedings of the ICSP-10, Tokio, Japan, 15–18 September 2008; p. 2008080.
47. Mlikota, M.; Schmauder, S.; Dogahe, K.; Bozic, Z. Influence of local residual stresses on fatigue crack initiation. *Procedia Struct. Integr.* **2021**, *31*, 3–7. [CrossRef]
48. Lindemann, J.; Buque, C.; Appel, F. Effect of shot peening on fatigue performance of a lamellar titanium aluminide alloy. *Acta Mater.* **2006**, *54*, 1155–1164. [CrossRef]
49. Bag, A.; Delbergue, D.; Levesque, M.; Bocher, P.; Brochu, M. Study of short crack growth in shot peened 300M steel. In Proceedings of the ICSP-13, Montreal, QC, Canada, 18–21 September 2017.

50. Mutoh, Y.; Fair, G.H.; Noble, B.; Waterhouse, R.B. The effect of residual stresses induced by shot-peening on fatigue crack propagation in two high strength aluminium alloys. *Fatigue Fract. Engng Mater. Struct.* **1987**, *10*, 261–272. [CrossRef]
51. Aviles, A.; Aviles, R.; Albizuri, J.; Pallares-Santasmartas, A.R. Effect of shot-peening and low-plasticity burnishing on the high-cycle fatigue strength of DIN 34CrNiMo6 alloy steel. *Int. J. Fatigue* **2019**, *119*, 338–354. [CrossRef]
52. De los Rios, E.R.; Walley, A.; Milan, M.T.; Hammersley, G. Fatigue crack initiation and propagation on shot-peened surfaces in A316 stainless steel. *Int. J. Fatigue* **1995**, *17*, 493–499. [CrossRef]
53. Berns, H.; Weber, L. Fatigue progress in shot-peened surface layers. In Proceedings of the ICSP-3, Garmisch-Partenkirchen, Germany, 12–16 October 1987.
54. Misumi, M.; Ohkubo, M. Deceleration of small crack growth by shot-peening. *Int. J. Mater. Prod. Technol.* **1987**, *2*, 36–47.
55. Gao, Y.K.; Wu, X.R. Experimental investigation and fatigue life prediction for 7475-T7351 aluminum alloy with and without shot peening-induced residual stresses. *Acta Mater.* **2011**, *59*, 3737–3747. [CrossRef]
56. Deng, X.; Lu, F.; Cui, H.; Tang, X.; Li, Z. Microstructure correlation and fatigue crack growth behavior in dissimilar 9Cr/CrMoV welded joint. *Mater. Sci. Eng. A* **2016**, *651*, 1018–1030. [CrossRef]
57. Li, S.; Zhu, G.; Kang, Y. Effect of substructure on mechanical properties and fracture behavior of lath martensite in 0.1C-1.1Si-1.7Mn steel. *J. Alloys Compd.* **2016**, *675*, 104–115. [CrossRef]
58. Krupp, U.; Düber, O.; Christ, H.-J.; Künkler, B.; Schick, A.; Fritzen, C.-P. Application of the EBSD technique to describe the initiation and growth behaviour of microstructurally short fatigue cracks in a duplex steel. *J. Microsc.* **2004**, *213*, 313–320. [CrossRef]
59. Farrahi, G.H.; Majzoobi, G.H.; Hosseinzadeh, F.; Harati, S.M. Experimental evaluation of the effect of residual stress field on crack growth behaviour in C(T) specimen. *Eng. Fract. Mech.* **2006**, *73*, 1772–1782. [CrossRef]
60. Tange, A.; Takamura, N. Relation between shot-peening residual stress distribution and fatigue crack propagation life in spring steel. *Trans. Jpn. Soc. Mech. Eng.* **1991**, *36*, 47–53. [CrossRef]
61. Hu, Y.; Cheng, H.; Yu, J.; Yao, Z. An experimental study on crack closure induced by laser peening in pre-cracked aluminum alloy 2024-T351 and fatigue life extension. *Int. J. Fatigue* **2020**, *130*, 105232. [CrossRef]
62. Liehr, A.; Zinn, W. Energy-dispersive residual stress analysis under laboratory conditions: Concept for a new type of diffractometer. *Adv. Mater. Res.* **2014**, *996*, 192–196. [CrossRef]
63. Liehr, A.; Zinn, W.; Degener, S.; Scholtes, B.; Niendorf, T.; Genzel, C. Energy resolved residual stress analysis with laboratory X-ray sources. *HTM J. Heat Treatm. Mat.* **2017**, *72*, 115–121. [CrossRef]
64. Apel, D.; Meixner, M.; Liehr, A.; Klaus, M.; Degener, S.; Wagener, G.; Franz, C.; Zinn, W.; Genzel, C.; Scholtes, B. Residual stress analysis of energy-dispersive diffraction data using a two-detector setup: Part I—Theoretical concept. *Nucl. Instrum. Methods Phys. Res. Sect. A Accel. Spectrometers Detect. Assoc. Equip.* **2018**, *877*, 24–33. [CrossRef]
65. Apel, D.; Meixner, M.; Liehr, A.; Klaus, M.; Degener, S.; Wagener, G.; Franz, C.; Zinn, W.; Genzel, C.; Scholtes, B. Residual stress analysis of energy-dispersive diffraction data using a two-detector setup: Part II—Experimental implementation. *Nucl. Instrum. Methods Phys. Res. Sect. A Accel. Spectrometers Detect. Assoc. Equip.* **2018**, *877*, 56–64. [CrossRef]
66. Wildeis, A.; Christ, H.-J.; Brandt, R.; Thimm, M.; Fritzen, C.-P. Prüftechnik zur Durchführung von in-situ-Ermüdungsversuchen. In *Werkstoffprüfung*; Langer, J.B., Wächter, M., Eds.; DVM e.V.: Berlin, Germany, 2020.
67. Cayron, C. ARPGE: A computer program to automatically reconstruct the parent grains from electron backscatter diffraction data. *J. Appl. Crystallogr.* **2007**, *40*, 1183–1188. [CrossRef]
68. Hornbogen, E.; Thumann, M. (Eds.) *Die Martensitische Phasenumwandlung und Deren Werkstofftechnische Anwendung*; DGM—Informationsgesellschaft: Oberursel, Germany, 1986.
69. Caron, R.N.; Krauss, G. The tempering of Fe-C lath martensite. *Metall. Mater. Trans.* **1972**, *3*, 2381–2389. [CrossRef]
70. Morris, J.W. On the ductile-brittle transition in lath martensitic steel. *ISIJ Int.* **2011**, *51*, 1569–1575. [CrossRef]

Article

A New Approach to Estimate the Fatigue Limit of Steels Based on Conventional and Cyclic Indentation Testing

David Görzen, Pascal Ostermayer, Patrick Lehner, Bastian Blinn *, Dietmar Eifler and Tilmann Beck

Institute of Materials Science and Engineering, TU Kaiserslautern, 67663 Kaiserslautern, Germany; goerzen@mv.uni-kl.de (D.G.); ostermayer@mv.uni-kl.de (P.O.); lehner@mv.uni-kl.de (P.L.); eifler@mv.uni-kl.de (D.E.); beck@mv.uni-kl.de (T.B.)
* Correspondence: blinn@mv.uni-kl.de; Tel.: +49-631-205-5288

Abstract: For a reliable design of structural components, valid information about the fatigue strength of the material used is a prerequisite. As the determination of the fatigue properties, and especially the fatigue limit σ_w, requires a high experimental effort, efficient approaches to estimate the fatigue strength are of great interest. Available estimation approaches using monotonic properties, e.g., Vickers hardness (HV), and in some cases the cyclic yield strength, only allow a rough estimation of σ_w. The approaches solely based on monotonic properties lead to substantial deviations of the estimated σ_w in relation to the experimentally determined fatigue limit as they do not consider the cyclic deformation behavior. In this work, an estimation approach was developed, which is based on a correlation analysis of the fatigue limit σ_w, HV, and the cyclic hardening potential obtained in instrumented cyclic indentation tests (CIT). For this, eleven conditions from five different low-alloy steels were investigated. The CIT enable an efficient and quantitative determination of the cyclic hardening potential, i.e., the cyclic hardening exponent$_\text{CHT}$ e_II, and thus, the consideration of the cyclic deformation behavior in an estimation approach. In this work, a strong correlation of σ_w with the product of HV and $|e_\text{II}|$ was observed. In relation to an existing estimation approach based solely on HV, considering the combination of HV and $|e_\text{II}|$ enables the estimation of σ_w with an enormously increased precision.

Keywords: fatigue limit estimation; cyclic indentation testing; cyclic hardening potential; hardness; low-alloy steels

Citation: Görzen, D.; Ostermayer, P.; Lehner, P.; Blinn, B.; Eifler, D.; Beck, T. A New Approach to Estimate the Fatigue Limit of Steels Based on Conventional and Cyclic Indentation Testing. *Metals* **2022**, *12*, 1066. https://doi.org/10.3390/met12071066

Academic Editors: Martin Heilmaier and Martina Zimmermann

Received: 17 May 2022
Accepted: 18 June 2022
Published: 22 June 2022

Publisher's Note: MDPI stays neutral with regard to jurisdictional claims in published maps and institutional affiliations.

Copyright: © 2022 by the authors. Licensee MDPI, Basel, Switzerland. This article is an open access article distributed under the terms and conditions of the Creative Commons Attribution (CC BY) license (https://creativecommons.org/licenses/by/4.0/).

1. Introduction

Materials used in engineering applications are oftentimes subjected to cyclic loadings. Therefore, the knowledge of cyclic properties, and especially the fatigue limit, is indispensable for a safe and reliable design. To determine the fatigue properties, a high experimental effort is required, i.e., a relatively large number of fatigue tests and specimens. Consequently, methods to reduce this effort are of great industrial and scientific interest. For this purpose, short-time procedures [1–5] and correlations between the fatigue limit and mechanical parameters that can be determined with a low experimental effort (e.g., hardness and ultimate tensile stress (UTS)) [6–8] were elaborated.

Hardness tests are one of the simplest methods to characterize the monotonic strength of a material. By investigating six different steels, Garwood et al. [6] showed that Rockwell hardness (HRC) roughly correlates with the fatigue limit σ_w. Note that the fatigue limit σ_w discussed in the presented work is defined as the stress amplitude σ_a that leads to an ultimate number of cycles of 10^6–10^7, depending on the material investigated and the test frequency used. According to Murakami [9,10], the linear relationship found by Garwood et al. [6] can be expressed based on the Vickers hardness (HV) using Equation (1) (compare Figure 1), with σ_w in MPa. This relationship is valid up to a material-specific hardness value, ranging between 400 and 500 HV [6,9,10]. Note that the fatigue limits given in Figure 1 were determined in rotation bending tests and the deviation observed for materials with a higher hardness is partially caused by residual stresses at the surface [11]. As this loading condition results in a stress gradient from the surface to the inner material, an increased influence of residual stresses in relation to uniaxial fatigue testing occurs. In addition to the residual stresses and the loading condition, it must be considered that a higher surface roughness significantly decreases the fatigue limit [9,12].

$$\sigma_w = 1.6\,\text{HV} \pm 0.1\,\text{HV} \tag{1}$$

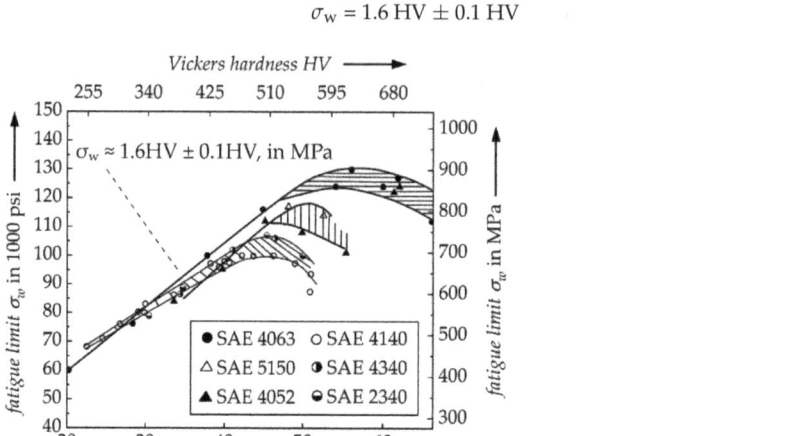

Figure 1. Fatigue limit as a function of hardness [6], based on [9,10].

In addition to Equation (1), the ASTM Handbook [13] provides a linear correlation of the fatigue limit and Brinell hardness (HB) for some steels, which is valid for a hardness up to 500 HB. The resulting correlation factor is 1.72, which is relatively close to Equation (1). However, this is only a rough estimation, as it is based on the assumption that the fatigue limit is half of the UTS [13]. Furthermore, the hardness can also be used to estimate the fatigue limit in case of defect-based failure, as shown by the investigations of Murakami [9,10] and Casagrande et al. [14].

In addition to hardness, the tensile properties, i.e., the yield strength σ_y or the UTS, were also used in different scientific works to assess the fatigue limit [7,8,15–17].

Although the abovementioned models provide sufficient estimations for some steels, a reliable prediction of the fatigue limit exclusively based on hardness or tensile properties is not achieved [15,18,19]. This is illustrated by the big scatter band shown in Figure 1, while the deviation between the estimated and experimentally determined fatigue limit increases with increasing hardness [10]. Furthermore, Fleck et al. [15] and Rennert et al. [17] show that the relation between the fatigue limit and yield strength, i.e., the ratio σ_w/σ_y, depends on the type and the strength of the respective steel. In general, in these works, an increase in σ_y leads to a less pronounced growth of σ_w, and thus, with an increase in σ_y, the ratio σ_w/σ_y decreases. This is explained by Fleck et al. [15] with the cyclic deformation behavior: While steels with a lower strength in the initial state show cyclic hardening, most high-strength steels exhibit cyclic softening. This is also underlined by the work of Lopez and Fatemi [19], who observed for two steels with the same hardness a different fatigue limit caused by a different cyclic deformation behavior. Consequently, for a reliable estimation of the fatigue limit σ_w, the cyclic deformation behavior must be considered.

Beyond the possibilities of conventional hardness measurement, instrumented indentation testing enables the analysis of the deformation behavior. Thus, other material properties, e.g., the Young's modulus E [20,21], can be determined, too. Moreover, the cyclic deformation behavior of a material can be characterized using instrumented cyclic indentation tests (CIT) [22,23]. The short-time procedure PhyBaL$_{\text{CHT}}$, which is based on CIT and enables determining the cyclic hardening potential, was used in [23] to describe the cyclic deformation behavior of differently heat-treated conditions of the low-alloy steel 42CrMo4. The cyclic hardening potential refers to the capacity of a metallic material to increase its strength during cyclic loading, especially via dislocation activities, which counteracts local stress concentrations, e.g., at microstructural inhomogeneities. As demonstrated in [23], the cyclic deformation behavior obtained with this testing approach corresponds to the cyclic deformation curves determined in uniaxial compressive fatigue tests. Furthermore, Kramer et al. [24] showed for the steel 18CrNiMo7-6 that the cyclic hardening potential determined in CIT correlates with the amount of cyclic hardening obtained in uniaxial push-pull fatigue tests.

In our various investigations, it was demonstrated that the cyclic hardening observed in CIT highly depends on different microstructural phenomena, e.g., the size and distribution of precipitates [25–27], the dislocation density [28], and the grain size [24]. Moreover, in [26,27], the results obtained in CIT showed a higher sensitivity to microstructural changes than conventional hardness measurements. Accordingly, in another of our own studies [29] on two differently heat-treated Cu-alloyed steels, X0.5CuNi2-2 and X21CuNi2-2, the ranking of the fatigue strength could only be explained by a combined consideration of the hardness and cyclic hardening potential, as conditions with nearly identical hardness showed a significant difference in σ_w, depending on their cyclic hardening potential.

Based on the results outlined above, PhyBaL$_{CHT}$ is a promising method to improve the estimation of the fatigue limit of steels as it enables consideration of the cyclic deformation behavior, i.e., the cyclic hardening potential. The high potential of this method is especially underlined by the results reported in [29]. Based on that, in the present work, the relationship between the hardness, the cyclic hardening potential obtained in CIT, and the fatigue limit σ_w was analyzed. For this purpose, different types of low-alloy steels with a bcc lattice structure were investigated. In addition to existing data from [29] for Cu-alloyed steels and [30,31] for different batches of C50E, hardness measurements, CIT, and fatigue tests were performed on differently heat-treated 42CrMo4 and a batch of bainitic 100CrMnSi6-4.

2. Materials and Methods

2.1. Materials

For the correlation analysis presented in this work, which is the basis for an improved estimation of the fatigue limit σ_w, three differently heat-treated batches of 42CrMo4, one batch of 100CrMnSi6-4, five batches of differently heat-treated Cu-alloyed steels with two different Cu contents, and two batches of C50E (railway steel R7) were used. In summary, the presented analyses are based on eleven conditions of five different low-alloy steels.

As a condition with a relatively high hardness, a bainitic 100CrMnSi6-4 with the chemical composition given in Table 1 was investigated. To generate a bainitic microstructure, the material was austenitized at 880 °C for 30 min and subsequently bainitized at 220 °C for 8 h, resulting in a retained austenite fraction of 20 vol.-%.

Table 1. Chemical compositions of 42CrMo4 [23] and 100CrMnSi6-4.

Element in wt.-%	C	Si	Mn	P	S	Cr	Mo	Ni	Cu	Al
100CrMnSi6-4	0.93	0.59	1.09	0.01	0.01	1.46	-	0.03	0.01	<0.001
42CrMo4	0.37	0.20	0.76	0.002	0.027	1.02	0.16	0.20	0.19	0.02

The 42CrMo4 conditions investigated comprise one normalized as well as two quenched and differently tempered conditions. As these variants are from the identical batch investigated in [23], a more detailed description of the microstructure and heat treatment parameters can be found in this preliminary publication. However, the chemical composition is given in Table 1 while the annealing temperatures T_a and times t_a are listed in Table 3.

For the Cu-alloyed steels, a detailed description of the chemical compositions, heat treatment parameters, and the resulting microstructure is given in [29,32]. The variants considered in this work differ in their C contents, i.e., 0.005 wt.-% (X0.5CuNi2-2) and 0.21 wt.-% (X21CuNi2-2), and in their annealing times t_a, respectively. Moreover, a quenched condition of X0.5CuNi2-2 was analyzed. Please note that these laboratory melts were named with our own labeling based on DIN EN 10027-1 [33] and constitute two low-alloy steels.

Moreover, two batches of C50E, which were extracted from a railway wheel and differ in extraction positions and thus, cooling conditions, were used. As these batches were extracted from a component, no defined heat treatment parameters can be provided. However, the resulting microstructures are characterized in detail in [27,28].

2.2. Mechanical Characterization

To analyze the correlation of the fatigue limit σ_w and the cyclic deformation behavior obtained in CIT, the respective data is required. Thus, for all material conditions, the cyclic hardening potential was determined using the short-time procedure PhyBaL$_{CHT}$, which is described in detail in [23,24]. For this, instrumented cyclic indentation tests (CIT) were performed using a Fischerscope H100C device and a Fischerscope HM2000 device (both from Helmut Fischer GmbH, Sindelfingen, Germany), which are both equipped with a Vickers indenter and enable a continuous measurement of the indentation force F and the indentation depth h. For the specimens of the high-strength steel 100CrMnSi6-4, a maximum indentation force of F_{max} = 2000 mN was used while the other steels were tested with F_{max} = 1000 mN. In CIT, a sinusoidal load function with a frequency of 1/12 Hz and, in total, 10 cycles were used. For each variant, 20 indentations were performed at 2 different polished sections, respectively. Thus, the results of each variant were determined based on 40 CIT, resulting in a high statistical reliability. To exclude any interference between the indentation points, they were placed at a distance of at least five times the indent diagonal, which is in accordance with [34].

From the second cycle on, the continuously measured signals result in an F-h hysteresis (compare Figure 2a). In analogy to the plastic strain amplitude obtained from a stress–strain hysteresis, the half width of the F-h hysteresis at mean loading is defined as the plastic indentation depth amplitude $h_{a,p}$. Similar to the cyclic deformation curve, the development of $h_{a,p}$ versus the number of cycles N is used to describe the cyclic deformation behavior (compare Figure 2b). The resulting $h_{a,p}$–N curve shows a stabilized slope from the fifth cycle on, which indicates saturation of the macro plastic deformation and a domination of microplasticity. This regime of the $h_{a,p}$–N curve can further be expressed by the power function $h_{a,p\,II}$ given in Equation (2) [24]:

$$5 \leq N \leq 10: h_{a,p\,II} = a_{II} \cdot N^{e_{II}} \tag{2}$$

The exponent e_{II} describes the slope of $h_{a,p\,II}$ and, thus, of the $h_{a,p}$–N curve in the stabilized regime. As a steeper slope of the $h_{a,p}$–N curve indicates a more pronounced cyclic hardening, e_{II} is used to quantify the cyclic hardening potential of the investigated material and, thus, is called the cyclic hardening exponent$_{CHT}$. A steeper slope of the curve, which is described by a higher absolute value of the exponent $|e_{II}|$, indicates a higher cyclic hardening potential, i.e., a higher capacity to counteract local stress concentrations [23,24,26,29]. As already discussed in the introduction, the fatigue strength relates to the cyclic hardening potential and, thus, $|e_{II}|$ is considered for the correlation analyses. Note that the cyclic hardening exponent$_{CHT}$ is not equivalent to the cyclic strain hardening exponent n' in the Ramberg–Osgood curve as e_{II} is, in contrast to n', determined in a multi-axial compressive stress state.

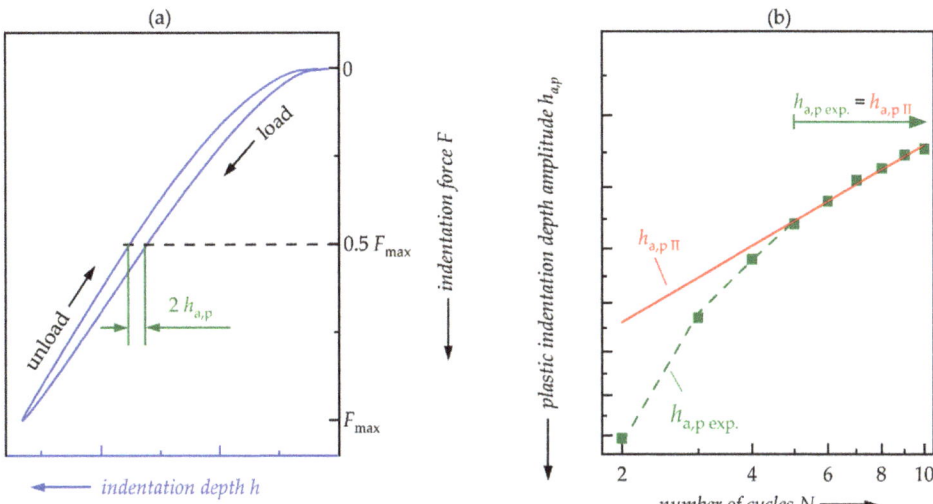

Figure 2. Schematic of the PhyBaL$_{CHT}$ procedure: (**a**) F–h hysteresis and (**b**) $h_{a,p}$–N curve [26].

In addition to e_{II}, the Vickers hardness was measured for all material variants, as Equation (1) is the starting point of the correlation analyses presented in this work. The Vickers hardness (HV) was measured with a ZwickRoell ZHU 250top universal hardness testing device (ZwickRoell GmbH & Co. KG, Ulm, Germany). In the case of the conditions of the steels X0.5CuNi2-2, X21CuNi2-2, and C50E, a test force of 98.07 N was used (HV 10) while the variants of steel 42CrMo4 and 100CrMnSi6-4 were tested with 294.12 N (HV 30). However, as the Vickers hardness is considered to be independent of the test force, the hardness values can be used adequately. Each HV value was determined by calculating the arithmetic mean of eight measurements. Please note that the values of HV of the steels X0.5CuNi2-2 and X21CuNi2-2 previously published in [32] differ slightly from the ones used in this work because additional hardness measurements were taken into account.

To correlate the values of hardness and cyclic hardening exponent$_{CHT}$ with the fatigue limits, valid data of σ_w is required for each condition. The fatigue limits of the different variants of X0.5CuNi2-2 and X21CuNi2-2 were obtained by extrapolating the power functions of the Woehler curves given in [29] to the ultimate number of cycles $N_L = 2 \times 10^6$, which is commonly used to determine the fatigue limit of these types of steels. Because of the small lifetime scatter observed for this material, this approach provides a sufficient reliability. For the steel C50E, σ_w was determined by VHCF experiments with $N_L = 2 \times 10^8$, as presented in [30]. However, because no failure occurred beyond ~10^6 cycles, the fatigue limit can be defined at $N_L = 2 \times 10^6$.

As for the 100CrMnSi6-4 variant investigated in this work, no fatigue data were available, a Woehler curve was determined for this condition. To this end, fatigue specimens with a polished gauge length and the geometry given in Figure 3a were used. Before polishing, the specimens were hard-turned, which could lead to pronounced residual stresses in the surface. Thus, the process-induced residual stresses were measured at a 40 kV tube voltage, 40 mA tube current and a scanning speed of 0.004°/s using an X-ray diffractometer of the type PANalytical Empyrean (Malvern Panalytical B.V., Almelo, the Netherlands) equipped with a Cu Kα_1 tube. The fatigue tests were stress-controlled with a stress ratio $R = -1$ and a frequency of 10 Hz. For this, a servo-hydraulic testing system of the type Schenck PSA 40 was utilized. The tests were performed at ambient temperature up to $N_L = 2 \times 10^6$.

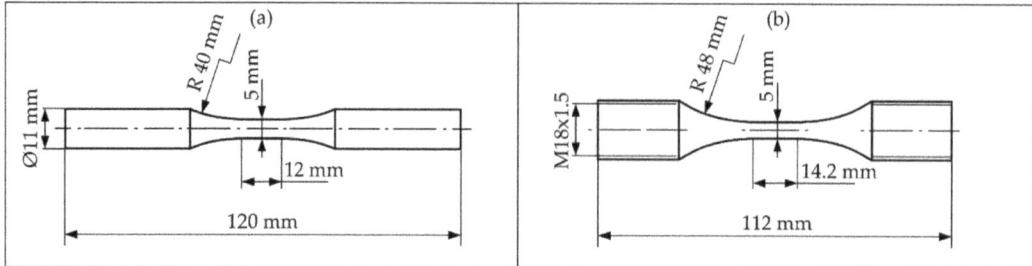

Figure 3. Fatigue specimen geometry of (**a**) 100CrMnSi6-4 and (**b**) 42CrMo4.

Similarly, for the variants of 42CrMo4, fatigue experiments were required, as in [23] only single fatigue tests were performed, which did not enable the estimation of σ_w. The fatigue specimens of 42CrMo4 had a geometry as illustrated in Figure 3b with a polished gauge length. For the quenched and tempered conditions, the fatigue tests were conducted in stress control with the resonant testing device RUMUL TESTRONIC 200 kN (RUSSENBERGER PRÜFMASCHINEN AG, Neuhausen am Rheinfall, Switzerland) and a frequency of approximately 86 Hz while for the normalized variant, the resonant testing device RUMUL TESTRONIC 100 kN (RUSSENBERGER PRÜFMASCHINEN AG, Neuhausen am Rheinfall, Switzerland) and a frequency of approximately 72 Hz were used. All tests were conducted at ambient temperature up to $N_L = 10^7$ cycles.

To determine the fatigue limits σ_w of the 42CrMo4 variants, the staircase method, which is described in detail in [35], was used. For each condition, the first experiment of this approach was performed at a stress amplitude $\sigma_{a,start}$, being in the range of σ_w. To roughly estimate $\sigma_{a,start}$, preliminary fatigue tests were performed for each condition. Each following step depended on the results of the prior experiment. If the specimen had reached the maximum number of cycles N_L (run-out), the following experiment was conducted at a higher stress amplitude, which was obtained by multiplying the former stress amplitude with the staircase factor d_{log}. If an experiment led to failure, the stress amplitude was reduced by dividing the previous stress amplitude by d_{log}. Therefore, all stress amplitudes σ_n were calculated as described in Equation (3) [35].

The staircase factor d_{log} depends on the scatter of the tested material and is calculated using the expected or known standard deviation of the whole data set (compare Equation (4)). For the tested conditions, the exponent s_{log} was estimated based on the preliminary fatigue experiments. For the normalized condition and the variant annealed at $T_a = 650\ °C$, s_{log} was set to 0.01 while for the variant annealed at $T_a = 550\ °C$, an s_{log} of 0.03 was chosen. After finishing the test series, a fictitious data point can be set, depending on the result of the last experiment. The fictitious point was set to the next smaller stress amplitude in case of failure or at the next higher stress amplitude if the experiment resulted in a run-out [35]. An exemplary fictitious test series of the staircase method is illustrated in Figure 4.

$$\sigma_n = \sigma_{a,start} \times (d_{log})^n \tag{3}$$

$$d_{log} = 10^{s_{log}} \tag{4}$$

The fatigue limit σ_w was calculated based on the number of stress levels i and the number of events f_i on each stress level. The lowest stress level was assigned to $i = 0$ and each higher stress level had the next larger number $i + 1$. Using Equations (5) and (6), the factors F and A were calculated, which were used to determine σ_w with Equation (7). Based on this, the fatigue limit σ_w was determined with a failure probability of 50% [35]:

$$F = \Sigma f_i \tag{5}$$

$$A = \Sigma(i \times f_i) \tag{6}$$

$$\sigma_w = \sigma_0 \times d_{\log}^{(A/F)} \qquad (7)$$

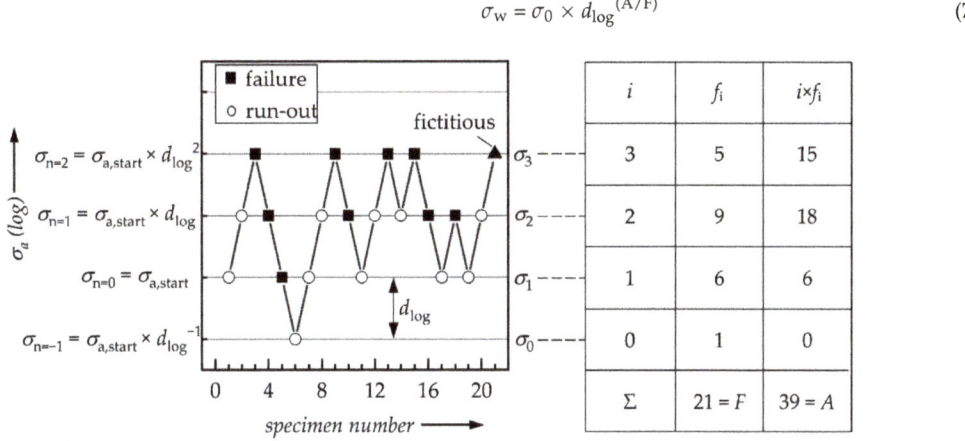

Figure 4. Schematic of the staircase method according to [35].

3. Results

3.1. Determination of the Fatigue Limit of Differently Heat-Treated 42CrMo4 and 100CrMnSi6-4

To obtain statically reliable fatigue limits σ_w of the different conditions of 42CrMo4, the staircase method was used, leading to the results summarized in Table 2 and Figure 5a. As expected, the normalized condition showed the lowest σ_w while the condition annealed at T_a = 550 °C yielded the highest fatigue limit.

Table 2. Parameters of the staircase method F, A, and fatigue limit σ_w of the 42CrMo4 conditions.

Material Condition	F	A	σ_w in MPa
normalized	10	11	261
T_a = 650 °C, t_a = 2 h	8	10	402
T_a = 550 °C, t_a = 2 h	10	19	479

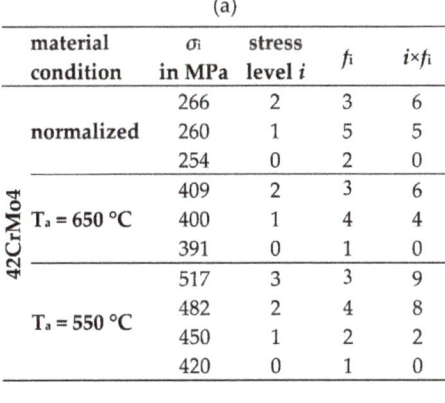
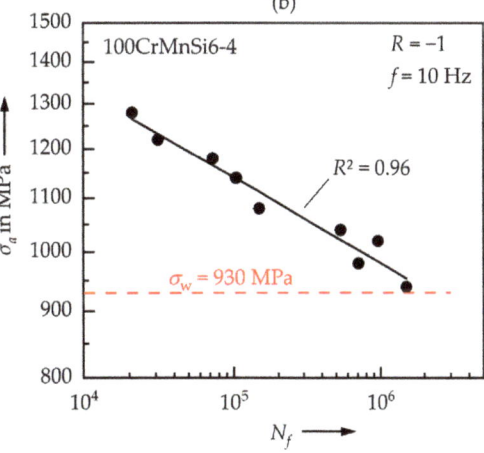

Figure 5. (**a**) Results obtained with the staircase method for different variants of 42CrMo4 and (**b**) the Woehler curve of 100CrMnSi6-4.

Considering the failure mechanisms that led to these σ_w, differences can be observed between the conditions. While for the normalized condition, crack initiation was only observed at surface slip bands, the quenched and tempered conditions also exhibited crack initiation at nonmetallic inclusions. In addition to crack initiation at defects, the condition annealed at $T_a = 650\ °C$ also showed crack initiation at slip bands at the surface. Note that for this condition, only one defect with $\sqrt{area} = 40\ \mu m$ was observed, which did not lead to a decrease in the number of cycles to failure N_f. Thus, no relevant influence of microstructural defects on the σ_w determined is expected for this condition. In contrast to the other conditions, the variant annealed at $T_a = 550\ °C$ only showed fatigue crack initiation at nonmetallic inclusions. However, apart from one specimen fractured at a stress amplitude of $\sigma_a = 450$ MPa, which was caused by a defect with $\sqrt{area} = 190\ \mu m$, only small defect sizes ($\sqrt{area} < 90\ \mu m$) were observed. As for the small defects, no correlation between the defect size and N_f was obtained, and as the one specimen with a bigger defect has only an insignificant influence on the value σ_w for $T_a = 550\ °C$, the influence of microstructural defects on σ_w could also be neglected.

Since for the 100CrMnSi6-4 a smaller number of specimens was available, for this condition, a conventional Woehler curve was determined, which is shown in Figure 5b. The results obtained show a, in comparison with typical high-strength steels, relatively low scatter, which is caused by the small crack-initiating defect sizes, i.e., $\sqrt{area} < 25\ \mu m$. Consequently, it is expected that also for this variant, the influence of crack-initiating defects on the fatigue limit is weak. Because of the small lifetime scatter, the Woehler curve shown in Figure 5b enables a valid determination of the fatigue limit of this material. For this purpose, the regression line was extended to $N_f = 2 \times 10^6$ and based on this, the fatigue limit σ_w is assessed to be approximately 930 MPa (compare Figure 5b).

The fatigue limits of the material conditions of Cu-alloyed steels and the C50E were determined in analogy to 100CrMnSi6-4. For these conditions, only fatigue crack initiation at the surface and no crack-initiating defects were observed and thus, for these variants, the influence of microstructural defects can be neglected, too.

3.2. Relationship between the Fatigue Limit, Hardness, and Cyclic Hardening Potential

To elaborate an improved approach for an estimation of the fatigue limit based on cyclic indentation testing, for all material conditions considered, the fatigue limit σ_w, the macro hardness (HV), and the cyclic hardening exponent$_{CHT}$ e_{II} were required. Thus, all conditions were characterized by hardness measurements and CIT. The results obtained in indentation testing and the experimentally determined fatigue limits are listed in Table 3, which is the data basis for the correlation analyses shown in the following. Table 3 also contains the standard deviation of HV and the 90% confidence interval of e_{II}, illustrating that the scatter of HV and $|e_{II}|$ is low for all materials.

Table 3. Fatigue limit σ_w, Vickers hardness (HV), and cyclic hardening exponent$_{CHT}$ $|e_{II}|$ of the investigated steels.

Material	Condition	σ_w in MPa	HV	$|e_{II}|$	References
	quenched	400	182 ± 3	0.453 ± 0.027	
X0.5CuNi2-2	$T_a = 550\ °C, t_a = 120$ s	460	205 ± 2	0.463 ± 0.022	[29,32]
	$T_a = 550\ °C, t_a = 2400$ s	510	207 ± 3	0.537 ± 0.023	
X21CuNi2-2	$T_a = 550\ °C, t_a = 120$ s	480	257 ± 3	0.383 ± 0.024	[29,32]
	$T_a = 550\ °C, t_a = 2400$ s	530	263 ± 3	0.429 ± 0.023	
C50E	position 1	370	275 ± 3	0.290 ± 0.007	[30]
	position 3	340	260 ± 4	0.289 ± 0.008	
42CrMo4	normalized	261	192 ± 6	0.292 ± 0.010	[23]
	$T_a = 650\ °C, t_a = 2$ h	402	286 ± 2	0.292 ± 0.009	
	$T_a = 550\ °C, t_a = 2$ h	479	364 ± 5	0.270 ± 0.007	
100CrMnSi6-4	-	930	709 ± 5	0.274 ± 0.007	

The σ_w values shown in Table 3 were determined using fatigue specimens with polished surfaces, leading to an elimination of the influence of the surface roughness. As the specimens were manufactured from bars with diameters significantly bigger than the gauge diameters using low feed rates and manual polishing, thermally and finishing-induced residual stresses are assumed to be relatively low. Additionally, only push-pull fatigue testing ($R = -1$) was applied and, thus, the whole material volume in the gauge length was loaded homogenously, leading to a less pronounced impact of the residual stresses and surface roughness in relation to other loading conditions, e.g., rotating bending. Consequently, the influence of the residual stresses was neglected.

Note that the specimens of 100CrMnSi6-4 were hard-turned and subsequently polished. Due to the hard-turning, for this condition, pronounced compressive residual stresses in the loading direction (−733 MPa) were measured in the surface-near area, which corresponds to [36]. However, the crack initiation started at defects located at a distance to the surface bigger than 800 µm, which, hence, significantly exceeds the depth of the residual stresses induced by hard-turning (compare [36]). Consequently, the influence of the residual stresses induced by hard-turning on σ_w can be neglected in this case, too.

In summary, the values of σ_w given in Table 3 are mainly related to the mechanical properties of the material volume. In this context, the properties of the material volume refer to the integral, global material properties in contrast to local ones, e.g., at the surface. More details about the microstructure and the properties of the material volume are given in the reference mentioned in Table 3. As the results determined in indentation testing are also dominated by the properties of the material volume, the material parameters listed in Table 3 constitute an excellent basis for the approach proposed.

Initially, for all material conditions considered in this work, the relation between the fatigue limit and the Vickers hardness was analyzed. For this purpose, the σ_w are plotted against the respective HV in Figure 6a. Moreover, in Figure 6a, the commonly used Equation (1) is represented by a straight line, showing a rough correlation with the experimental data. However, some conditions exhibit a rather large deviation from the relation given in Equation (1), especially in the case of 100CrMnSi6-4 and the Cu-alloyed steels. In accordance with [6,9], the relatively hard 100CrMnSi6-4 (706 HV) yields lower σ_w than expected from Equation (1). However, for the softer Cu-alloyed steels, a significantly higher σ_w can be observed in relation to Equation (1). It should, thus, be noted that the deviation from this estimation approach is not limited to high-strength conditions.

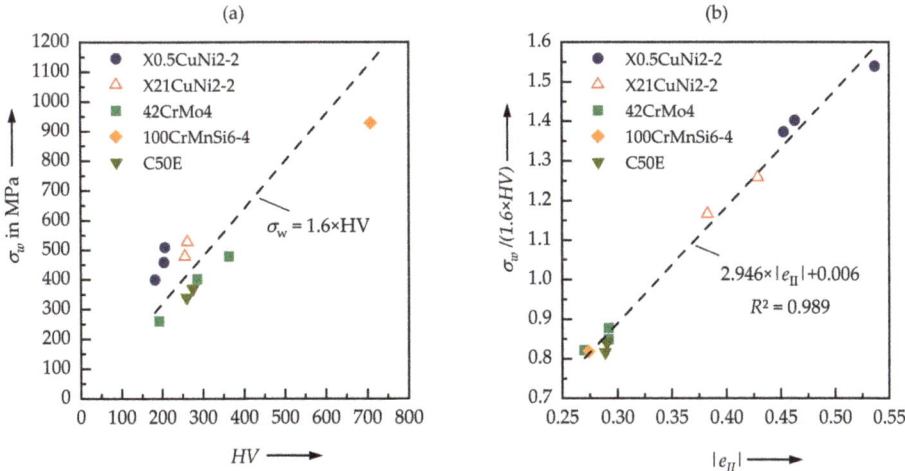

Figure 6. (a) Relationship of σ_w as a function of HV and (b) the deviation of the experimental determined to the estimated fatigue limit $\sigma_w / (1.6 \times HV)$ plotted versus $|e_{II}|$.

Comparing e_{II} of these materials, the overestimated 100CrMnSi6-4 exhibits low $|e_{II}|$ while the underestimated Cu-alloyed steels all show strong cyclic hardening and, thus, a high $|e_{II}|$ (see Table 3). Consequently, the relative deviation between the experimentally determined σ_w and the estimation based on Equation (1) was quantified by $\sigma_w/(1.6 \times HV)$. These relative deviations were related to $|e_{II}|$, which is illustrated in Figure 6b. Applying a linear regression of $\sigma_w/(1.6 \times HV)$ as a function of $|e_{II}|$ results in an impressively good correlation, which can be expressed by Equation (8). Note that Equation (8) contains all three material parameters considered in this work. By rearranging Equation (8), the fatigue limit σ_w can be expressed as the dependency of HV and $|e_{II}|$ (see Equation (9)):

$$\sigma_w/(1.6 \times HV) = 2.946 \times |e_{II}| + 0.006 \tag{8}$$

$$\sigma_w = 4.7136 \times |e_{II}| \times HV + 0.0096 \times HV \tag{9}$$

The summand $0.0096 \times HV$ of Equation (9) results from the mathematical regression analysis and represents the section point with the σ_w-axis. This summand is considered as a mathematical artifact. Note that even for a very high hardness of 1000 HV, this summand would shift the estimated fatigue limit by only 9.6 MPa and consequently, it can be neglected.

Considering Equation (9), the fatigue limit correlates with the product of HV and $|e_{II}|$. However, the factor (4.7136) given in Equation (9) was calculated based on the estimation from Equation (1), which yields pronounced deviations. Thus, a regression analysis of σ_w and $HV \times |e_{II}|$ was examined only based on the data given in Table 3, which is illustrated in Figure 7.

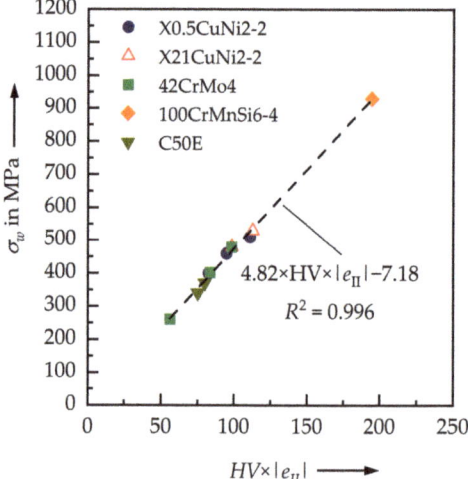

Figure 7. Relationship of σ_w as a function of $HV \times |e_{II}|$.

The diagram shown in Figure 7 demonstrates an excellent correlation between the fatigue limit σ_w and the product $HV \times |e_{II}|$, which can be expressed mathematically by Equation (10). Thus, this equation can be used to estimate σ_w based on the Vickers hardness (HV) and the cyclic hardening exponent$_{CHT}$ $|e_{II}|$. Note that this relation covers data with a high range of $HV \times e_{II}$ (56 to 194) and σ_w (261 MPa to 930 MPa):

$$\sigma_w = 4.82 \times HV \times |e_{II}| - 7.18 \tag{10}$$

However, the factor (4.82) and the subtrahend (7.18) of this equation result from regression analysis and, hence, highly depend on the data basis used. Although the data used in the presented work is sufficient to verify the fundamental applicability of the relation, the explicit quantification requires a substantially bigger data basis, especially in the range between σ_w = 500–900 MPa. Consequently, from the available data, only the general form of this relation can be derived, which is given in Equation (11):

$$\sigma_w = C_1 \times HV \times |e_{II}| + C_2 \quad (11)$$

4. Discussion

The regression analysis conducted in Section 3.2 results in a new estimation approach of the fatigue limit σ_w based on the hardness and the cyclic hardening potential determined in CIT (see Equations (10) and (11)). To analyze the improvement of the estimation achievable with this approach, an additional regression analysis was performed on a reduced data set, without the conditions X0.5CuNi2-2 (t_a = 2400 s), X21CuNi2-2 (t_a = 2400 s), C50E (position 3), and 42CrMo4 (T_a = 650 °C). As illustrated in Figure 8a, the regression based on Equation (11) and the reduced data set leads to C_1 = 4.82 and C_2 = −2.34. These coefficients were used to estimate the fatigue limit of the omitted material conditions. Figure 8b demonstrates that the experimentally determined fatigue limits $\sigma_{w, exp.}$ are very close to the estimated fatigue limit $\sigma_{w, est.}$. To quantitatively compare the estimation based on HV × |e_{II}| with the estimation based solely on HV, the relative differences ($\Delta\sigma_{w, est.}$) between $\sigma_{w, exp.}$ and $\sigma_{w, est.}$ estimated with Equations (1) and (11), respectively, were determined for each condition by Equation (12). The resulting deviations are illustrated in Figure 8c:

$$\Delta\sigma_{w, est.} = (\sigma_{w, exp.} - \sigma_{w, est.})/\sigma_{w, est.} \quad (12)$$

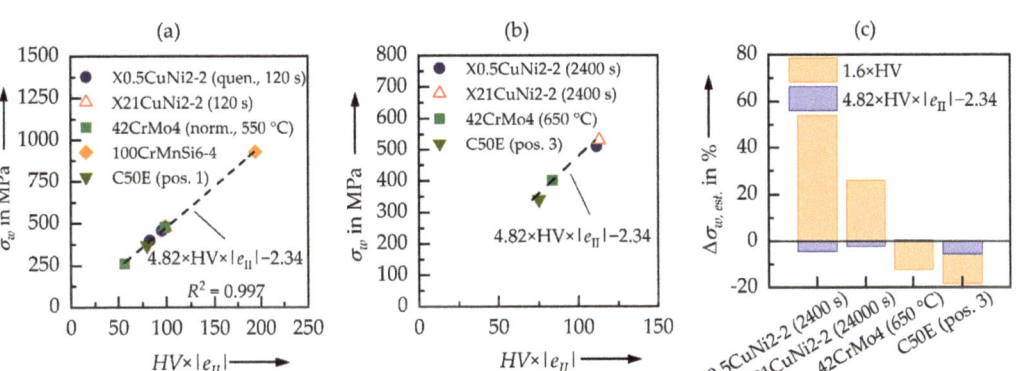

Figure 8. (a) Regression analysis of σ_w as a function of HV × |e_{II}| for selected material conditions; (b) verification of the regression analysis by selected material conditions; (c) relative deviation of σ_w experimentally determined to the estimated fatigue limit based on HV and HV × |e_{II}|, respectively.

As demonstrated by the results shown in Figure 8c, a strong improvement of the fatigue limit estimation is realized by considering the cyclic hardening potential in addition to the hardness. In accordance with the results shown by Fleck et al. [15] and Fatemi and Lopez [19], this improvement is caused by the integration of the cyclic deformation behavior, i.e., the cyclic hardening exponent$_{CHT}$ e_{II}.

Although e_{II} is determined based on a relatively small number of cycles (N = 10), it was shown in preliminary work that this material parameter strongly corresponds to the cyclic deformation curves [23,24] as well as the fatigue strength and lifetime [29,32,37] determined in uniaxial fatigue tests. Moreover, the limited number of cycles enables a fast (about 2 h) analysis of the cyclic deformation characteristics. Consequently, the cyclic hardening exponent$_{CHT}$ obtained with PhyBaL$_{CHT}$ allows the consideration of the cyclic deformation behavior in an efficient approach to estimate the fatigue limit σ_w.

Note that the estimation approach presented here was only validated for low-alloy steels with several carbon contents and a bcc lattice structure. Moreover, only push-pull fatigue experiments with $R = -1$ were considered and, thus, this approach needs to be extended to other loading conditions, e.g., rotational bending testing. In addition to this, the data basis is limited to polished surfaces and materials whose fatigue limit is not affected by microstructural defects and residual stresses.

Despite the abovementioned constraints, this approach yields promising results for a wide range of construction materials and, thus, has high potential for industrial and scientific applications. However, a bigger data basis is required, which is an objective of future research.

5. Conclusions and Prospects

In this work, a new approach to estimate the fatigue limit σ_w of low-alloy steels is presented. For this purpose, the correlation of the fatigue limit σ_w with the Vickers hardness (HV) and the cyclic hardening potential was analyzed based on eleven material conditions, which comprise five different low-alloy steels. To determine the cyclic hardening potential, instrumented cyclic indentation tests (CIT) were performed under each material condition. By analyzing the cyclic deformation behavior in CIT, the cyclic hardening exponent e_{CHT} $|e_{II}|$, which represents the cyclic hardening potential of the material, was determined.

The results obtained demonstrate that an estimation of σ_w simply based on hardness measurement is insufficient. However, by considering the combination of HV and the cyclic deformation behavior, i.e., the cyclic hardening potential, the estimation of the fatigue limit σ_w can be improved enormously. Consequently, the cyclic hardening potential, which represents the cyclic deformation behavior, and can be easily determined by cyclic indentation testing, was used for the first time to obtain a valid estimation of σ_w, requiring only a small volume of material and a short testing time. Note that for this testing approach, the microstructure and residual stress state of the samples should be equivalent to the specimens or components considered. Thus, this approach has high potential for industrial and scientific applications.

Although the data considered in the presented work is sufficient for verification of the fundamental viability, an extended data base is required for a valid quantification. Moreover, the transferability of this approach to other materials, e.g., aluminum alloys, austenitic steels, or titanium alloys, needs to be investigated [38,39]. Furthermore, the cyclic hardening potential might also be used to improve the estimation of the fatigue limit in the case of defect-based failure. In addition to this, the scatter of the hardness, cyclic hardening potential, and resulting fatigue limit must be integrated [40]. Finally, an analysis of the influence of the loading condition, surface roughness, residual stress state, and microstructural defects, which were excluded in this work by the experimental design, is required, too. These aspects will be objectives of future work.

Author Contributions: Conceptualization, D.G., P.O., P.L., B.B., D.E. and T.B.; methodology, D.G., P.O., P.L., B.B., D.E. and T.B.; formal analysis, D.G.; investigation, D.G. and P.O.; writing—original draft preparation, D.G., P.O. and B.B.; writing—review and editing, P.L., D.E. and T.B.; visualization, D.G., P.O. and P.L.; supervision, B.B. and T.B.; project administration, B.B. and T.B.; funding acquisition, T.B. All authors have read and agreed to the published version of the manuscript.

Funding: This research was funded by the German Research Foundation (DFG), grant numbers "BE 2350/9-2" (project number 335746905) and "BE 2350/14-1" (project number 420401443), as well as by the priority research activity of Rhineland Palatinate "Advanced Materials Engineering (AME)".

Data Availability Statement: Not applicable.

Acknowledgments: The authors want to the thank all the co-authors from preliminary work, especially Hendrik Kramer, Marcus Klein, and Peter Starke, which was the basis for the presented approach.

Conflicts of Interest: The authors declare no conflict of interest.

References

1. Blinn, B.; Beck, T.; Jost, B.; Klein, M.; Eifler, D. PhyBaL$_{SL}$—Short-time procedure for the determination of the fatigue lifetime of metallic materials under service loading. *Int. J. Fatigue* **2021**, *144*, 106060. [CrossRef]
2. Jost, B.; Klein, M.; Beck, T.; Eifler, D. Out-of-Phase TMF lifetime calculation of EN-GJS-600 (ASTM 80-55-06) ductile cast iron based on strain increase tests and evaluation of cyclic deformation behavior in isothermal measuring intervals. *Int. J. Fatigue* **2018**, *117*, 274–282. [CrossRef]
3. Starke, P.; Walther, F.; Eifler, D. "PHYBAL" a Short-Time Procedure for a Reliable Fatigue Life Calculation. *Adv. Eng. Mater.* **2010**, *12*, 276–282. [CrossRef]
4. Starke, P.; Bäumchen, A.; Wu, H. SteBLife—A new short-time procedure for the calculation of S-N curves and failure probabilities. *Mater. Test.* **2018**, *60*, 121–127. [CrossRef]
5. Wu, H.; Bäumchen, A.; Engel, A.; Acosta, R.; Boller, C.; Starke, P. SteBLife—A new short-time procedure for the evaluation of fatigue data. *Int. J. Fatigue* **2019**, *124*, 82–88. [CrossRef]
6. Garwood, M.F.; Zurburg, H.H.; Erickson, M.A. Correlation of laboratory tests and service performance. In *Interpretation of Tests and Correlation with Service*; Garwood, M.F., Zurburg, H.H., Erickson, M.A., Gensamer, M., Burwell, J.T., La Que, F.L., Eds.; American Society for Metals: Cleveland, OH, USA, 1951; pp. 1–77.
7. Pang, J.C.; Li, S.X.; Wang, Z.G.; Zhang, Z.F. General relation between tensile strength and fatigue strength of metallic materials. *Mater. Sci. Eng. A* **2013**, *564*, 331–341. [CrossRef]
8. Pang, J.C.; Li, S.X.; Wang, Z.G.; Zhang, Z.F. Relations between fatigue strength and other mechanical properties of metallic materials. *Fatigue Fract. Eng. Mater. Struct.* **2014**, *37*, 958–976. [CrossRef]
9. Murakami, Y. *Metal Fatigue: Effects of Small Defects and Nonmetallic Inclusions*, 1st ed.; Elsevier: Amsterdam, The Netherland, 2002; ISBN 0080440649.
10. Murakami, Y. Material defects as the basis of fatigue design. *Int. J. Fatigue* **2012**, *41*, 2–10. [CrossRef]
11. Thompson, S.W.; Parthasarathi, V.; Findley, K.O. A comparison of bending-fatigue properties of surface-induction-hardened SAE 1045 bar steels with and without vanadium and the influence of comparable low-temperature induction-tempering and furnace-tempering treatments. *Mater. Sci. Eng. A* **2021**, *807*, 140812. [CrossRef]
12. Novovic, D.; Dewes, R.C.; Aspinwall, D.K.; Voice, W.; Bowen, P. The effect of machined topography and integrity on fatigue life. *Int. J. Mach. Tools Manuf.* **2004**, *44*, 125–134. [CrossRef]
13. Mitchell, M.R. Fundamentals of Modern Fatigue Analysis for Design. In *ASM Handbook: Volume 19 Fatigue and Fracture*; Lampman, S.R., Ed.; ASM International: Materials Park, OH, USA, 1996; pp. 227–249. ISBN 978-1-62708-193-1.
14. Casagrande, A.; Cammarota, G.P.; Micele, L. Relationship between fatigue limit and Vickers hardness in steels. *Mater. Sci. Eng. A* **2011**, *528*, 3468–3473. [CrossRef]
15. Fleck, N.A.; Kang, K.J.; Ashby, M.F. Overview no. 112. *Acta Metall. Mater.* **1994**, *42*, 365–381. [CrossRef]
16. Liu, J.; Zenner, H. Dauerschwingfestigkeit und zyklisches Werkstoffverhalten. *Mater. Werkst.* **1989**, *20*, 327–332. [CrossRef]
17. Rennert, R.; Kullig, E.; Vormwald, M.; Esderts, A.; Luke, M. *Analytical Strength Assessment of Components: Made of Steel, Cast Iron and Aluminium Materials: FKM Guideline*; 7th Revised Edition 2020; VDMA Verlag GmbH: Frankfurt am Main, Germany, 2021; ISBN 978-3-8163-0745-7.
18. Lee, K.; Song, J. Estimation methods for strain-life fatigue properties from hardness. *Int. J. Fatigue* **2006**, *28*, 386–400. [CrossRef]
19. Lopez, Z.; Fatemi, A. A method of predicting cyclic stress–strain curve from tensile properties for steels. *Mater. Sci. Eng. A* **2012**, *556*, 540–550. [CrossRef]
20. Hay, J.L. Introduction to instrumented indentation testing. *Exp. Tech.* **2009**, *33*, 66–72. [CrossRef]
21. Oliver, W.C.; Pharr, G.M. Measurement of hardness and elastic modulus by instrumented indentation: Advances in understanding and refinements to methodology. *J. Mater. Res.* **2004**, *19*, 3–20. [CrossRef]
22. Saraswati, T.; Sritharan, T.; Mhaisalkar, S.; Breach, C.D.; Wulff, F. Cyclic loading as an extended nanoindentation technique. *Mater. Sci. Eng. A* **2006**, *423*, 14–18. [CrossRef]
23. Blinn, B.; Görzen, D.; Klein, M.; Eifler, D.; Beck, T. PhyBaL$_{CHT}$—Influence of indentation force on the results of cyclic hardness tests and investigations of comparability to uniaxial fatigue loading. *Int. J. Fatigue* **2019**, *119*, 78–88. [CrossRef]
24. Kramer, H.S.; Starke, P.; Klein, M.; Eifler, D. Cyclic hardness test PHYBAL$_{CHT}$—Short-time procedure to evaluate fatigue properties of metallic materials. *Int. J. Fatigue* **2014**, *63*, 78–84. [CrossRef]
25. Blinn, B.; Görzen, D.; Fischer, T.; Kuhn, B.; Beck, T. Analysis of the Thermomechanical Fatigue Behavior of Fully Ferritic High Chromium Steel Crofer®22 H with Cyclic Indentation Testing. *Appl. Sci.* **2020**, *10*, 6461. [CrossRef]
26. Schwich, H.; Görzen, D.; Blinn, B.; Beck, T.; Bleck, W. Characterization of the precipitation behavior and resulting mechanical properties of copper-alloyed ferritic steel. *Mater. Sci. Eng. A* **2020**, *772*, 138807. [CrossRef]
27. Blinn, B.; Winter, S.; Weber, M.; Demmler, M.; Kräusel, V.; Beck, T. Analyzing the influence of a deep cryogenic treatment on the mechanical properties of blanking tools by using the short-time method PhyBaLCHT. *Mater. Sci. Eng. A* **2021**, *824*, 141846. [CrossRef]
28. Klein, M.W.; Blinn, B.; Smaga, M.; Beck, T. High cycle fatigue behavior of high-Mn TWIP steel with different surface morphologies. *Int. J. Fatigue* **2020**, *134*, 105499. [CrossRef]
29. Görzen, D.; Schwich, H.; Blinn, B.; Bleck, W.; Beck, T. Influence of different precipitation states of Cu on the quasi-static and cyclic deformation behavior of Cu alloyed steels with different carbon contents. *Int. J. Fatigue* **2020**, *136*, 105587. [CrossRef]

30. Wagner, V. *Wechselverformungsverhalten Hochbeanspruchter Eisenbahnradstähle im Very High Cycle Fatigue (VHCF-) Bereich*; Dissertation; TU Kaiserslautern, Lehrstuhl für Werkstoffkunde (WKK): Kaiserslautern, Germany, 2011; ISBN 3-932066-26-X.
31. Wagner, V.; Starke, P.; Kerscher, E.; Eifler, D. Cyclic deformation behaviour of railway wheel steels in the very high cycle fatigue (VHCF) regime. *Int. J. Fatigue* **2011**, *33*, 69–74. [CrossRef]
32. Görzen, D.; Schwich, H.; Blinn, B.; Song, W.; Krupp, U.; Bleck, W.; Beck, T. Influence of Cu precipitates and C content on the defect tolerance of steels. *Int. J. Fatigue* **2021**, *144*, 106042. [CrossRef]
33. DIN EN 10027-1; DIN Deutsches Institut für Normung e. V. Designation Systems for Steels—Part 1: Steel Names. Beuth Verlag GmbH: Berlin, Germany, 2017.
34. DIN EN ISO 14577-1; DIN Deutsches Institut für Normung e. V. Metallic Materials—Instrumented Indentation Test for Hardness and Materials Parameters—Part 1: Test Method. Beuth Verlag GmbH: Berlin, Germany, 2015.
35. DIN 50100; DIN Deutsches Institut für Normung e. V. Load Controlled Fatigue Testing: Execution and Evaluation of Cyclic Tests at Constant Load Amplitudes on Metallic Specimens and Components. Beuth Verlag GmbH: Berlin, Germany, 2021.
36. Ostermayer, P.; Ankener, W.; Blinn, B.; Smaga, M.; Eifler, D.; Beck, T. Analysis of the Subsurface Volume of Differently Finished AISI 52100 by Cyclic Indentation and X-Ray Diffraction. *Steel Res. Int.* **2021**, *92*, 2100253. [CrossRef]
37. Blinn, B.; Krebs, F.; Ley, M.; Teutsch, R.; Beck, T. Determination of the influence of a stress-relief heat treatment and additively manufactured surface on the fatigue behavior of selectively laser melted AISI 316L by using efficient short-time procedures. *Int. J. Fatigue* **2020**, *131*, 105301. [CrossRef]
38. Niu, X.; Zhu, S.-P.; He, J.-C.; Liao, D.; Correia, J.A.; Berto, F.; Wang, Q. Defect tolerant fatigue assessment of AM materials: Size effect and probabilistic prospects. *Int. J. Fatigue* **2022**, *160*, 106884. [CrossRef]
39. Niu, X.-P.; Wang, R.-Z.; Liao, D.; Zhu, S.-P.; Zhang, X.-C.; Keshtegar, B. Probabilistic modeling of uncertainties in fatigue reliability analysis of turbine bladed disks. *Int. J. Fatigue* **2021**, *142*, 105912. [CrossRef]
40. Chausov, M.; Pylypenko, A.; Maruschak, P.; Menou, A. Phenomenological Models and Peculiarities of Evaluating Fatigue Life of Aluminum Alloys Subjected to Dynamic Non-Equilibrium Processes. *Metals* **2021**, *11*, 1625. [CrossRef]

Article

In-Situ Characterization of Microstructural Changes in Alloy 718 during High-Temperature Low-Cycle Fatigue

Sebastian Barton *, Maximilian K.-B. Weiss and Hans Jürgen Maier

Institut für Werkstoffkunde (Materials Science), Leibniz Universität Hannover, An der Universität 2, 30823 Garbsen, Germany
* Correspondence: barton@iw.uni-hannover.de

Abstract: Components made of nickel-based alloys are typically used for high-temperature applications because of their high corrosion resistance and very good creep and fatigue strength, even at temperatures around 1000 °C. Corrosive damage can significantly reduce the mechanical properties and the expected remaining service life of components. In the present study, a new method was introduced to continuously determine the change in microstructure occurring as a result of exposure to high temperature and cyclic mechanical loading. For this purpose, the conventional low-cycle fatigue test procedure was modified and a non-destructive, electromagnetic testing technique was integrated into a servohydraulic test rig to monitor the microstructural changes. The measured values correlate with the magnetic material properties of the specimen, allowing the microstructural changes in the specimen's subsurface zone to be analyzed upon high-temperature fatigue. Specifically, it was possible to show how different loading parameters affect the maximum chromium depletion as well as the depth of chromium depletion, which influences the magnetic properties of the nickel-based material. It was also observed that specimen failure is preceded by a certain degree of microstructural change in the subsurface zone. Thus, the integration of the testing technology into a test rig opens up new possibilities for improved prediction of fatigue failure via the continuous recording of the microstructural changes.

Keywords: alloy 718; fatigue; oxidation; chromium depletion; non-destructive testing

Citation: Barton, S.; Weiss, M.K.-B.; Maier, H.J. In-Situ Characterization of Microstructural Changes in Alloy 718 during High-Temperature Low-Cycle Fatigue. *Metals* **2022**, *12*, 1871. https://doi.org/10.3390/met12111871

Academic Editors: Martin Heilmaier and Martina Zimmermann

Received: 7 October 2022
Accepted: 25 October 2022
Published: 2 November 2022

Publisher's Note: MDPI stays neutral with regard to jurisdictional claims in published maps and institutional affiliations.

Copyright: © 2022 by the authors. Licensee MDPI, Basel, Switzerland. This article is an open access article distributed under the terms and conditions of the Creative Commons Attribution (CC BY) license (https://creativecommons.org/licenses/by/4.0/).

1. Introduction

Components in aircraft, aerospace, industrial, and automotive engineering are frequently subjected to severe loads due to thermal fluctuations and cyclic mechanical stresses. In order to ensure a sufficient service life under these loading conditions, these components are often manufactured using nickel-based alloys, the so-called superalloys. These alloys exhibit very high resistance to corrosion, as well as high creep and fatigue strength, even at temperatures around 1000 °C. Their excellent properties at these high temperatures are achieved by alloying up to 15 elements with nickel as the base element. High mechanical strength is reached in particular by the formation of ordered precipitates and carbides. Corrosion protection is provided by alloying with sufficiently high contents of the coating-forming elements chromium and aluminum [1–4].

A component's service life is often determined by mechanical and corrosive factors. The high temperatures and an oxygen-containing atmosphere lead to oxidation of the component surface. If the chromium and aluminum contents are sufficiently high, protective coating layers are formed that drastically slow down further oxidation and typically, parabolic oxide layer growth occurs. As the protective layer forming elements diffuse from the component subsurface to the component surface, the component subsurface becomes depleted in these elements. If the subsurface is excessively depleted in these elements, the oxide layer can no longer be re-established in the event of local damage. As a result, the protective effect of the oxide layer is finally lost. Cyclic mechanical stress acting on the component can lead to such local damage in form of cracks in the protective oxide layers. Due to the constant reformation of the oxide layer, the element depletion in the component subsurface increases rapidly. If the oxide layers no longer provide sufficient protection, oxidation products may form in the material and on grain boundaries. This can lead to notch effects, crack initiation, and component failure [5–8].

The depletion of alloying elements can also lead to a change in the magnetic material properties. In fact, an initially paramagnetic material behavior can change to a ferromagnetic one. This effect makes it possible to detect microstructural changes as a result of high-temperature corrosion using non-destructive testing (NDT) techniques [9–14]. Harmonic analysis of eddy current signals (EC-HA) is a non-destructive testing technique suitable for characterizing such changes in the magnetic material properties. This testing technique is mostly used for non-destructive characterization of steel alloys with ferromagnetic properties [15–17]. For detecting microstructural changes in nickel-based superalloys by EC-HA, the fact can be exploited that the chromium depletion correlates with the Curie temperature and the amplitude of the third harmonic at 25 °C [18–20]. In the context of the present study, EC-HA was integrated into a servohydraulic test rig in order to record the change of the specimen subsurface zone under high-temperature fatigue conditions in a non-destructive manner.

There are various test techniques for assessing the behavior of high-temperature materials under thermal and mechanical loading conditions. These can be divided into isothermal, thermo-mechanical, or thermal fatigue tests [21–24]. The objective of the present study was to evaluate an EC-HA system designed to monitor the microstructural degradation in the subsurface region under such conditions. Although the present study was limited to isothermal high-temperature low-cycle fatigue (LCF) test, the approach is applicable to other loading scenarios as well.

2. Materials and Methods

The experiments were carried out using the nickel-based alloy 718 in annealed condition. The alloy composition was 18.6 wt.% Fe, 18.4 wt.% Cr, 5.1 wt.% Nb, 2.9 wt.% Mo, 0.9 wt.% Ti, 0.5 wt.% Al, 0.3 wt.% Co, and balance Ni. For testing, cylindrical specimens with a diameter in the gauge section of 8 mm were employed. The samples were tested in the as-machined condition (surface roughness of R_z = 2 µm).

The modified isothermal LCF tests were performed on an MTS Landmark 370.10 servohydraulic universal test system (MTS Systems Corp., Eden Prairie, MN, USA). The key components of the test set-up are presented in Figure 1. The LCF parts of the tests were conducted in total strain control with triangular wave shape and a test frequency of 0.0083 Hz. The actual strain was measured by a high-temperature extensometer (MTS Systems Corp., Eden Prairie, MN, USA) attached to the specimen. The specimen was heated inductively via an induction coil enclosing the specimen. The temperature of the specimen was determined without contact via a pyrometer. To ensure a constant emission coefficient of the specimen surface, the specimen was painted black with heat-resistant paint. The temperature measurements were calibrated using thermocouples attached to companion specimens.

The EC-HA system was integrated into the fatigue test rig, allowing in-situ electromagnetic characteristics to be monitored as shown in Figure 1. Specifically, a temperature-resistant EC-HA tactile sensor (in-house development) was used to record the magnetic properties. Protection against high specimen temperatures was realized by a thin-walled housing made of a paramagnetic, low-conductive metal. The sensor was operated at a frequency of 1.6 kHz and had a measuring spot diameter of 15 mm. Therefore, not the entire sample was probed, but only a selected part of the sample. The penetration depth and the volume probed are essentially dependent on the electrical and magnetic material properties in addition to the selected test frequency. Since alloy 718 has paramagnetic properties in its initial state as well as a low electrical conductivity (approx. 1 MS/m), high penetration depths result. In fact, for the geometry used it would also be possible to detect the formation of ferromagnetic phases on the backside of the sample. However, this changes when a thin, ferromagnetic surface layer forms as a result of high-temperature oxidation. Then the effective penetration depth decreases to a few μm. To position the sensor on the specimen, a linear axis with a spindle drive and a traverse path of 250 mm was used. With the traversing unit, the sensor was moved from a position outside the effective field of the inductor (= parking position) to the specimen (= measuring position). By means of an integrated spring mechanism, the sensor was pressed onto the specimen surface with a constant force when the sensor was in the measuring position.

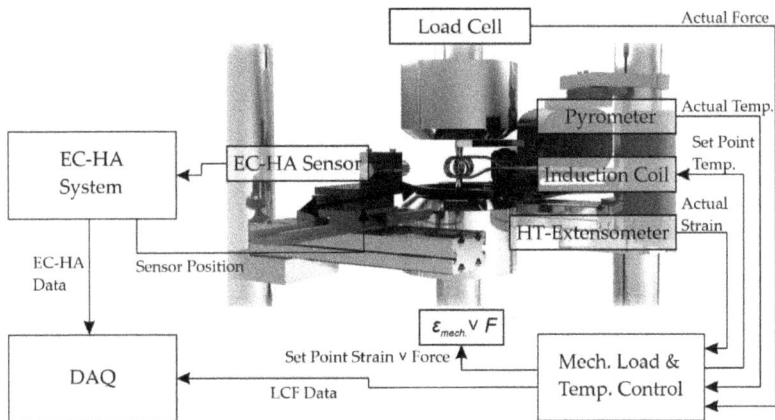

Figure 1. Schematic illustration of the LCF test rig and the integrated non-destructive EC-HA testing system.

As described in ref. [18], the Curie temperature and the amplitude of the third harmonic can be used to evaluate the microstructural state of the subsurface zone. In order to reduce the number of influencing factors in the test sequence, only isothermal, strain-controlled LCF tests were carried out in the present study. The actual test sequence is shown in Figure 2. It consists of a conventional high-temperature LCF part and a cooling/measuring period. At the beginning of each test cycle, the specimen was heated to the target temperature of 800 °C in the gauge section while the stress in the specimen was controlled at 0 MPa, allowing the specimen to expand freely during heating. At the end of heating, a holding phase of 120 s was used to minimize temperature gradients in the gauge length. Thereafter, the specimen strain was set to 0%. Then strain-controlled mechanical loading commenced, using a triangular wave shape. After a defined number of load cycles, heating was turned off and the samples was cooled down in stress control to allow specimen stress-free thermal contraction. When the specimen temperature dropped below 450 °C, the EC-HA sensor was moved from the parking position to the measuring position, Figure 2b. The EC-HA signal was then recorded until the sample temperature dropped below 25 °C. The duration of the cooling of the samples to a value < 25 °C was about 10 min. Thereafter, the EC-HA sensor was retracted and heating to the next LCF cycle began.

The entire test was terminated when the peak stress at maximum strain dropped to a level below 50% of the one in the saturation regime. At such a large stress drop clearly detectable crack growth had occurred. The experimental parameters used in the fatigue tests are summarized in Table 1. The temperature specified in Table 1 is the set temperature. In all three tests, an average temperature of 798 °C was measured during the high-temperature LCF part, with temperature variations within the gage length remaining within the permissible limits according to ASTM E 606.

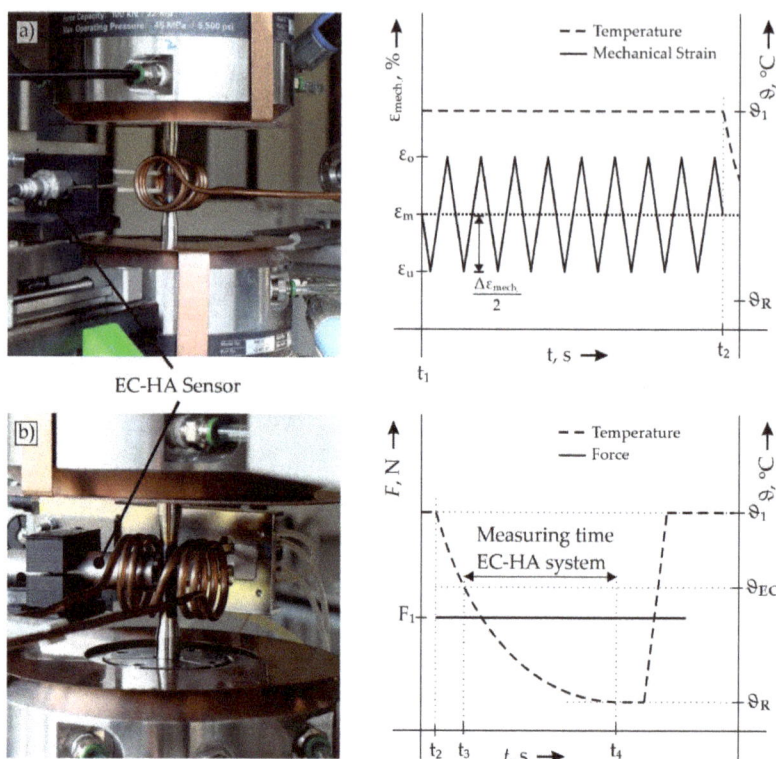

Figure 2. Modified LCF test procedure with combined EC-HA measurement: (**a**) strain-controlled LCF at a constant temperature of ϑ_1 with EC-HA sensor in park position, (**b**) EC-HA sensor in measurment position recording the electromagnetic characteristics of the specimen between ϑ_{EC} and ϑ_R under stress-free conditions ($F_1 = 0$).

Table 1. Test parameters of the modified isothermal LCF tests.

Specimen	Temperature ϑ_1, °C	Total Strain Amplitude $\Delta\varepsilon_{mech.}/2$, %	Duration per Load Cycle, s	Load Cycles per Test Cycle, -
1	800	0.3	120	20
2	800	0.5	120	20
3	800	0.7	120	20

The magnet-inductive testing method of harmonic analysis of eddy current signals was designed such as to be most sensitive for the detection of ferromagnetic phases and for the evaluation of the changes caused by high-temperature oxidation. Specifically, the microstructural changes in the specimen subsurface zone were determined by evaluating the amplitude of the third harmonic as a function of the specimen temperature. The details of the NDT technique are given in Ref. [18].

Ferromagnetic phases show a ferromagnetic behavior until the so-called Curie Temperature reached. Once this temperature is exceeded, a former ferromagnetic phase becomes paramagnetic. According to the procedure described in ref. [18], the Curie temperature of the sample can be determined by defining a threshold value in the amplitude of the third harmonic. This threshold value can be correlated with the maximum chromium depletion in the subsurface zone, see Figure 3a, since a reduction in chromium content leads to a ferromagnetic behavior of the nickel-based alloy 718. Thus, the value of the amplitude of the third harmonic at a temperature of 25 °C can be correlated with the depth of chromium depletion, cf. Figure 3b. By monitoring the amplitude of the third harmonic over the specimen temperature after each test cycle, the change in microstructure in the specimen subsurface zone up to specimen failure can be non-destructively detected. Since in the described experimental setup the minimum sample temperature corresponds to the room temperature (≈25 °C), no Curie temperatures below 25 °C were detected in the experiments conducted in the present study.

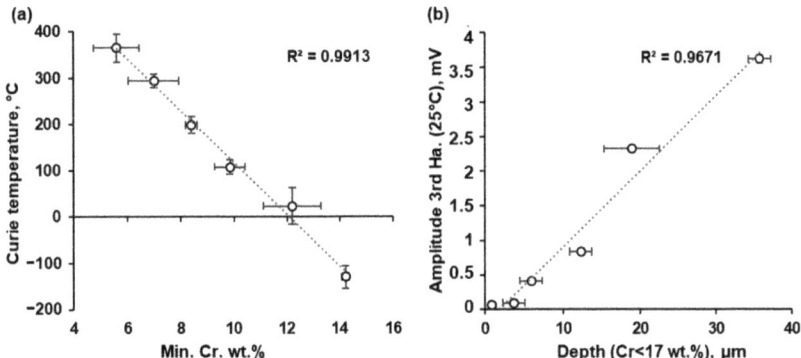

Figure 3. (a) Minimum Cr content vs. Curie temperature and (b) size of the chromium-depleted subsurface region vs amplitude of the third harmonic Reprinted with permission from ref. [18]. Copyright 2022 Taylor & Francis.

In conjunction with the recording of the electromagnetic measurements, the specimens were examined with regard to the microstructural changes occurring in the subsurface region. For this purpose, the specimens were cut along the longitudinal axis, with the cut being aligned such that its plane passed through the measuring spot of the EC-HA sensor. The specimens were then embedded in electrically conductive resin and polished. Element mappings by energy dispersive spectroscopy using X-rays (EDX) were obtained using a XV-scanning electron microscope (SEM) Mira VP (VisiTec Mikrotechnik GmbH, Grevesmühlen, Germany) operated at an acceleration voltage of 20 kV to determine the element distribution around the crack initiation site.

3. Results

In Figure 4 the change of the Curie temperature, observed via the change in the amplitude of the third harmonic as a function of the specimen temperature for different loading cycles is displayed. After the first test cycle, i.e., 20 individual load cycles, the amplitude signal cannot be distinguished from the baseline noise, which shows that the Curie temperature is lower than 25 °C. At 374 cycles, which corresponded to 27% of the fatigue life, the pre-defined amplitude threshold value of 1.5 times the baseline noise for determining the current Curie temperature was exceeded. As seen in Figure 4, a clear increase in the amplitude of the third harmonic, i.e., an increase in the Curie temperature can be seen after 520 load cycles. The 520 load cycles correspond approximately to 37.5 % of fatigue life. As cycling progresses, both the amplitude of the third harmonic, and thus the Curie temperature continued to increase. After 1380 cycles, shortly before reaching the end of the fatigue test a 1382 cycles, the amplitude of the third harmonic showed a value of 0.43 mV and the Curie temperature amounted to 110 °C.

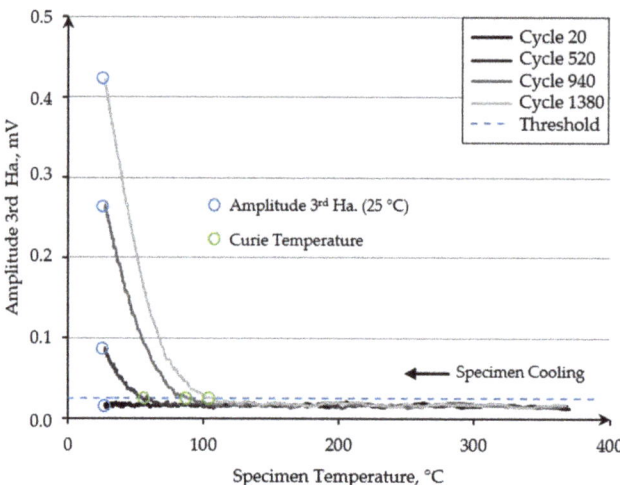

Figure 4. Amplitude of the third harmonic versus specimen temperature for selected cycles in the modified isothermal LCF test with a strain amplitude of $\varepsilon_{mech.} = 0.5\%$, a strain ratio of $R_\varepsilon = -1$ and a specimen test temperature of $\vartheta_1 = 800\,°C$.

Figure 5 shows selected stress-strain hysteresis loops from the specimen used to monitor the changes in the third harmonic in Figure 4. Whereas the latter has already revealed a detectable microstructural change at 374 cycles (27% of fatigue life), the effect of crack growth on apparent stress–strain response is only apparent clearly close to the end of the test in Figure 5e), where the maximum stress was seen to decrease substantially, and the hysteresis loop began to show an inclination in the compressive part. Both effects demonstrated that substantial crack formed at this stage.

The potential to detect fatigue damage early on via EC-HA is further evaluated in Figure 6, which shows the course of the amplitude of the third harmonic and the Curie temperature as a function of fatigue life. Here, a clearly detectable increase of the amplitude of the third harmonic can be seen at about 25% fatigue progress, which corresponds to the microstructural change in the probed subsurface layer. The calculated change in Curie temperature corresponds to a substantial depletion in local chromium content.

By tracking the electromagnetic parameters during fatigue, insight can be gained into damage evolution. Figure 7 shows the progress of the Curie temperatures of the three samples considered during the fatigue tests. As expected, the specimen strained with the highest amplitude fails at the lowest number of cycles. The curves of the specimens with strain amplitudes of 0.3% and 0.7% revealed a sharp increase in the Curie temperature in the cycles shortly before failure occurred. A better comparison is obtained plotting the Curie temperatures vs. the normalized fatigue life, cf. Figure 7b. Between a Curie temperature of 85 °C and 105 °C all specimens reached about 80% of fatigue life, irrespective of their actual strain amplitude.

In addition to the Curie temperatures, the amplitudes of the third harmonic curves at 25 °C were evaluated. As the Curie temperature reflects a different part of the data provided by tracking the third harmonic, the curves shown in Figure 8 progress differently.

Similar to Figure 7, the curves in Figure 8a are steeper at higher strain amplitudes. However, normalization, cf. Figure 8b, does not collapse the data onto a common master curve. This difference can be explained based on the total duration of the fatigue test. At lower strain amplitude, the samples are exposed to the high-temperature phase for more cycles, allowing changes in the microstructure as a result of high-temperature oxidation to take place over a longer duration. Thus, the increase in the amplitude of the third harmonic, normalized to the fatigue life, is greatest with a strain of 0.3%, i.e., for the sample with longest test period. A longer testing time leads to greater chromium depletion and thus more ferromagnetic material, which is reflected by the increase in the amplitude of the third harmonic.

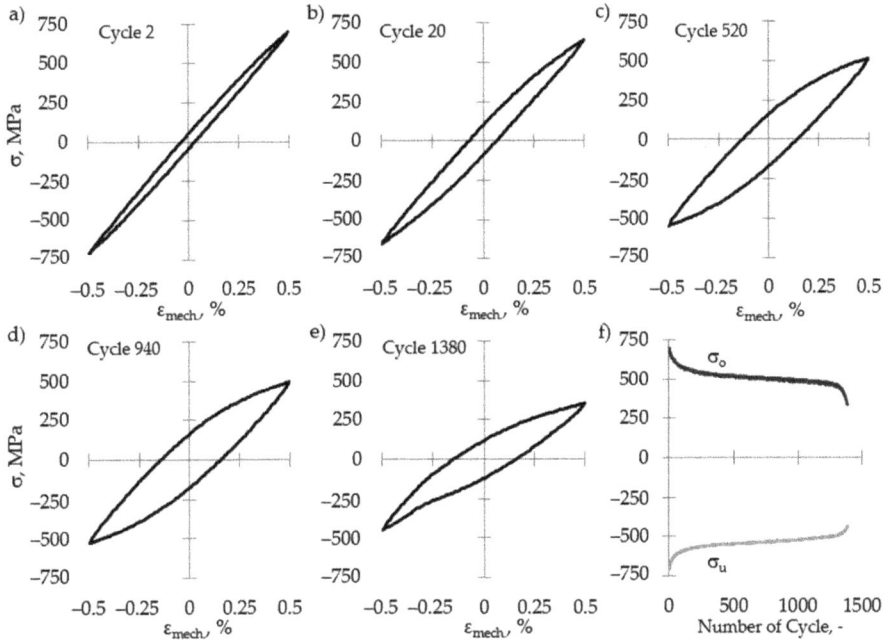

Figure 5. Stress–strain hysteresis loops after (**a**) 2 cycles, (**b**) 20 cycles, (**c**) 520 cycles, (**d**) 940 cycles, (**e**) 1380 cycles, and (**f**) maximum and minimum stress vs number of cycles for the fatigue test corresponding to Figure 4.

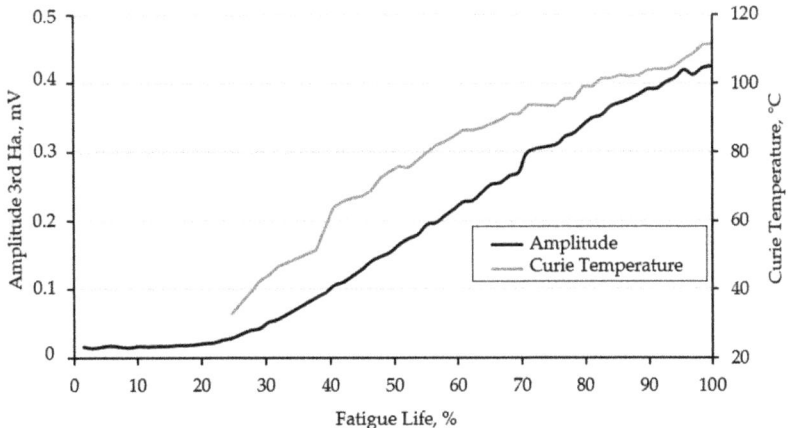

Figure 6. Amplitude of the third harmonic and Curie temperature versus fatigue life of the specimen subjected to 0.5% strain amplitude at 800 °C.

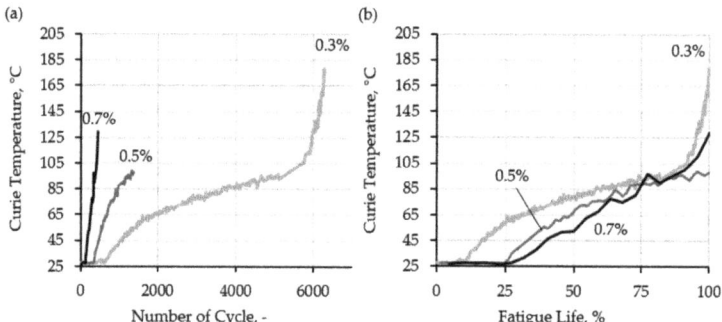

Figure 7. Evolution of Curie temperature in the fatigue tests with different strain amplitudes as a function of (**a**) the number of cycles and (**b**) normalized fatigue life; recompiled from ref. [19].

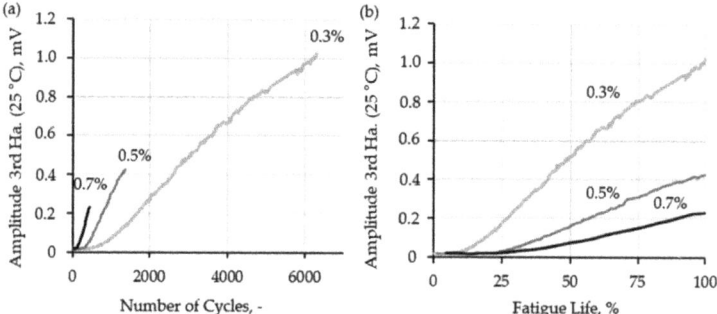

Figure 8. Evolution of the amplitude of the third harmonic at 25 °C in fatigue tests with different strain amplitudes as a function of (**a**) the number of cycles and (**b**) normalized fatigue life; recompiled from ref. [19].

To investigate the influence of high-temperature oxidation in more detail, micrographs were taken from the fatigued specimens around the cracked area. Optical microscope images showed that in addition to the primary crack which led to the specimen failure, further smaller cracks were induced, Figure 9. In the specimen with the longest test duration, Figure 9a, these secondary, smaller cracks were most pronounced.

Figure 9. Optical microscope images of the fatigued specimens in the region of crack initiation: (**a**) 0.3%, (**b**) 0.5%, and (**c**) 0.7% strain amplitude.

EDX mappings for the element chromium showed the formation of a chromium-rich oxide layer, cf. Figure 10. Increased chromium content at the cracks surfaces, although less pronounced, was also seen. This is confirmed by the in the mapping of oxygen. High concentrations of oxygen were found at the surface and the crack flanks. This indicates the formation of oxides on the crack flanks. In addition, a chromium-depleted near surface zone was visible. This is most evident in the specimens that were subjected to 0.3% (1) and 0.7% (3) strain. In the specimen subjected to 0.5% strain (2), chromium depletion on the crack flank was very weak. The mappings for the elements nickel and iron did not show any special peculiarities.

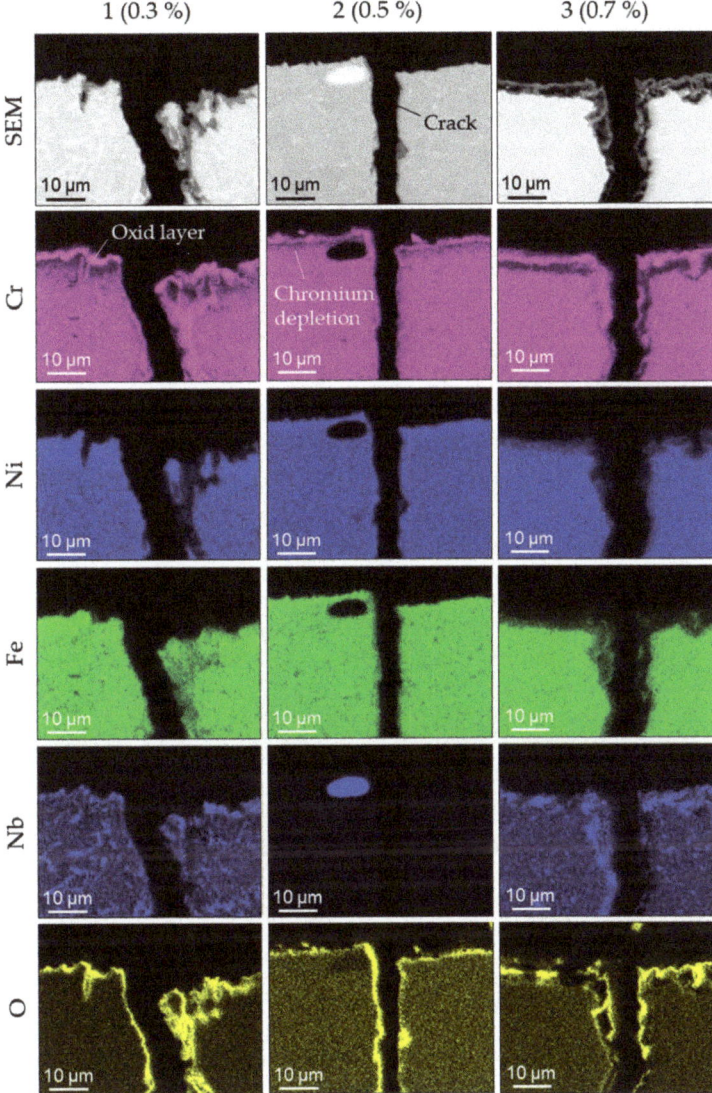

Figure 10. SEM images and EDX mappings for the elements chromium, nickel, iron, niobium, and oxygen.

Subsequently, as shown in Figure 11a, EDX line scans were taken from the specimen surface towards the center of the specimen and from the crack start diagonally into the interior of the specimen. Upon initial crack formation, the exposed substrate no longer features a protective coating. Subsequently, the oxides form, which increases the chromium depletion. This became more pronounced with increasing test duration as this process is controlled by diffusion. Thus, EDX analysis demonstrated that the depth affected by chromium depletion was larger in the case of the specimen cycled at a strain amplitude of 0.3%, cf. Figure 11c. By contrast, the chromium concentration profiles measured at the un-cracked surface, Figure 11b, were much more similar. This difference can be explained if it is assumed that material transport is more rapid in the crack wake because of pipe-diffusion in the plastically deformed region.

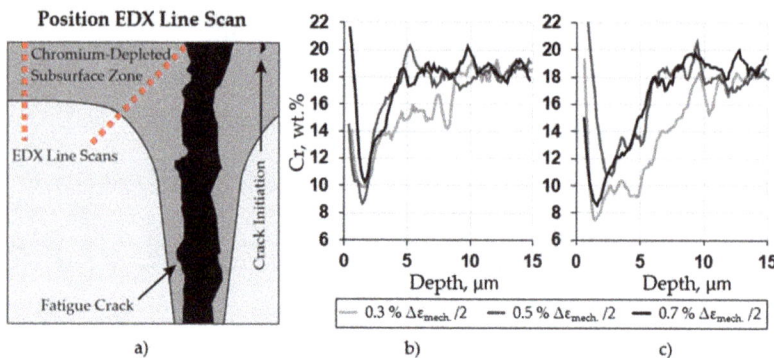

Figure 11. (**a**) Schematic illustration of the chromium-depleted zones and traces of the EDX line scans marked in red for scans from (**b**) the un-cracked surface to the center and (**c**) at the crack initaion site.

4. Discussion

The fatigue life of a cyclically loaded material can be divided into two parts: (i) the fraction of cycles required to initiate cracking and (ii) the fraction of cycles responsible for crack propagation [25]. In the crack initiation phase, a material can cyclically harden, soften, or exhibit an almost constant stress−strain response under cyclic loading. This depends, in part, on the material, the induced stress states, temperature, and type of fatigue test [26–28]. Corrosion-induced damage can have a substantial influence on crack initiation and early crack growth, which subsequently leads to fatigue failure. For example, many studies on crack growth during high-temperature fatigue of alloy 718 show that crack growth rates in tests conducted in air are significantly higher than those conducted in vacuum. Especially in tests with low test frequencies, as in the present study, high-temperature oxidation has a great influence on the crack growth rate [29–33].

The high-temperature oxidation of alloy 718 manifests itself, in particular, in a pronounced depletion of the subsurface zone in chromium. Chromium depletion as a result of high-temperature oxidation in alloy 718 has been studied using metallography by many authors, e.g., refs. [34–36]. In the present study, the amplitude of the third harmonic of EC-HA testing was used, which senses the volume of material affected by chromium depletion, and the Curie temperature determined, correlates with the maximum depth of chromium depletion. In this way, the progress of high-temperature corrosion can be evaluated non-destructively [18]. By implementing the EC-HA testing system into an isothermal LCF test rig, an estimate of the remaining fatigue life is possible, cf. Figure 7b. From a mechanistic point of view, this is based on the newly created surfaces upon crack initiation and early crack growth and hence accelerated chromium depletion.

It was shown that chromium depletion is not only dependent on time and temperature but also on cyclic straining. The electromagnetically measured data have shown that both the maximum chromium depletion and the depth of chromium depletion increase more rapidly with cycle number at larger strain amplitude, cf. Figures 7 and 8. The EDX line scans also show that the maximum chromium depletion occurs at the site of crack initiation, Figure 11. During crack growth, an unprotected surface forms in the already depleted subsurface zone. This leads to significantly accelerated chromium depletion. This is also evident in the Curie temperature curves. This explains the strong increase in the Curie temperature for sample 1 ($\Delta\varepsilon_{mech.}/2 = 0.3$ %) that occurred at approx. 80% of the test duration, when more and/or larger cracks form, cf. Figure 7. For sample 3 ($\Delta\varepsilon_{mech.}/2 = 0.7$ %), the same trend was seen close to the end of fatigue life, but the effect is less pronounced. This is to be expected as at the higher mechanical strain amplitude crack growth is faster, and thus exposure time of the unprotected surfaces is shorter. Yet, sample 2 ($\Delta\varepsilon_{mech.}/2 = 0.5$ %) did not show such a sharp increase close to failure. In this sample, crack initiation did not take place in the area probed by the measuring spot of the sensor. The crack only grew into the measuring area just before final failure. Consequently, high-temperature oxidation could only act briefly on the open crack flank in the area probed by the sensor. Thus, there was only a small additional decrease of the minimum chromium content in this area and no further increase of the Curie temperature was seen in the data. In summary, the change in the signal depends both on the strain amplitude and the crack initiation site with respect to the location of the sensor.

The integration of electromagnetic testing technology into loading devices has already been reported, but the focus was on detecting martensitic transformations [37,38]. The application in high temperature LCF tests has not been described so far. The present results showed that by applying the NDT technique of harmonic analysis of eddy current signals at regular intervals, an estimation of the remaining service life can be made. Interestingly, normalization of the recorded electromagnetic data in terms of relative fatigue life, cf. Figure 7, shows that for the material studied at a Curie temperature of approx. 90 °C, about 20% residual service life remains independent of the mechanical loading. Such high Curie temperatures are straightforward to detect. Thus, by continuously monitoring the magnetic properties, it is possible to detect and evaluate a change in the material even at low damage states. This characteristic makes the non-destructive testing technique presented here potentially interesting as a test method for components in order to estimate their remaining service life.

As can be seen in Figure 9, the thermal and mechanical loadings induced a large number of cracks. Clearly, the first crack formed is not necessarily the one that leads to unstable crack growth, which then determines fatigue failure [30,39]. It is rather the case that several small incipient cracks occur, which then do not go beyond the stable crack growth regime. As with the primary crack, high-temperature corrosion acts on the open flanks that form in these cracks. This accelerates chromium depletion in these regions. These microstructural changes caused by high-temperature fatigue are also detected by EC-HA.

In order to determine and evaluate the microstructure of the subsurface layer by means of the presented method, the Curie temperature and the amplitude of the third harmonic must be determined repeatedly. For this purpose, phases in which the specimen cools down from the test temperature must be integrated into the test sequence. For actual components, shutdown periods may be exploited for such measurements. With respect to modeling fatigue life, it should be noted that cooling and reheating can cause stresses in the protective oxide layers, which can also lead to cracks. This can accelerate the environmental attack. Thus, the exact influence of the modified test procedure on the results of the fatigue tests have to be investigated in further studies. It should also be possible to transfer the findings to tests on thermomechanical or thermal fatigue tests. Provided that phases in which the specimen cools down to temperatures < 100 °C can be integrated, the current test procedure is straightforward to apply.

The investigations presented were carried out on samples of alloy 718. Transferability to other nickel- or cobalt-based alloys is likely. Specifically, these alloys must feature substantial changes in magnetic material properties as a result of high-temperature corrosion. This is particularly the case for alloys with high chromium contents. Chromium strongly reduces the Curie temperature in nickel-based alloys [40] and cobalt-based alloys [41]. Thus, a significant change in the local alloy composition can be expected in these alloys as a result of high-temperature corrosion, which in turn results in a change in the magnetic properties.

5. Conclusions

By using an electromagnetic non-destructive testing (NDT) technique, integrated into a modified, isothermal low-cycle fatigue (LCF) test rig, it was possible to estimate the residual specimen lifetime by measuring the change of the magnetic sample properties.

The change of the magnetic properties is caused by a chromium depletion, which changes the magnetic properties of the nickel alloy from paramagnetic to ferromagnetic. Dependent on the chromium content, the Curie temperature individually changes. In the present study, the change in the Curie temperature was detected by the NDT technique in-situ. Once a crack is formed, the newly created surface accelerates the chromium depletion and hence increases the Curie temperature. If the remaining chromium content is too low, no further corrosion protection is given and hence the service lifetime is reduced significantly in a high-temperate corrosive environment. This can be detected earlier in the electromagnetic testing signals as compared with monitoring the stress−strain response in the LCF test. Thus, the NDT data can be used for modeling residual service life as well as for gaining deeper insights into the crack propagation at high-temperature fatigue in high-temperature corrosion environments.

Author Contributions: Conceptualization, S.B., M.K.-B.W. and H.J.M.; methodology, S.B., M.K.-B.W. and H.J.M.; validation, S.B., M.K.-B.W. and H.J.M.; formal analysis, S.B., M.K.-B.W. and H.J.M.; investigation, S.B., M.K.-B.W. and H.J.M.; writing—original draft preparation, S.B. and M.K.-B.W.; writing—review and editing, S.B. and H.J.M.; visualization, S.B., M.K.-B.W. and H.J.M.; supervision, S.B. and H.J.M.; project administration, S.B.; funding acquisition, H.J.M. All authors have read and agreed to the published version of the manuscript.

Funding: This research was funded by the Deutsche Forschungsgemeinschaft (DFG, German Research Foundation) –SFB 871/3 –119193472.

Institutional Review Board Statement: Not applicable.

Informed Consent Statement: Not applicable.

Data Availability Statement: Data are available upon request.

Acknowledgments: The authors thank Deutsche Forschungsgemeinschaft for financial support. The publication of this article was funded by the Open Access Fund of the Leibniz Universität Hannover.

Conflicts of Interest: The authors declare no conflict of interest.

References

1. Xia, W.; Zhao, X.; Yue, L.; Zhang, Z. A review of composition evolution in Ni-based single crystal superalloys. *J. Mater. Sci. Technol.* **2020**, *44*, 76–95. [CrossRef]
2. Campbell, F.C. Superalloys. In *Manufacturing Technology for Aerospace Structural Materials*; Elsevier: Amsterdam, The Netherlands, 2006; pp. 211–272, ISBN 9781856174954.
3. Reed, R.C. *The Superalloys*; Cambridge University Press: Cambridge, UK, 2009; ISBN 9780521859042.
4. Pettit, F.S.; Meier, G.H. Oxidation and Hot Corrosion of Superalloys. In Proceedings of the Superalloys 1984 (Fifth International Symposium), Superalloys, Champion, PA, USA, 7–11 October 1984; pp. 651–687.
5. Li, X.; Li, W.; Imran Lashari, M.; Sakai, T.; Wang, P.; Cai, L.; Ding, X.; Hamid, U. Fatigue failure behavior and strength prediction of nickel-based superalloy for turbine blade at elevated temperature. *Eng. Fail. Anal.* **2022**, *136*, 106191. [CrossRef]
6. Stinville, J.C.; Martin, E.; Karadge, M.; Ismonov, S.; Soare, M.; Hanlon, T.; Sundaram, S.; Echlin, M.P.; Callahan, P.G.; Lenthe, W.C.; et al. Fatigue deformation in a polycrystalline nickel base superalloy at intermediate and high temperature: Competing failure modes. *Acta Mater.* **2018**, *152*, 16–33. [CrossRef]
7. Kofstad, P. *High Temperature Corrosion*; Elsevier Applied Science: London, UK, 1988; ISBN 1-85166-154-9.
8. Rajabinezhad, M.; Bahrami, A.; Mousavinia, M.; Seyedi, S.J.; Taheri, P. Corrosion-Fatigue Failure of Gas-Turbine Blades in an Oil and Gas Production Plant. *Materials* **2020**, *13*, 900. [CrossRef] [PubMed]
9. Guo, J.; Cao, T.; Cheng, C.; Meng, X.; Zhao, J. The Relationship Between Magnetism and Microstructure of Ethylene Pyrolysis Furnace Tubes after a Long-term Service. *Microsc. Microanal.* **2018**, *24*, 478–487. [CrossRef]
10. Takahashi, S.; Sato, Y.; Kamada, Y.; Abe, T. Study of chromium depletion by magnetic method in Ni-based alloys. *J. Magn. Magn. Mater.* **2004**, *269*, 139–149. [CrossRef]
11. Rahmani, K. Magnetic Property Changes of CoNiCrAlY Coating Under Cyclic Oxidation and Hot Corrosion. *Oxid. Met.* **2020**, *93*, 75–86. [CrossRef]
12. Mook, G.; Simonin, J.; Feist, W.D.; Hinken, J.H.; Perrin, G. Detection and characterization of magnetic anomalies in gas turbine disks. In Proceedings of the 9th European Conference on NDT, Berlin, Germany, 25–29 September 2006.
13. Aspden, R.G.; Economy, G.; Pement, F.W.; Wilson, I.L. Relationship between magnetic properties, sensitization, and corrosion of incoloy alloy 800 and inconel alloy 600. *Metrics* **1972**, *3*, 2691–2697. [CrossRef]

14. Schnell, A.; Germerdonk, K.; Antonelli, G. A Non-destructive Testing Method of Determining the Depletion of a Coating. U.S. Patent US7175720B2, 13 February 2007.
15. Fricke, L.V.; Thürer, S.E.; Jahns, M.; Breidenstein, B.; Maier, H.J.; Barton, S. Non-destructive, Contactless and Real-Time Capable Determination of the α'-Martensite Content in Modified Subsurfaces of AISI 304. *J. Nondestruct. Eval.* **2022**, *41*, 72. [CrossRef]
16. Mercier, D.; Lesage, J.; Decoopman, X.; Chicot, D. Eddy currents and hardness testing for evaluation of steel decarburizing. *Mater. Sci.* **2006**, *39*, 652–660. [CrossRef]
17. Stegemann, D.; Reimche, W.; Feiste, K.L.; Heutling, B. Determination of Mechanical Properties of Steel Sheet by Electromagnetic Techniques. In *Nondestructive Characterization of Materials VIII*; Green, R.E., Ed.; Springer: Boston, MA, USA, 1998; pp. 269–275, ISBN 978-1-4613-7198-4.
18. Barton, S.; Zaremba, D.; Maier, H.J. Microstructural degradation in the subsurface layer of the nickel base alloy 718 upon high-temperature oxidation. *Mater. High Temp.* **2021**, *38*, 147–157. [CrossRef]
19. Barton, S. Zerstörungsfreie Bewertung des Randzonenzustands und Schädigungsgrads in Nickelbasislegierungen Infolge von Hochtemperaturkorrosion. Ph.D. Thesis, Gottfried Wilhelm Leibniz Universität Hannover, Hannover, Germany, TEWISS—Technik und Wissen GmbH, Garbsen, Germany, 2022.
20. Burkhardt, G.L.; Kwun, H. Nonlinear harmonics method and system for measuring degradation in protective coatings. U.S. Patent US19980168185 19981007, 7 October 1998.
21. Guth, S. *Schädigung und Lebensdauer von Nickelbasislegierungen unter Thermisch-Mechanischer Ermüdungsbeanspruchung bei Verschiedenen Phasenlagen*; KIT Scientific Publishing: Karlsruhe, Germany, 2016; ISBN 978-3-7315-0445-0.
22. Deng, W.; Xu, J.; Hu, Y.; Huang, Z.; Jiang, L. Isothermal and thermomechanical fatigue behavior of Inconel 718 superalloy. *Mater. Sci. Eng. A* **2019**, *742*, 813–819. [CrossRef]
23. Fournier, D.; Pineau, A. Low cycle fatigue behavior of inconel 718 at 298 K and 823 K. *Metall. Mater. Trans. A* **1977**, *8*, 1095–1105. [CrossRef]
24. Fissolo, A.; Gourdin, C.; Ancelet, O.; Amiable, S.; Demassieux, A.; Chapuliot, S.; Haddar, N.; Mermaz, F.; Stelmaszyk, J.M.; Constantinescu, A. Crack initiation under thermal fatigue: An overview of CEA experiencePart II (of II): Application of various criteria to biaxial thermal fatigue tests and a first proposal to improve the estimation of the thermal fatigue damage. *Int. J. Fatigue* **2009**, *31*, 1196–1210. [CrossRef]
25. Sangid, M.D. The physics of fatigue crack initiation. *Int. J. Fatigue* **2013**, *57*, 58–72. [CrossRef]
26. Kirka, M.M.; Greeley, D.A.; Hawkins, C.; Dehoff, R.R. Effect of anisotropy and texture on the low cycle fatigue behavior of Inconel 718 processed via electron beam melting. *Int. J. Fatigue* **2017**, *105*, 235–243. [CrossRef]
27. Droste, M.; Henkel, S.; Biermann, H.; Weidner, A. Influence of Plastic Strain Control on Martensite Evolution and Fatigue Life of Metastable Austenitic Stainless Steel. *Metals* **2022**, *12*, 1222. [CrossRef]
28. Smaga, M.; Boemke, A.; Daniel, T.; Skorupski, R.; Sorich, A.; Beck, T. Fatigue Behavior of Metastable Austenitic Stainless Steels in LCF, HCF and VHCF Regimes at Ambient and Elevated Temperatures. *Metals* **2019**, *9*, 704. [CrossRef]
29. Wagenhuber, E.-G.; Trindade, V.B.; Krupp, U. The Role of Oxygen-Grain-Boundary Diffusion During Intercrystalline Oxidation and Intergranular Fatigue Crack Propagation in Alloy 718. In Proceedings of the Symposium on Superalloys 718, 625, 706 and Derivattives, Pittsburgh, PA, USA, 2–5 October 2005; pp. 591–600, ISBN 978-0-87339-602-8.
30. Gustafsson, D.; Moverare, J.; Johansson, S.; Hörnqvist, M.; Simonsson, K.; Sjöström, S.; Sharifimajda, B. Fatigue crack growth behaviour of Inconel 718 with high temperature hold times. *Procedia Eng.* **2010**, *2*, 1095–1104. [CrossRef]
31. Ghonem, H.; Nicholas, T.; Pineau, A. Elevated temperature fatique crack growth in alloy 718 part II: Effects of environmental and material variables. *Fatigue Fract. Eng. Mater. Struct.* **1993**, *16*, 577–590. [CrossRef]
32. Viskari, L.; Hörnqvist, M.; Moore, K.L.; Cao, Y.; Stiller, K. Intergranular crack tip oxidation in a Ni-base superalloy. *Acta Mater.* **2013**, *61*, 3630–3639. [CrossRef]
33. Leo Prakash, D.G.; Walsh, M.J.; Maclachlan, D.; Korsunsky, A.M. Crack growth micro-mechanisms in the IN718 alloy under the combined influence of fatigue, creep and oxidation. *Int. J. Fatigue* **2009**, *31*, 1966–1977. [CrossRef]
34. Delaunay, F.; Berthier, C.; Lenglet, M.; Lameille, J.-M. SEM-EDS and XPS Studies of the High Temperature Oxidation Behaviour of Inconel 718. *Mikrochim. Acta* **2000**, *132*, 337–343. [CrossRef]
35. Sanviemvongsak, T.; Monceau, D.; Desgranges, C.; Macquaire, B. Intergranular oxidation of Ni-base alloy 718 with a focus on additive manufacturing. *Corros. Sci.* **2020**, *170*, 108684. [CrossRef]
36. Garat, V.; Deleume, J.; Cloue, J.-M.; Andrieu, E. High Temperature Intergranular Oxidation of Alloy 718. In Proceedings of the Symposium on Superalloys 718, 625, 706 and Derivattives, Pittsburgh, PA, USA, 2–5 October 2005; pp. 559–569, ISBN 978-0-87339-602-8.
37. Celada-Casero, C.; Kooiker, H.; Groen, M.; Post, J.; San-Martin, D. In-Situ Investigation of Strain-Induced Martensitic Transformation Kinetics in an Austenitic Stainless Steel by Inductive Measurements. *Metals* **2017**, *7*, 271. [CrossRef]
38. Cao, B.; Iwamoto, T.; Bhattacharjee, P.P. An experimental study on strain-induced martensitic transformation behavior in SUS304 austenitic stainless steel during higher strain rate deformation by continuous evaluation of relative magnetic permeability. *Mater. Sci. Eng. A* **2020**, *774*, 138927. [CrossRef]
39. Jones, R. Fatigue crack growth and damage tolerance. *Fatigue Fract. Eng. Mater. Struct.* **2014**, *37*, 463–483. [CrossRef]

40. Yang, Z.; Lu, S.; Tian, Y.; Gu, Z.; Mao, H.; Sun, J.; Vitos, L. Assessing the magnetic order dependent γ-surface of Cr-Co-Ni alloys. *J. Mater. Sci. Technol.* **2021**, *80*, 66–74. [CrossRef]
41. Bolzoni, F.; Leccabue, F.; Panizzieri, R.; Pareti, L. Magnetic properties and anisotropy of Co-Cr alloy. *J. Magn. Magn. Mater.* **1983**, *31–34*, 845–846. [CrossRef]

Article

VHCF Behavior of Inconel 718 in Different Heat Treatment Conditions in a Hot Air Environment

Sebastian Schöne [1,2,*], Sebastian Schettler [1], Martin Kuczyk [2], Martin Zawischa [1] and Martina Zimmermann [1,2]

1 Fraunhofer Institute for Material and Beam Technology IWS, 01277 Dresden, Germany; sebastian.schettler@iws.fraunhofer.de (S.S.); martin.zawischa@iws.fraunhofer.de (M.Z.); martina.zimmermann@tu-dresden.de (M.Z.)
2 Institute of Materials Science, Technische Universität Dresden, 01069 Dresden, Germany; martin.kuczyk1@tu-dresden.de
* Correspondence: sebastian.schoene@iws.fraunhofer.de

Abstract: The very high cycle properties of Inconel 718 in two different heat treatment conditions were investigated at a test temperature of 500 °C. One condition was optimized for fatigue strength and displayed a finer-grained microstructure, while the second batch had a more coarse-grained microstructure. For the high-temperature ultrasonic fatigue testing, a new test concept was developed. The method is based on the principle of a hot-air furnace and thus differs from the conventionally used induction heaters. The concept could be successfully evaluated in the course of the investigations. The materials' microstructure was analyzed before and after fatigue testing by means of metallographic and electron backscatter diffraction (EBSD)analysis. The results show a significant influence of the heat treatment on the fatigue strength caused by the specific microstructure. Further, a difference in crack propagation behavior due to microstructural influences and non-metallic precipitations was observed.

Keywords: Inconel 718; ultrasonic fatigue testing; VHCF; high-temperature; hot air; crack growth

1. Introduction

Moving parts in power plant turbines or aerospace applications are subject to very high numbers of load cycles over their operating lifetime. In addition to cyclic mechanical stress, also high thermal loads occur in such greatly stressed application areas. Therefore, high-temperature materials such as nickel-based superalloys are commonly used. However, the fatigue behavior of metallic materials at elevated temperatures fundamentally differs from the behavior at room temperature [1]. In order to ensure safe and reliable service, both the influence of cyclic and mechanical stress and the exposition to high temperatures have to be evaluated.

A frequently used and already fundamentally investigated high-temperature resistant material is Inconel 718. Several studies have already analyzed the fatigue behavior at room temperature, including ultrasonic fatigue investigations up to the very high cycle fatigue (VHCF) region. Particularly relevant for use in safety-relevant components, failure up to the range of 1×10^8 cycles could be detected [2,3]. However, the focus of previous investigations was on fatigue behavior at elevated temperatures. Compared to room temperature, fatigue results showed deviating fatigue mechanisms such as an oxide-induced crack closure effect which resulted in increased fatigue life at 500 °C [4,5]. In addition, some results from ultrasonic fatigue testing show a continuous decrease in fatigue strength well into the VHCF range [4,6]. In most investigations, specimen material was conditioned by appropriate heat treatment strategies to achieve the highest possible fatigue strength. Therefore, the influence of a deviation from the established heat treatment condition of Inconel 718 on fatigue strength and fatigue mechanism has not yet been studied in detail. Studies on other nickel-based superalloys, however, demonstrated that the heat treatment condition could have a significant influence on the fatigue behavior at elevated temperatures, in particular in the VHCF regime. Differences in fatigue behavior based on the interaction between dislocations and precipitations were found [1,7,8].

As shown by previous work, the use of special ultrasonic fatigue test equipment is appropriate to investigate the material behavior in the VHCF range. Further, numerous concepts have been developed to investigate the VHCF fatigue behavior of materials at elevated temperatures, mainly based on ultrasonic fatigue systems combined with induction furnaces. Even though this method allows high test temperatures, the setup is very intricate. Considerable effort is required to exclude the possible influences of the electromagnetic field on measurement and test technology. For example, the use of non-contact measuring devices that are insensitive to electromagnetic fields, such as pyrometers or fiber optic sensors for displacement measurement, becomes necessary [9–13]. The present study shows a new test concept for ultrasonic fatigue testing at temperatures of at least 500 °C. The concept is based on the principle of a hot-air furnace and thus differs from the typically used induction heaters.

This test setup was used to conduct high-temperature fatigue tests on age-hardenable nickel-based superalloy Inconel 718. The fatigue strength of Inconel 718 in two different heat treatment conditions was compared at a test temperature of 500 °C and up to 5×10^9 cycles. The investigations aimed to study fatigue mechanisms at elevated temperatures as a function of the heat treatment condition and thus the microstructure. Therefore, one batch was analyzed and tested in a heat treatment condition that causes a very coarse-grained microstructure and, therefore, a good impact strength. A second batch was heat-treated in a way that the δ-phase precipitations were preserved together with a finer grain. Based on the significant difference in morphology and grain structure, deviation of fatigue strength is expected. This study aims to investigate differences in fatigue mechanisms caused by different microstructures under the influence of a hot air environment.

2. Materials and Methods

2.1. Material, Specimen Preparation, and Microstructural Characterisation Methods

The present study was performed on Inconel 718 with a chemical composition that is given in mass% in Table 1. Inconel 718 is an age-hardenable nickel-based superalloy, where the strengthening effect relates to the precipitation of the secondary γ''-phase in the γ metal matrix. In addition, typical intermetallic phases are formed. The material was investigated in two different age-hardened heat treatment conditions, which differ in solution temperature as well as aging temperature and time. Details are shown in Figure 1a. Due to the lower solution temperature of batch 2, the primary δ-phase (Ni_3Nb) remains at the grain boundaries. This leads to a significantly finer-grained microstructure after the following two-stage precipitation-hardening process. As a result, notch sensitivity and fatigue strength improve compared to the heat treatment of batch 1. Due to the coarse grain structure, batch 1 shows a high transverse ductility and impact strength. In addition to the different potential applications based on the heat treatment used, the significant difference in the resulting microstructure provides a basis to investigate fatigue mechanisms in Inconel 718 [14].

Table 1. Chemical composition (mass%) of the material studied in both batches.

Element	Ni	Cr	Fe	Nb	Mo	Ti	Al
Inconel 718	52.0	18.6	17.9	6.0	3.6	0.8	0.6

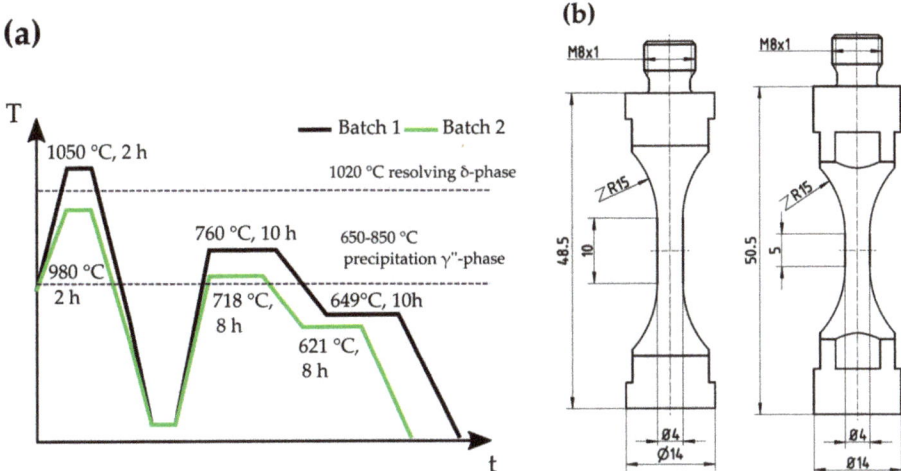

Figure 1. T-t plot of the two heat treatments applied (**a**) and specimen geometry in mm (**b**) with different gauge length of batch 1 and batch 2.

Fatigue specimens were machined from 16 mm bars into hourglass-shaped ultrasonic fatigue specimens with cylindrical gauge lengths of 4 mm in diameter. The gauge length of batch 1 was 10 mm (test volume 126 mm^2), while the gauge length of batch 2 was limited to 5 mm (test volume 63 mm^2) to enable ultrasonic fatigue tests at higher stress amplitudes. Both specimen geometries are displayed in Figure 1b. The influence of the different test volumes on the fatigue test results is negligible since the statistical distribution of crack-inducing defects is ensured in both test volumes. After heat treatment, the specimens were mechanically ground and polished to eliminate the influence of surface roughness. Metallographic cross-sections of both batches were prepared for microstructural and electron backscatter diffraction (EBSD) examination. For EBSD and EDX analysis, a scanning electron microscope of the type JSM-IT700HR (JEOL Ltd., Tokyo, Japan) in combination with a Symmetry S2 EBSD detector and an Ultim Max EDX detector (Oxford Instruments plc, Abingdon, UK) was used.

2.2. Fatigue Testing in Hot Air Environment

Fatigue tests were performed on an ultrasonic fatigue test stand (University of Natural Resources and Life Sciences, Vienna, Austria) at a stress ratio of R = −1 in laboratory air at a test temperature of 500 °C. In order to enable ultrasonic fatigue testing at elevated temperatures, a novel heating system was developed (Figure 2).

The concept is based on specimen heating with hot air, where a hot air stream is generated by a heating element supplied by a blower. In contrast to induction heating, the sample is not heated intrinsically but by convection. This distinction opens up the possibility of investigating various materials, even non-electrically conductive. Additionally, tests in a hot-air environment close to the application scenario are possible. In general, the concept enables fatigue testing at high temperatures without complex measurement technology. In contrast to systems based on inductive heating with resulting strong electromagnetic fields, for example, the use of inductive displacement sensors is possible. In the developed device, the specimen and the lower end of the load string are located in an insulated heating chamber. In order to optimize the energy utilization, the hot airstream is focused only on the specimen's gauge length and gets recycled through a closed circulation system. In order to prevent overheating of the whole testing system, the load string was cooled using compressed air.

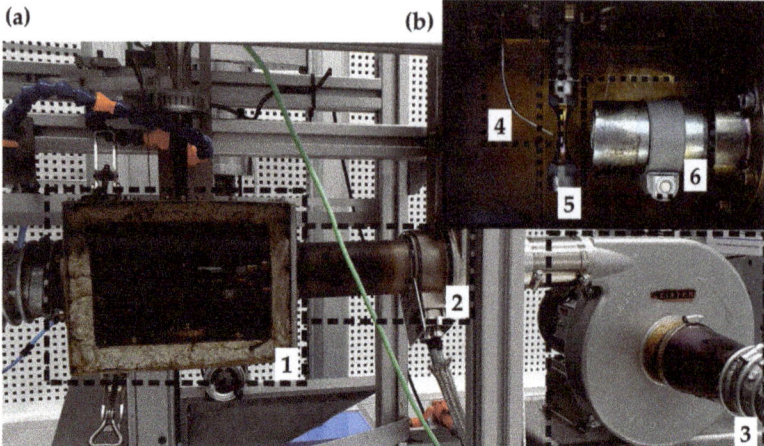

Figure 2. Hot air furnace for high-temperature ultrasonic fatigue testing. (**a**) Overview. (**b**) Heating chamber inside: (1) insulated heating chamber, (2) heating element, (3) blower, (4) thermocouple, (5) specimen and (6) hot air nozzle.

Specimen temperature is measured contactless by a thermocouple mounted behind and in the slipstream of the specimen. Temperature calibration is performed using a specimen with a central hole that allows measurement of the real core temperature due to a thermocouple. Due to the small specimen diameter in the measuring range, sufficiently uniform heating for practical applications can be achieved despite the one-sided heat input. Temperature regulation is carried out by a PID controller (Wachendorff Prozesstechnik GmbH & Co. KG, Geisenheim, Germany) through a thyristor power actuator (Thermokon Sensortechnik GmbH, Mittenaar, Germany). In order to limit uncontrolled self-heating of the specimens during the high-frequency fatigue test, experiments were performed in pulse–pause mode. Pulse and pause were individually adapted to each load amplitude. For high load amplitudes, pulse–pause ratios of 150 ms pulse time and 2000 ms pause time were used. For low load amplitudes, ratios of 200 ms pulse and 250 ms pause were used without any significant temperature rise detectable by a change in resonant frequency. Short pulse and long pause times reduce the effective test frequency at high-stress amplitudes, whereas the ratio selected for the low-stress amplitudes allows a short test for the VHCF range. In this paper, we report on the very high cycle fatigue properties of Inconel 718 in two different heat treatment conditions at a test temperature of 500 °C. In addition, the adjacent airflow dissipates the intrinsically generated heat, as in a compressed air cooling system.

2.3. Young's Modulus Determination

To determine reliable strength parameters with an ultrasonic fatigue system, precise knowledge of Young's modulus is crucial. Since materials' Young's modulus decreases with increasing test temperature, the resonance length of ultrasonic fatigue specimen will be shorter and the resonance frequency lower than that at ambient temperature.

Therefore, it is essential to determine Young's modulus at the test temperature. For this purpose, the fatigue specimens were investigated by laser-induced surface acoustic wave spectroscopy using a "LAwave" measurement system (Fraunhofer IWS, Dresden, Germany) [15]. This method allows the non-destructive and very precise determination of the elastic properties directly in the test area of the specimens, even at elevated temperatures. Typically, this method is used to characterize coatings [16] and effects due to other surface treatments such as machining [17] or hardening [18], but the measurement of Young's modulus within the upper 1–1.5 mm of a homogeneous material surface is also possible. The measuring principle is based on the measurement of the propagation velocity of surface acoustic waves, which depends on the elastic properties and density of a material. The waves were generated by a short laser pulse on the specimen surface, and their running time was measured by a piezoelectric transducer in distance of about 8 mm. To determine Young's modulus, density values of 8.26 g/cm^3 at room temperature, 8.20 g/cm^3 at 200 °C, and 8.09 g/cm^3 at 500 °C were used. Furthermore, a Poisson's ratio of 0.29 was assumed.

3. Results

3.1. Microstructural and Mechanical Analysis

The microstructures of the conditions investigated are shown in Figure 3. In addition to the clear difference in grain size (Figure 4), non-metallic inclusions in the form of TiN and NbC can be found through EDX and EBSD analysis. TiN precipitates are characterized by their angular shape, while NbC precipitates assume an irregular, smaller shape. Fine lenticular particles of the δ-phase can only be observed along the grain boundaries in the microstructure of batch two.

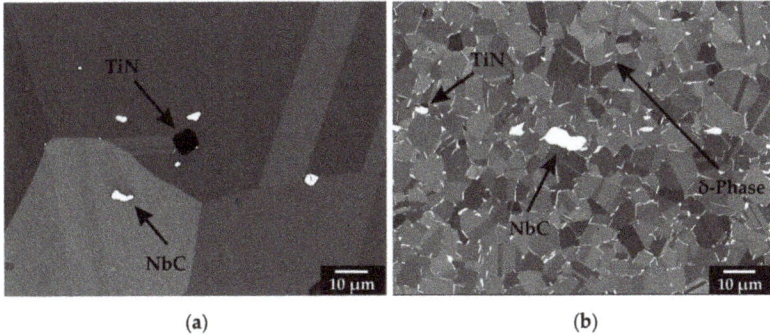

Figure 3. SEM microscopy of the microstructure; (**a**) Batch 1; (**b**) Batch 2.

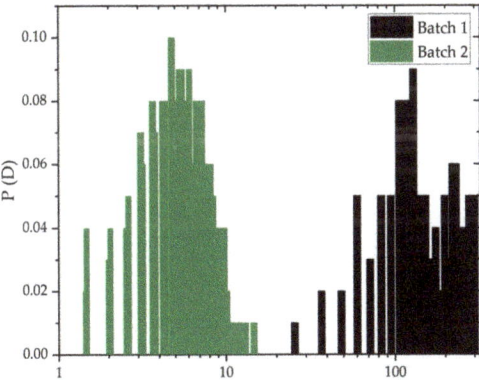

Figure 4. Comparison of the grain size distribution.

In order to investigate the influences of the microstructural differences found on mechanical parameters, measurements of hardness and Young's modulus were conducted. In order to take into account the significant difference in grain size of both batches, microhardness (HV0.3) and macrohardness measurements (HV 10) were performed comparatively. As shown by the results in Table 2, in conjunction with the standard deviation σ, there is no significant difference in hardness between the two batches. The Young's modulus was determined in the range of room temperature up to the test temperature of 500 °C using the surface acoustic wave spectroscopy. The obtained data were extrapolated using linear polynomial fitting for the whole measurement range. As with the hardness measurement, no significant deviation can be found in Young's modulus between the two batches. The difference found can be attributed mainly to fluctuations in the microstructure, which have a greater influence on the measurement results for the significantly coarser-grained batch one than for batch two. This is also evident from the higher deviation of the measured values by surface acoustic wave spectroscopy. Greater confidence in Young's modulus determined is possible with a high number of measurements. Nevertheless, a significantly increased attenuation behavior of batch one could be observed during the measurements.

Table 2. Hardness and Young's Modulus for both batches.

	HV0.3	σ	HV10	σ	Young's Modulus (GPa)		
					20 °C	200 °C	500 °C
Batch 1	454	9	450	3	206	194	171
Batch 2	446	6	459	3	206	195	177

3.2. High-Temperature Fatigue Behavior in the VHCF Regime

Figure 5 shows the S-N plot for both batches of Inconel 718 obtained at 500 °C by ultrasonic fatigue testing under a load ratio of R = −1. Batch two shows significantly higher fatigue strength compared to batch one above 10^5 cycles. In the range below 10^5 cycles, the difference in fatigue strength does not turn out so obviously. While a significant decrease in fatigue strength due to high temperature is obvious for batch one, batch two reaches values comparable to fatigue results of Inconel 718 at room temperature [2,6]. Both batches show failure even beyond 10^6 cycles, while no failures can be observed beyond 10^8 cycles. All tested stress amplitudes are well below the typical 0.2% proof stress of about 1000 MPa and, therefore, in the range of fully elastic deformation, which can be reliably tested by ultrasonic fatigue technology [5].

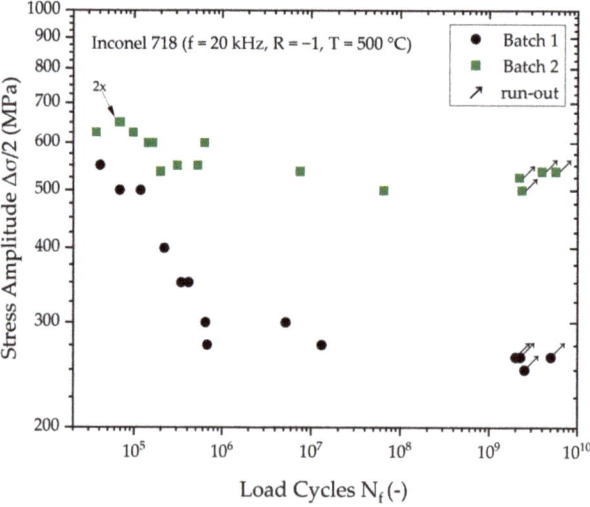

Figure 5. Results of the fatigue tests obtained at 500 °C for batch 1 and batch 2.

Following fatigue tests, fractographic analyses were performed using scanning electron microscopy. In Figure 6, the significant differences in fracture surfaces between the investigated batches become obvious. In Figure 6b, a typical smooth fatigue fracture surface of batch two with concentric beach lines is shown. Fracture surfaces of batch one, on the other hand, are characterized by fissured surfaces with marks of intercrystalline cracking.

In all examined specimens, the crack origin is located in areas next to the sample surface. However, no precise microstructural feature could be identified as a crack-initiation origin for batch two. Here, locations of increased plastic deformation such as grain or twin boundaries can have a crack-inducing effect. On the fracture surfaces of batch one, TiN particles could be observed in the dark area of the crack initiation point. These observations are consistent with results from previous studies [2,3].

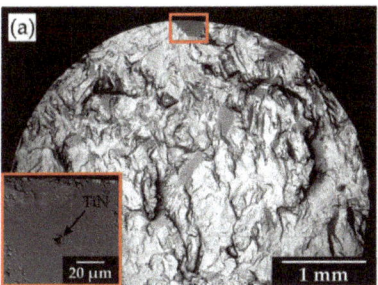 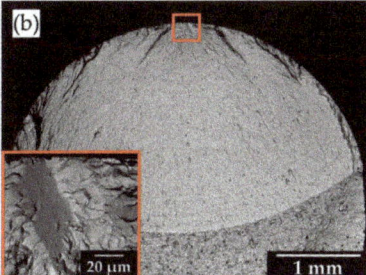

Figure 6. SEM microscopy of fracture surface (**a**) Batch 1, σ_a = 275 MPa, N_f = 1.35 × 10^7 cycles; (**b**) Batch 2, σ_a = 500 Mpa, N_f = 6.5 × 10^7 cycles.

3.3. Crack Growth Behavior

Based on the results of fatigue tests and fractographic analyses, a microstructural influence on crack growth can be assumed. It has already been shown repeatedly that the crack growth phase takes up only a small part of the total life of materials in the VHCF range and that this is largely determined by crack initiation. Nevertheless, conclusions on mechanisms of very early crack growth can be drawn from findings on crack growth behavior, and important effects in fatigue behavior at elevated temperatures can be identified. To analyze this, in addition to the fracture surface analyses, the resonance frequency during fatigue testing was investigated. The resonance frequency is very sensitive to fatigue crack growth and is therefore widely used for damage detection in ultrasonic fatigue testing technology [3]. The fatigue crack growth causes a decreasing stiffness of the specimen in the test area and thus a significant decrease in the resonance frequency. Figure 7 shows that the crack growth phase in the coarse-grained batch one is about ten times longer than in the finer-grained batch two. This behavior is exhibited by all specimens with a relevant high fracture load cycle number.

Due to the differences in fracture surfaces and crack growth rate, additional EBSD investigations were carried out in the area around the crack to identify any differences in the crack growth mechanisms. Therefore, metallographic sections of the test area were prepared parallel to the specimen axis. Figure 8 shows exemplary crack paths of the investigated batches starting at the surface inside the gauge length. The fine microstructure of batch two leads to a much straighter crack path and thus to a smoother fracture surface shown in Figure 6. Furthermore, EBSD data show that, especially in batch one, the crack runs both at the grain boundaries and through the grains. Upon further investigation, it becomes clear that grains with a low Schmid factor are preferably bypassed via the grain boundaries, while grains with a higher Schmid factor are cut by the crack (Figure 9). In some cases, directional changes of 90° can be observed within such grains. In the coarse microstructure of batch one, the crack is hindered by the found precipitations. In the fine structure of batch two, these phenomena could not be observed in detail. Here, mainly transcrystalline crack growth is present.

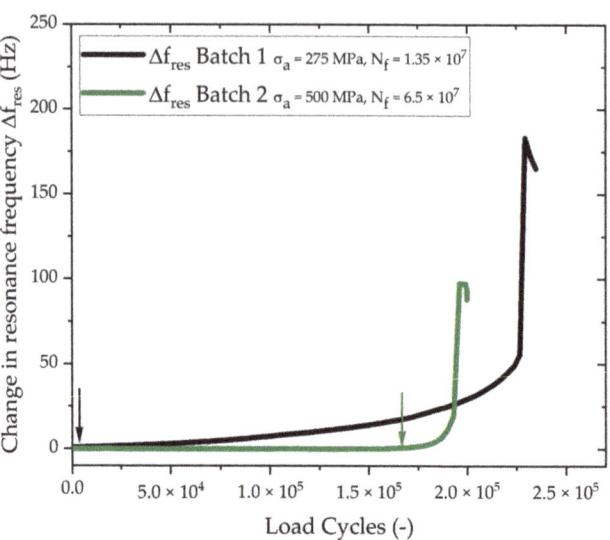

Figure 7. Exemplary comparison of the resonant frequency change over the last 2×10^5 cycles of batch 1 and batch 2 with start of crack growth phase ($\Delta f > 1$ Hz) marked by arrows.

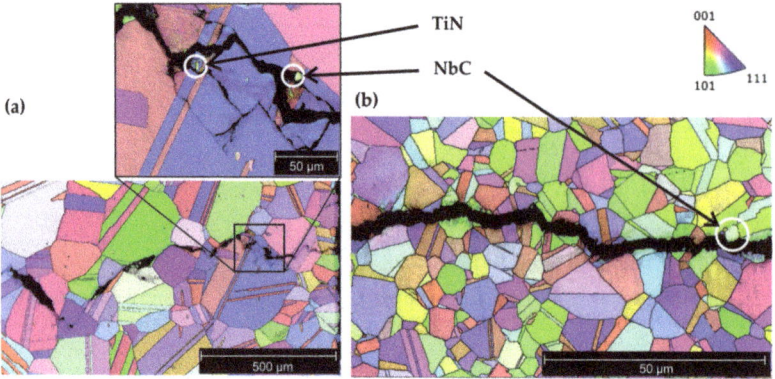

Figure 8. EBSD analysis of crack tip and crack paths with non-metallic inclusions. (**a**) Batch 1, (**b**) Batch 2.

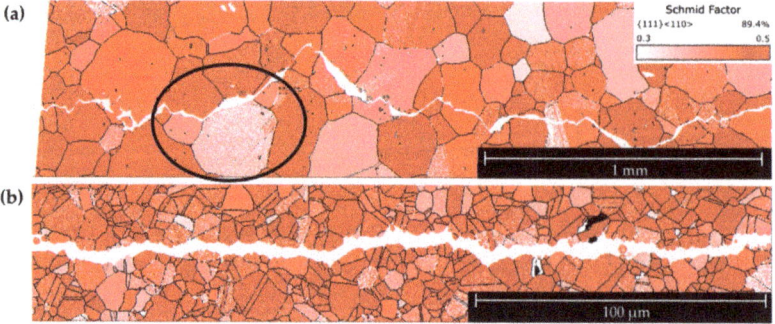

Figure 9. Analysis of Schmid factors for grains along the crack path starting at the surface. (**a**) Batch 1 with bypassed grains (circle), (**b**) Batch 2.

4. Discussion

In order to investigate the effect of two heat treatments on the fatigue behavior of Inconel 718 in a hot air environment, microstructural analysis, as well as ultrasonic fatigue tests at 500 °C and up to 5×10^9 cycles, were performed. The results show significant differences in the microstructure and thus an influence on fatigue mechanism and consequently on the fatigue strength.

Differences in the heat treatments affect the microstructure mainly by grain size and content of δ-phase precipitations but do not significantly impact the hardness or Young's modulus. The absence of difference in the measured hardness is remarkable since the different grain boundary densities, as well as the finely dispersed δ-phase precipitates, should imply a higher hardness in batch two by precipitation hardening effects. The influence of the finely dispersed γ''-phase in the metal matrix apparently outweighed the influences of the investigated precipitations and grain boundaries on this quasi-static behavior.

Since the material was subjected to a temperature of 500 °C during the fatigue test, temperature-induced microstructural changes are of particular interest. Neither in samples with long nor with short test times could such changes be detected based on the investigations carried out. At test temperature below 0.4 times the melting temperature of around 1300 °C and at the mainly short testing time of ultrasonic fatigue testing, changes based on recrystallization and dislocation movement were not to be expected. Nevertheless, the combination of elevated temperature and high-cycle plastic deformation can lead to phase transformations, especially at long-term test times in the VHCF regime. In the investigated range of high to very high cycle fatigue, plastic deformations do not occur globally. Typically, areas around brittle precipitates in a ductile matrix are critical points of local plastic deformation. Under such conditions, the transformation of the γ''- phase into δ-phase particles was observed by Jambor et al. [19]. Due to their location in the grain interior, these precipitates significantly reduce the fatigue strength, especially in the coarse-grained microstructure, compared to δ-phase particles in the range of the grain boundaries of batch two. Since the research program in this project did not foresee any detailed analyses of the γ''-precipitates evolution during elevated temperature VHCF tests, likely influences are still an open question.

As in the formation of the microstructure, the investigated batches also differ in fatigue strength. The fatigue results obtained at 500 °C for both heat treatment conditions show a typical one-step progression of S-N curves and therefore suggest a true fatigue strength. Contrary to the results obtained here, previous studies on Inconel 718 have assumed a constant decrease in fatigue strength at elevated temperatures. Batch two achieves the expected strength values in the range of literature data at room temperature. Batch one, on the other hand, shows significantly lower values. Such differences are mainly attributable to the different grain boundary densities and the presence of non-metallic inclusions to a considerable extent. Due to their brittleness and stress concentration, the non-metallic inclusions are potential initiation sites for fatigue cracks, both on the specimen surface and inside the material. Grain boundaries, on the other hand, inhibit crack growth due to their barrier effect.

Nevertheless, since fatigue life in the VHCF range is dominated by crack initiation and very early crack growth, the behavior of fatigue cracks as well as temperature-induced interaction of micro-cracks with the environment must be considered to explain the fatigue behavior of Inconel 718 at elevated temperature. As described above, the reason for the significantly different fatigue strength of the two batches is the optimized heat treatment and the resulting microstructure. The lower fatigue strength of batch one can be explained not only by the lower grain boundary density but also by the location of these precipitates. As can be seen, by SEM and EBSD examinations, an increased number of precipitates is located in the interior of the grains. Compared to precipitations in the area of the grain boundaries, these have a more critical effect as crack initiation sites and can lead to the formation of critical crack lengths at lower stress amplitudes.

In contrast to the crack initiation site's effect, precipitations are also able to slow down or even stop early crack propagation by obstacle effect, as well as grain boundaries can do [20]. These effects, which counteract crack growth, are particularly noticeable in the case of batch two. This is reflected by the significantly increased fatigue strength. Thus, the formation of isolated slip bands responsible for crack initiation due to increased activity of dislocations at elevated temperatures are hindered or maybe even suppressed. As a result, fatigue strength values in the range of the results at room temperature can be achieved in batch two. The unchanged fatigue strength of nickel-based alloys at temperatures in the range of 400–600 °C can be explained by a more pronounced homogeneous dislocation motion based on the specific interaction between dislocations and particles. In investigations on Nimonic 80A, strength-reducing processes were only observed above 800 °C [1]. Extensive studies on the influence of precipitates and their interaction with dislocations in the fatigue of nickel-base alloys have been carried out by Stöcker et al. [7,8].

Since the difference in fatigue strength depicted in Figure 5 is very pronounced, not only differences in crack initiation but also the influence of early crack growth shall be considered. This is also supported by the difference in crack growth between the two batches, as shown in Figure 6, and the findings from fractographic and EBSD analysis. The fracture surfaces found indicate a long crack initiation phase following a fast crack growth. For the more fissured fracture surface of batch one, intercrystalline crack growth is evident, while the fracture surface of batch two shows a smoother surface due to the smaller grain size. EBSD data show that the crack in the coarse microstructure of batch one grows preferentially through grains with high Schmid factor, whereas grains with low Schmid factor are bypassed. Grain boundaries, twins, and precipitations form obstacles, hinder crack growth and lead to a longer crack growth phase. This effect has been proven for EN-AW 6082, where microstructural features such as large precipitates and grain boundaries had a significant effect on the crack propagation at growth rates close to the threshold region [20]. In the finer microstructure of batch two, mainly transcrystalline crack growth was observed. Due to the high grain boundary density and the additionally intercalated delta precipitates, early crack growth is thus more strongly suppressed than in batch one. This is a likely explanation for the increased fatigue strength and the shorter crack growth phase. Furthermore, in the coarser-grained microstructure of batch one, crack initiation at precipitates inside the grain can lead to a critical crack length being exceeded unhindered already within one grain. Additionally, a high damping behavior of batch one became evident during Young's modulus measurement using surface acoustic wave spectroscopy, which is due to the significantly coarser-grained microstructure. A high damping capacity may indicate an increased energy absorption capacity that significantly slows crack growth, which in turn leads to the long crack growth phase. However, this cannot be proven with certainty since grain-size-dependent effects, such as scattering at the grain boundaries, can also cause increased damping.

If the load amplitudes are small enough and a fatigue crack propagates at growth rates close to the threshold region, the crack is stopped in the early stage. If the applied loads exceed a threshold, the crack quickly grows to a critical size under an additional effect of matrix softening. This circumstance explains the absence of fractures at load cycles above 6×10^7 cycles, even though fatigue tests were performed up to 5×10^9 cycles. The interaction of all effects is particularly evident in batch two at a stress level of 538 MPa. Here, fractures occurred at about 2×10^5, 7×10^7, as well as two runouts at about 4×10^9 [4,6].

In addition to crack initiation and early crack growth, the interaction between the microcrack and the environment plays a decisive role in high-temperature fatigue. Such temperature-induced interactions of microcracks with the environment are oxide-induced crack closure effects. In some cases, crack closure effects in Inconel 718 lead to higher fatigue strength at 500 °C compared to room temperature. Kawagoishi et al. [5] observed that cracks can grow to a length of 20–30 μm at the surface before they start non-propagating. In the work of Yan et al. [4], significantly higher strength values at 500 °C were achieved at a test frequency of 55 Hz than in the tests carried out in this work. This suggests that at test frequencies of 20 kHz, the oxide-induced crack closure effect no longer has a crucial impact. The duration of crack opening and the time period of crack growth in the oxidizing environment is too short at a test frequency of 20 kHz to allow a significant oxide layer to grow. Rather, already discussed microstructural influences may be a reason for a significant reduction in crack propagation at very low amplitudes and high frequencies. Nevertheless, once the early crack growth has been stopped temporarily by microstructural obstacles, the oxide-induced crack closure effect can lead to the permanent end of crack growth at constant amplitude due to the high ambient temperature.

5. Conclusions

In this study, fatigue tests on Inconel 718 were performed at 500 °C in the VHCF regime under fully reverse loading up to 5×10^9 cycles. In the course of this study, a new test concept for ultrasonic fatigue testing up to 500 °C was designed. It is based on the principle of a hot-air furnace and thus stands out from the induction heating commonly used. In the context of the investigations, the influence of two heat treatments on the fatigue strength at elevated operating temperatures was investigated. The two batches differed significantly in their grain size caused by the remaining δ-precipitations in one batch. The most important conclusions obtained are summarized as follows:

1. The developed testing concept allows reliable ultrasonic fatigue testing up to at least 500 °C and 5×10^9 cycles. At the same time, it requires less technical effort and is more economical than previous developments.
2. The Young's modulus could be determined up to the test temperature of 500 °C by laser-induced surface acoustic wave spectroscopy using an "LAwave" measurement system.
3. The differences in the microstructure of the two batches are not reflected in the hardness and Young's modulus.
4. The fatigue test results show a significant difference in cyclic life caused by different heat treatments. The finer-grained microstructure permits fatigue strength in the range of results at room temperature and significantly exceeds the values of the more coarse-grained batch.
5. In both batches, no fractures at load cycles above 6×10^7 cycles were found. This is attributed to the suppression of early crack growth at growth rates near the threshold region by microstructural obstacles. Whether oxide-induced crack closure effects also contributed to the lack of failure beyond 6×10^7 cycles cannot be fully ruled out.
6. Fractographic investigations and EBSD analysis show differences in crack propagation, which also result in different periods of fatigue crack propagation. By means of EBSD investigations, it was observed that the fatigue cracks grew along the boundaries of grains with low Schmid factor, while grains with high Schmid factor were more likely to pass through. In addition, the crack deflection was observed at non-metallic inclusions.

Based on the successfully applied methodology, further ultrasonic fatigue investigations in the VHCF area should mainly focus on the influence of microstructure development, γ'''-precipitation, and oxide-induced crack closure effects.

Author Contributions: Conceptualization, S.S. (Sebastian Schettler) and M.Z. (Martina Zimmermann); methodology, S.S. (Sebastian Schöne), S.S. (Sebastian Schettler), M.K. and M.Z. (Martin Zawischa); validation, S.S. (Sebastian Schöne), S.S. (Sebastian Schettler) and M.Z. (Martina Zimmermann); formal analysis, S.S. (Sebastian Schöne); investigation, S.S. (Sebastian Schöne), M.K. and M.Z. (Martin Zawischa); resources, S.S. (Sebastian Schettler) and M.Z. (Martina Zimmermann); writing—original draft preparation, S.S. (Sebastian Schöne); writing—review and editing, M.Z. (Martina Zimmermann), S.S. (Sebastian Schöne); visualization, S.S. (Sebastian Schöne); supervision, S.S. (Sebastian Schettler) and M.Z. (Martina Zimmermann); project administration, S.S. (Sebastian Schöne) and S.S. (Sebastian Schettler). All authors have read and agreed to the published version of the manuscript.

Funding: This research received no external funding.

Institutional Review Board Statement: Not applicable.

Informed Consent Statement: Not applicable.

Data Availability Statement: The data presented in this study are available on request from the corresponding author.

Conflicts of Interest: The authors declare no conflict of interest.

References

1. Zimmermann, M.; Stöcker, C.; Christ, H.-J. On the effects of particle strengthening and temperature on the VHCF behavior at high frequency. *Int. J. Fatigue* **2011**, *33*, 42–48. [CrossRef]
2. Ma, X.; Duan, Z.; Shi, H.; Murai, R.; Yanagisawa, E. Fatigue and fracture behavior of nickel-based superalloy Inconel 718 up to the very high cycle regime. *J. Zhejiang Univ. Sci. A* **2010**, *11*, 727–737. [CrossRef]
3. Chen, Q.; Kawagoishi, N.; Wang, Q.; Yan, N.; Ono, T.; Hashiguchi, G. Small crack behavior and fracture of nickel-based superalloy under ultrasonic fatigue. *Int. J. Fatigue* **2005**, *27*, 1227–1232. [CrossRef]
4. Yan, N.; Kawagoishi, N.; Chen, Q.; Wang, Q.Y.; Nisitani, H. Fatigue Properties of Inconel 718 in Long Life Region at Elevated Temperature. *KEM* **2003**, *243*, 321–326. [CrossRef]
5. Kawagoishi, N.; Chen, Q.; Nisitani, H. Fatigue strength of Inconel 718 at elevated temperatures. *Fatigue Fract. Eng. Mater. Struct.* **2000**, *23*, 209–216. [CrossRef]
6. Amanov, A.; Pyun, Y.-S.; Kim, J.-H.; Suh, C.-M.; Cho, I.-S.; Kim, H.-D.; Wang, Q.; Khan, M.K. Ultrasonic fatigue performance of high temperature structural material Inconel 718 alloys at high temperature after UNSM treatment. *Fatigue Fract. Eng. Mater. Struct.* **2015**, *38*, 1266–1273. [CrossRef]
7. Stöcker, C.; Zimmermann, M.; Christ, H.-J. Localized cyclic deformation and corresponding dislocation arrangements of polycrystalline Ni-base superalloys and pure Nickel in the VHCF regime. *Int. J. Fatigue* **2011**, *33*, 2–9. [CrossRef]
8. Stöcker, C.; Zimmermann, M.; Christ, H.-J. Effect of precipitation condition, prestrain and temperature on the fatigue behaviour of wrought nickel-based superalloys in the VHCF range. *Acta Mater.* **2011**, *59*, 5288–5304. [CrossRef]
9. Wagner, D.; Cavalieri, F.J.; Bathias, C.; Ranc, N. Ultrasonic fatigue tests at high temperature on an austenitic steel. *Propuls. Power Res.* **2012**, *1*, 29–35. [CrossRef]
10. Smaga, M.; Boemke, A.; Daniel, T.; Skorupski, R.; Sorich, A.; Beck, T. Fatigue Behavior of Metastable Austenitic Stainless Steels in LCF, HCF and VHCF Regimes at Ambient and Elevated Temperatures. *Metals* **2019**, *9*, 704. [CrossRef]
11. Furuya, Y.; Kobayashi, K.; Hayakawa, M.; Sakamoto, M.; Koizumi, Y.; Harada, H. High-temperature ultrasonic fatigue testing of single-crystal superalloys. *Mater. Lett.* **2012**, *69*, 1–3. [CrossRef]
12. Schmiedel, A.; Henkel, S.; Kirste, T.; Morgenstern, R.; Weidner, A.; Biermann, H. Ultrasonic fatigue testing of cast steel G42CrMo4 at elevated temperatures. *Fatigue Fract. Eng. Mater. Struct.* **2020**, *43*, 2455–2475. [CrossRef]
13. Smalcerz, A.; Przylucki, R. Impact of Electromagnetic Field upon Temperature Measurement of Induction Heated Charges. *Int. J. Thermophys.* **2013**, *34*, 667–679. [CrossRef]
14. Oradei-Basile, A.; Radavich, J.F. A Current T-T-T Diagram for Wrought Alloy 718. *Superalloys* **1991**, *718*, 325–335.
15. Schneider, D.; Schwarz, T. A photoacoustic method for characterising thin films. *Surf. Coat* **1997**, *91*, 136–146. [CrossRef]
16. Schneider, D.; Schultrich, B.; Scheibe, H.J.; Ziegele, H.; Griepentrog, M. A laser-acoustic method for testing and classifying hard surface layers. *Thin Solid Films* **1998**, *332*, 157–163. [CrossRef]
17. Pähler, D.; Schneider, D.; Herben, M. Nondestructive characterization of sub-surface damage in rotational ground silicon wafers by laser acoustics. *Microelectron. Eng.* **2007**, *84*, 340–354. [CrossRef]
18. Schneider, D.; Hofmann, R.; Schwarz, T.; Grosser, T. Evaluating surface hardened steels by laser-acoustics. *Surf. Coat. Technol.* **2012**, *206*, 8–9. [CrossRef]
19. Jambor, M.; Bokůvka, O.; Nový, F.; Trško, L.; Belan, J. Phase Transformations in Nickel base Superalloy Inconel 718 during Cyclic Loading at High Temperature. *Prod. Eng. Arch.* **2017**, *15*, 15–18. [CrossRef]
20. Kirsten, T.; Kuczyk, M.; Wicke, M.; Brückner-Foit, A.; Bülbül, F.; Christ, H.-J.; Zimmermann, M. Influence of Microstructural Inhomogeneities on the Fatigue Crack Growth Behavior Under Very Low Amplitudes for Two Different Aluminum Alloys. In *Mechanical Fatigue of Metals*; Correia, J.A., Jesus, A.M., de Fernandes, A.A., Calçada, R., Eds.; Springer International Publishing: Cham, Switzerland, 2019; pp. 303–310. ISBN 978-3-030-13979-7.

Article

Effect of Wall Thickness and Surface Conditions on Creep Behavior of a Single-Crystal Ni-Based Superalloy

Selina Körber [1], Silas Wolff-Goodrich [2], Rainer Völkl [1] and Uwe Glatzel [1,*]

[1] Metals and Alloys, University of Bayreuth, Prof.-Rüdiger-Bormann-Str. 1, 95447 Bayreuth, Germany; selina.koerber@uni-bayreuth.de (S.K.); rainer.voelkl@uni-bayreuth.de (R.V.)
[2] Max-Planck-Institut für Eisenforschung GmbH, Max-Planck-Str. 1, 40237 Duesseldorf, Germany; swolffgoodrich@gmail.com
* Correspondence: uwe.glatzel@uni-bayreuth.de; Tel.: +49-921-55-6600

Abstract: The influence of wall thickness and specimen surface on the creep behavior of the single-crystal nickel-based superalloy MAR M247LC is studied. Specimens with wall thicknesses of 0.4, 0.8, 1 and 2 mm, with and without casting surface, are compared to specimens of the same wall thickness prepared from bulk material. Creep behavior turned out to be independent from surface conditions even for the thinnest specimens. The thickness debit effect is not pronounced for short creep rupture times ($\leq$100 h at 980 °C), whereas it is significant for creep rupture times longer than ~200 h at 980 °C. The thickness debit effect is time-dependent and caused by oxidation and diffusion-controlled mechanisms.

Keywords: creep; thin-walled; casting surface; single crystal; MAR M247LC

1. Introduction

Decreasing wall thicknesses and ever filigree partitioning walls in turbine blades promise better aerodynamics, higher cooling efficiencies and weight savings. However, previous investigations have shown that creep properties may deteriorate with decreasing wall thicknesses. This so-called thickness debit effect can affect the service life of turbine components.

Gibbons [1] gives an overview of the influence of specimen thickness on the relative service life of different conventionally cast (CC) and directionally solidified (DS) nickel-based superalloys. A comparison by Duhl [2] between conventionally cast (CC), directionally solidified (DS) and single crystal (SX) nickel-based superalloys shows that wall thicknesses below 4.0 mm lead to a strong deterioration in creep properties. The thickness debit effect is less pronounced in directionally solidified (DS/SX) specimens than in conventionally cast polycrystalline (CC) specimens [2]. In the case of polycrystalline and directionally solidified specimens, the grain size and morphology play important roles, since failure is mainly caused by crack formation at grain boundaries [3–7]. Furthermore, casting defects such as pores lead to a deterioration in creep properties, since these significantly reduce the load-bearing cross-section in thin-walled specimens [4,6,7].

Pandey et al. [8,9] compare creep behavior of round and hollow specimens of the nickel-based superalloy IN X750 and show that the service life of hollow specimens is significantly lower than that of round specimens. A critical wall thickness of 3.2 mm is defined for round specimens and 2.0 mm for hollow specimens [8,9]. Doner and Heckler [3,6] illustrate that 0.5 mm thin flat samples of the nickel-based superalloy CMSX-3 show significantly shorter life times than round specimens with a diameter of 6.4 mm.

Seetharaman and Cetel [10] study the creep behavior of the single-crystal nickel-based superalloy PWA1484. They show that reducing the wall thickness from 1.0 to 0.3 mm leads to a reduction in the fracture time to 30%. A pronounced thickness debit effect is observed in strong alumina-forming nickel–based superalloys due to the formation of γ'-depleted or γ'-free regions, respectively [11–17].

The presented literature generally reports results from creep tests on specimens having a much smaller size in one cross-sectional dimension than in the perpendicular dimension. The term "thickness", how it is used here, means the size in the smaller dimension in mm. Thus, differences in creep behavior are inherently attributed to cross-sectional areas. A question to be answered, therefore, is at what thickness does the cross-sectional area become a critical factor determining the creep properties.

Previous investigations were almost exclusively carried out on thin-walled specimens that were prepared from bulk material [13,16,17]. The focus of this work is the creep behavior of "thin-cast" nickel-based superalloy MAR M247LC single-crystals, with single-crystals having wall thicknesses down to 0.4 mm. In addition, the influence of the specimen surface on creep properties is investigated.

2. Materials and Methods

2.1. Material

Thin-walled single-crystals of the nickel-based superalloy MAR M247LC were cast via Bridgman process in an induction furnace under vacuum at a pressure of 0.1 Pa. The composition of MAR M247LC is given in Table 1.

Table 1. Composition of MAR M247LC in wt.%.

Ni	Cr	Co	Mo	W	Al	Ti	Ta	Hf	C	B	Zr
61.5	8.1	9.3	0.5	9.4	5.7	0.7	3.3	1.4	0.07	0.017	0.007

Batches of the master alloy were placed in a ceramic crucible and hereafter induction melted. The molten material was poured into a ceramic mold that was preheated to a temperature of 1450 °C and positioned on a water-cooled Cu plate. After that, the ceramic shell mold was withdrawn through a water-cooled Cu baffle with a rate of 3 mm/min, as described in the work of Konrad et al. [18]. The ceramic shell molds for thin-walled casting were applied on 3D-printed positive polymer models instead of conventional wax models manufactured via injection molding [19].

Thin-walled single-crystal specimens with wall thicknesses of 0.4, 0.8, 1.0 and 2.0 mm were cast. Detailed investigations of the as-cast dendritic structure of these thin-cast specimens were already published [20]. Casting geometry is shown in Figure 1 and consisted of two thin-walled windows (30 × 8 mm^2), having wall thicknesses of 0.4, 0.8, 1.0 or 2.0 mm, respectively.

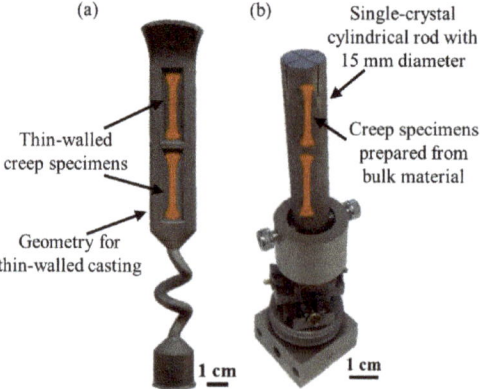

Figure 1. Casting geometry and schematic illustration of creep specimen preparation for (**a**) thin-walled castings and (**b**) single-crystal cylindrical rod.

Orientation of the single-crystal castings with respect to the [001] crystal direction was determined using electron backscattered diffraction (GeminiSEM Sigma VP 300, Zeiss, Jena, Germany). The deviation of the thin-walled castings was less than 5° in all cases. In addition, a cylindrical single crystal with a diameter of 15 mm and a length of 120 mm was cast for the preparation of thin creep specimens from bulk material. The cylindrical single crystal was oriented to the [001] direction within 2°.

All castings were subjected to a heat treatment under vacuum (10^{-3}–10^{-4} Pa) consisting of a solution heat treatment at 1220, 1240 and 1250 °C for 2 h each. After cooling down to 1000 °C with cooling rates higher than 30 °C/min, a two-stage ageing was carried out at 1080 °C for 4 h and 900 °C for 8 h.

2.2. Creep Testing

Flat miniature creep specimens (overall length ~28.5 mm, gauge length ~4.00 mm; gauge width ~2.90 mm) were prepared from single-crystal castings using electrical discharge machining as schematically shown in Figure 1. For the here-called creep tests with casting surface, specimen surfaces were left untreated after heat treatment. For the here-called creep tests without casting surface, the specimen surface was ground with SiC paper with grits of 1000 after heat treatment. For the so-called creep tests on specimens prepared from bulk, the recast layer formed through electrical discharge machining on the heat-treated samples was removed by grinding with SiC paper.

The thickness of the gauge section of every individual miniature creep specimen was determined with a micrometer screw (precision ± 0.01 mm) and the width of the gauge section was determined with a caliper gauge (precision ± 0.01 mm), respectively.

For constant engineering stress creep tests, the load was applied to the sample through a steel pull rod by means of calibrated weights. In the worst cases of an approximately 0.4 mm thick specimen, a nominal engineering stress of 230 MPa was guaranteed within ±6 MPa. Creep tests were carried out at a temperature of 980 °C.

Creep specimens were fixed in alumina clamps and heated in a radiation furnace in air or under vacuum. Strain was measured with an error of $\Delta\varepsilon \approx \pm 0.07\%$ with a non-contacting video extensometer [21,22], avoiding any grooves on the miniature specimens. Temperature regulation was realized with type-S thermocouples. Table 2 gives an overview of the creep tests carried out.

Table 2. Overview of creep tested specimens. Creep tests with conditions highlighted in bold were carried out twice.

Wall Thickness	In Air			Under Vacuum		
	With Casting Surface	Without Casting Surface	Prepared from Bulk Material	With Casting Surface	Without Casting Surface	Prepared from Bulk Material
0.4 mm	150 MPa 190 MPa 230 MPa	230 MPa	150 MPa **230 MPa**	150 MPa 230 MPa	**230 MPa**	230 MPa
0.8 mm	150 MPa 230 MPa	230 MPa	**230 MPa**	150 MPa 230 MPa	230 MPa	230 MPa
1.0 mm	**230 MPa**	230 MPa	230 MPa	**230 MPa**	230 MPa	230 MPa
2.0 mm	150 MPa 230 MPa	230 MPa	230 MPa	**230 MPa**	230 MPa	230 MPa

Creep tests with 150 MPa, 190 MPa and 230 MPa engineering stress were carried out on selected samples for the determination of the stress exponent n in a power law relation between the minimum creep rate and the engineering stress.

$$\dot{\varepsilon}_{min} = c \cdot \sigma^n \tag{1}$$

2.3. Microstructural Characterization

Microstructural investigations were carried out with a scanning electron microscope (SEM) 1540 EsB Cross Beam (Zeiss, Jena, Germany) operating at 15 kV accelerating voltage. Secondary electron imaging was carried out with an Everhart–Thornley detector. The chemical composition of surface and oxide layers was determined using energy-dispersive X-ray spectroscopy (EDS). Cross and longitudinal sections of each specimen after heat treatment and creep deformation were embedded, ground with SiC paper, polished with diamond slurry and, finally, polished with colloidal silica. Thereafter, the specimens were etched with a solution of 3 g Mo-(VI)-oxide, 100 mL H_2O, 100 mL HNO_3 and 100 mL HCl, preferentially dissolving the γ' phase.

Samples for TEM analysis were prepared using a Scientific SCIOS DualBeam SEM (Thermo Fisher, Waltham, MA, USA) with a Ga ions FIB system. Cross-section specimens were removed with milling trenches using an ion beam accelerating voltage of 30 kV and currents between 30 nA and 300 pA. Thinning of TEM lamellas to <100 nm thickness was carried out using 2 kV ion beam accelerating voltage and 35 pA ion beam current. TEM analysis was conducted with a JEOL 2100Plus (Jeol, Freising, Germany) at 200 kV acceleration voltage. Dark field images to reveal an ordered structure were taken with a $\vec{g}$ = <001> diffraction vector.

3. Results

3.1. Microstructure

An approximately 3–5 μm thick surface layer was observed after the heat treatment of the thin-cast MAR M247LC under vacuum, see Figure 2a. Beneath this layer, there was a homogeneous matrix/γ' structure with cuboidal γ' precipitates.

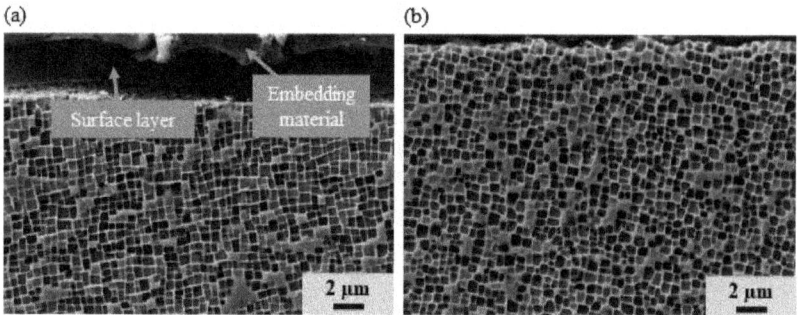

Figure 2. Microstructure in the near-surface area after heat treatment under vacuum of thin-cast specimens (**a**) with surface layer, (**b**) surface layer removed by grinding with SiC paper with grits of 1000.

Figure 3 gives the chemical composition of the surface layer. An enrichment of the γ'-forming element aluminum alongside a nickel enrichment was detected, while chromium and cobalt were depleted within this layer.

The surface layer observed after the heat treatment under vacuum was further investigated with TEM. An electron transparent lamella was prepared with the FIB technique described above so that both the single-phase surface layer (area 1 in Figure 3) and the two-phase structure (area 2 in Figure 3) could be analyzed in the TEM (see Figure 4a). Figure 4b illustrates the TEM diffraction patterns within the surface layer and the two-phase structure. Figure 4c shows a TEM dark field diffraction contrast image obtained with superlattice reflection. Here, γ' appears in light shading and the matrix phase in dark. The superlattice diffraction spots of an $L1_2$ ordered phase together with the chemical composition given in Figure 3 verified that the surface layer was single-phase γ'. Hence, for the so-called creep tests with casting surface, the initial surface at the beginning of the test consisted of single-phase γ'.

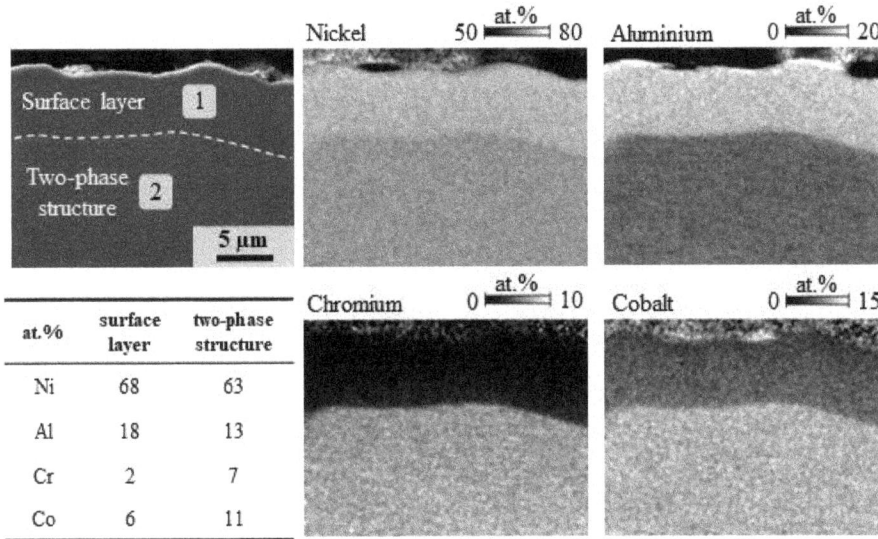

Figure 3. Chemical composition of the surface layer formed during heat treatment under vacuum.

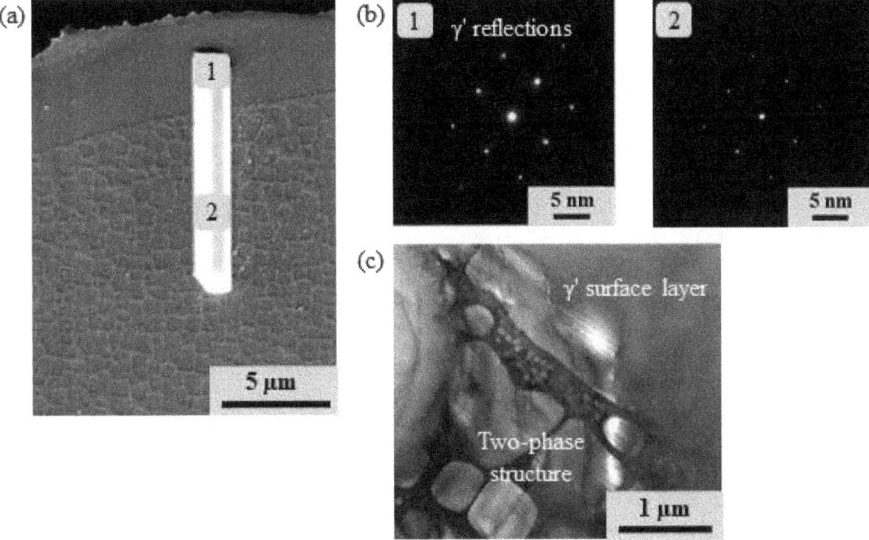

Figure 4. TEM lamella preparation and TEM analysis: (a) position of TEM lamella prepared with focused ion beam; (b) TEM diffraction patterns in the surface layer (area 1) and the two-phase structure; (c) dark field image with γ' reflection of the transition area.

Surface grinding for the so-called creep tests without casting surface removed the single-phase γ'-scale, so that the homogeneous matrix/γ' structure reached up to the creep specimen surface (see Figure 2b).

Figure 5 shows typical SEM images of near surface areas of single-crystal MAR M247LC after creep deformation in air (Figure 5a) and under vacuum (Figure 5b). During creep deformation in air, a γ'-free zone was formed beneath an oxide layer. Below these two zones, a rafted two-phase structure was seen (Figure 5a). In contrast, creep tests under vacuum prevented oxidation and a γ'-free zone was not detected. The rafted microstructure was present up to the specimen surface. In both cases, rafting of the γ' precipitates took place perpendicular to the load direction, anticipated for a nickel-based superalloy with a negative misfit under tensile load [23,24]. The bright areas correspond to the matrix, whereas the darker areas are γ' phase.

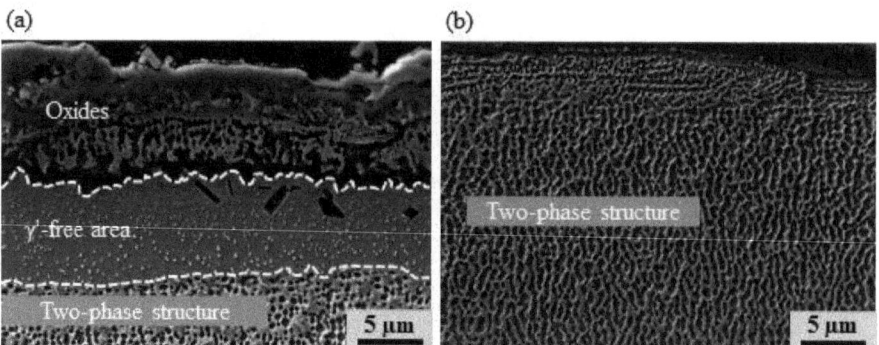

Figure 5. Microstructure after creep deformation at 980 °C and 230 MPa (**a**) in air after 50 h and (**b**) under vacuum after 75 h.

3.2. Creep Behavior

Figure 6 shows the strain versus time and strain rate versus strain creep curves of the single-crystal MAR M247LC specimens with various thicknesses in air at 980 °C and under an engineering tensile stress of 230 MPa. Rupture of the specimens was marked in the diagrams with a cross. Creep diagrams were divided according to three different surface conditions, i.e., with casting surface (Figure 6a,b), without casting surface (Figure 6c,d) and prepared from bulk material (Figure 6e,f).

The rupture times of all the creep tests carried out in air at 230 MPa and 980 °C, independent from the thickness, were in the same range, i.e., in between 40 and 75 h. Except for one outlier with a 0.4 mm thickness, the strains at rupture were in a single range of 20–25%. In addition, the creep curves all showed very similar progressions. A short primary creep was present with decreasing creep rates. Minimum creep rates were generally reached after just 1–2.5% engineering strain. The creep accelerated steadily after minimum creep rates were reached, which could be attributed to microstructural changes due to rafting. No pronounced secondary creep with an approximately constant creep rate was observed, but rather, a tertiary creep started immediately after the primary creep.

Figure 7 shows the creep curves of specimens with casting surface of different wall thicknesses tested in air at 980 °C under an engineering stress of 150 MPa. A creep rupture for wall thicknesses of 0.4 and 0.8 mm took place after 200 h. These two creep curves showed sharp increases in the creep rate near the ends of their live times, hence, very much different progressions than all creep curves from tests under 230 MPa at 980 °C in air. The creep test of the specimen with a wall thickness of 2 mm was interrupted after approximately 180 h, because an almost constant creep rate was reached.

Figure 8 shows the creep curves of tests under vacuum at 980 °C and an engineering stress of 230 MPa again carried out on specimens with various thicknesses and surface conditions. With the single exception of a 2 mm thick specimen prepared from the bulk, the rupture times of all the creep tests carried out in vacuum at 230 MPa and 980 °C were in the same range of 25–100 h. Strains at rupture in all creep tests were also in a single range of 25–30%. All creep curves showed very similar progressions reaching minimum creep rates after 1–2.5% engineering strain. Thereafter, creep accelerated steadily, typical for tertiary creep.

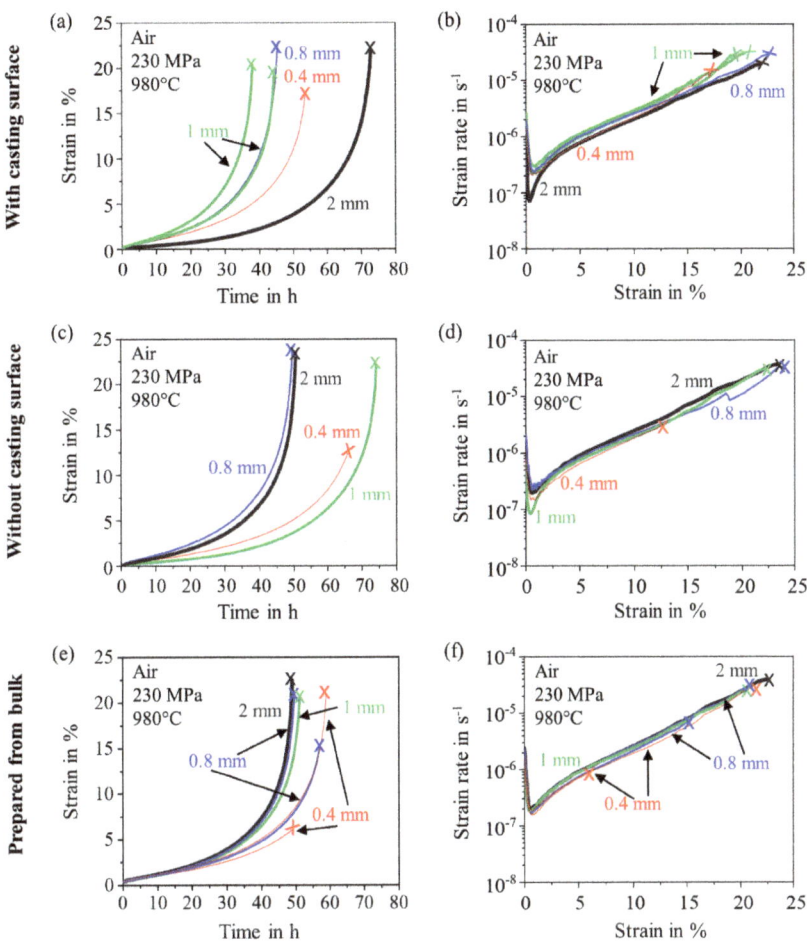

Figure 6. Creep curves of specimens with various thickness tested in air at 980 °C under a high stress of 230 MPa: (**a,b**) with casting surface, (**c,d**) without casting surface, (**e,f**) prepared from bulk material.

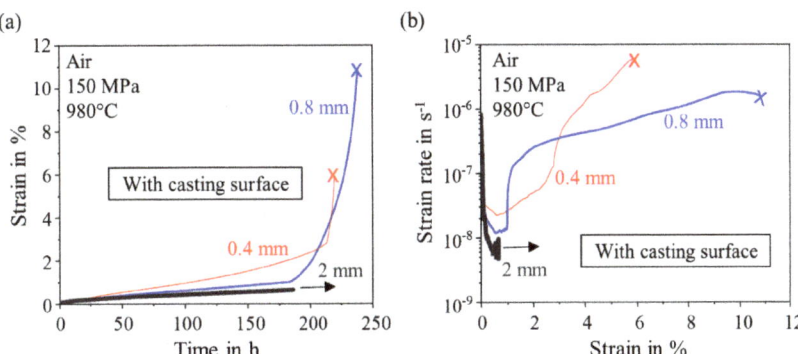

Figure 7. Creep curves of thin-cast specimens with casting surface of different wall thickness in air at 980 °C under a low stress of 150 MPa: (**a**) strain versus time; (**b**) strain rate versus strain.

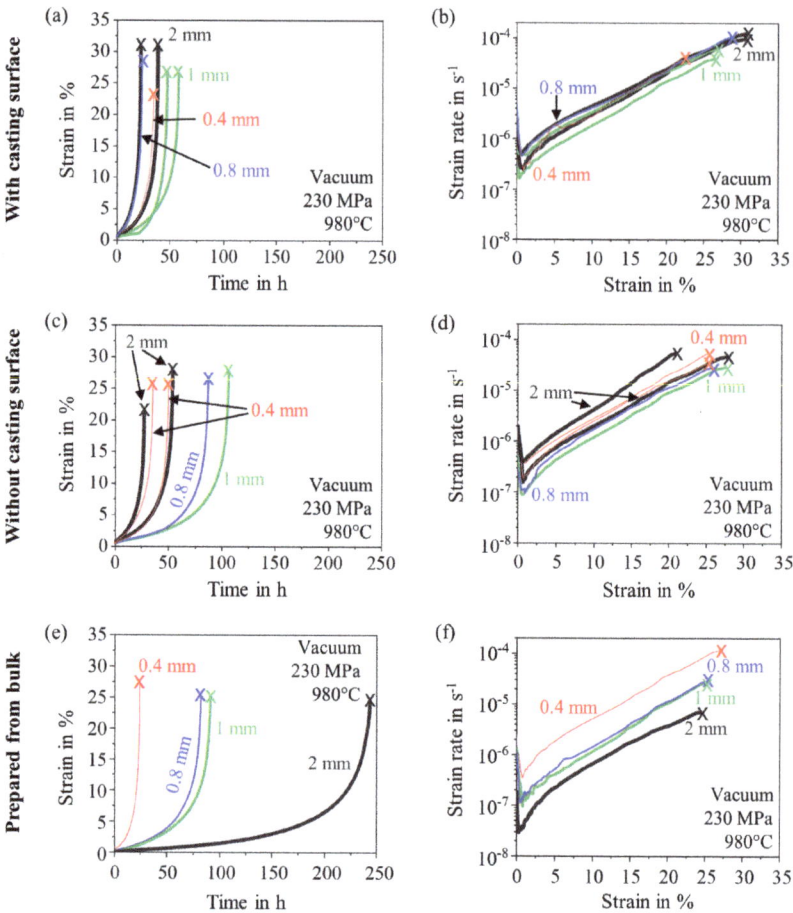

Figure 8. Creep curves of thin-walled specimens for the investigation of the influence of the wall thickness under vacuum at a stress of 230 MPa at 980 °C: (**a**,**b**) with casting surface, (**c**,**d**) without casting surface, (**e**,**f**) prepared from bulk material.

4. Discussion

The formation of a single-phase γ′ layer during a heat treatment under vacuum was caused by the high vapor pressures of chromium and cobalt. Chromium and cobalt are elements that stabilize the matrix. If these elements evaporate during a heat treatment, the matrix phase dissolves and the γ′ phase becomes stabilized [25–27].

The stress exponent in the power law relation between the minimum creep rate and the engineering stress of the single-crystal MAR M247LC in the [001] orientation was in the range of 5–7 in all circumstances. Hence, it was independent from the surface condition, the specimen thickness and the creep test atmosphere. It may be concluded that the dominant creep deformation mechanism was a dislocation climb [23] in all circumstances. Stress exponents in this work at a temperature of 980 °C were in good agreement with the literature, which indicated a stress exponent of 8.4 in creep tests under vacuum with stresses of 150 MPa and 230 MPa [17]. Bensch et al. [13] examined the stress exponent n of the single-crystal nickel-based superalloy MAR M247LC at 980 °C too. The stress exponent was expected to be 5.6 with a γ′ volume fraction of 49%, while it was 8.7 with a γ′ volume fraction of 62%.

Cast nickel-based superalloy MAR M247LC single crystals with various thickness were creep tested at 980 °C in air and vacuum and in two distinct load regimes, i.e., under a high engineering load of 230 MPa and a low engineering load of 150 MPa. Different surface conditions (with or without casting surface, prepared from bulk), tested both in air and under vacuum, showed that specimen surface conditions do not play a major role. Previous investigations on thin-cast polycrystalline specimens of the nickel-based superalloy IN100 with wall thicknesses of 0.9 and 1.3 mm confirmed this observation [28].

In addition, at high engineering loads, leading to generally short creep live times, creep properties turned out to be independent from the wall thickness. The scatter of the creep live times was within the usual deviation for creep experiments [29]. In contrast, at the low engineering load level and, therefore, longer creep live times (>200 h), a clear thickness debit effect was observed.

In the literature it is commonly observed that the thickness debit effect is more pronounced at higher temperatures and lower stress levels, i.e., longer creep times. Krieg et al. [16] observed a significant reduction in creep life with decreasing wall thickness (0.3/0.5/1.0 mm) on thin-walled specimens of the nickel-based superalloy PWA 1484 prepared from bulk single crystals (70/80 MPa, 1100 °C). In contrast to that, no significant thickness debit effect was detected for 170 MPa at 980 °C [16].

Brunner et al. [17] illustrated a corresponding behavior in thin specimens prepared from bulk material with a wall thickness of 0.3 and 1 mm of the single-crystal nickel-based superalloy MAR M247LC. At a low stress level (150 MPa, 980 °C) or at higher temperatures (i.e., 1100 °C), creep strength decreases more with decreasing wall thickness compared to higher stress levels (230 MPa, 980 °C) [17].

Bensch et al. [11,30] identified oxide layer formation and the associated growth of γ'-free and γ'-depleted areas as a decisive factor influencing the creep behavior of thin-walled specimens of nickel-based superalloys. Bensch et al. [11,30] demonstrated experimentally and with model simulations that the strength of the thickness debit effect also depends on the chemical composition of a single-crystal nickel-based superalloy. Simulations estimated that creep properties of MAR M247LC deteriorate by 50% as soon as the wall thickness is inferior by 0.4 mm. The determined critical wall thickness for René N5 was approximately 0.6 mm. Compared to MAR M247LC, René N5 showed a more pronounced thickness debit effect due to a stronger Al_2O_3 layer formation [11,30]. This also corresponded to experimental investigations of Hüttner et al. [14,15] regarding the thickness debit effect in the single-crystal nickel-based superalloy René N5. Figure 9 summarizes the rupture times or creep live times, respectively, as a function of the wall thickness of the single-crystal nickel-based superalloys MAR M247LC and René N5.

The thickness debit effect was obviously governed by diffusion-controlled mechanisms. The formation of oxide layers and γ'-free/depleted areas played the decisive role. Hence, the thickness debit effect itself was notably time dependent. If creep deformation lasted long enough (>200 h, 980 °C), a significant proportion of the cross-sectional area was affected by these diffusion-controlled mechanisms, leading to deteriorating creep properties.

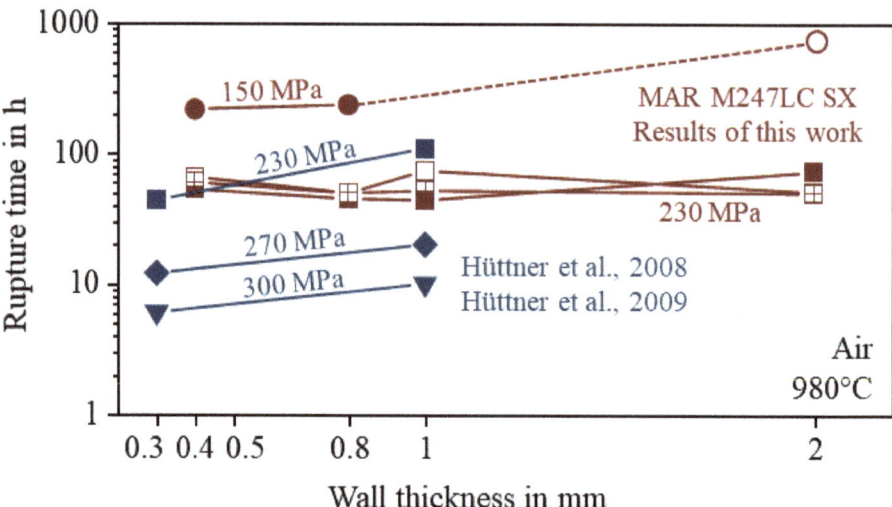

Figure 9. Summary of the results of this work in comparison to literature: rupture time over wall thickness at 980 °C in air; Data from [14,15].

5. Conclusions

Creep tests at 980 °C on thin nickel-based superalloy MAR M247LC single crystals were carried out in air and under vacuum. The influence of the surface condition and wall thickness (0.4/0.8/1.0/2.0 mm) on the creep behavior was investigated.

- Heat treatment under vacuum at an ambient pressure of 10^{-4} Pa led to a single-phase γ' surface layer.
- Surface condition did not affect the creep behavior of thin specimens.
- Creep tests at 980 °C under 230 MPa leading to creep rupture times <100 h showed no significant thickness debit effect.
- Creep tests at 980 °C under 150 MPa, leading to significantly longer creep rupture times (>200 h), showed a significant thickness debit effect, i.e., a strong deterioration of creep properties with wall thicknesses of 0.8–0.4 mm.
- Thickness debit effect was time-dependent, i.e., it was governed by the formation of oxide layers and γ'-free or γ'-depleted areas. With increasing creep times, these diffusion-controlled mechanisms affected larger proportions of the cross-section, thus, deteriorating the creep properties.

Author Contributions: Conceptualization, S.K., U.G. and R.V.; methodology, S.K.; software, S.K. and R.V.; validation, S.K.; formal analysis, S.K.; investigation, S.K. and S.W.-G.; resources, S.K.; data curation, S.K. and S.W.-G.; writing—original draft preparation, S.K.; writing—review and editing, U.G. and R.V.; visualization, S.K.; supervision, U.G.; project administration, U.G. and R.V.; funding acquisition, U.G. All authors have read and agreed to the published version of the manuscript.

Funding: This research was funded by Deutsche Forschungsgemeinschaft (DFG), research project number GL 181/52-1.

Institutional Review Board Statement: Not applicable.

Informed Consent Statement: Not applicable.

Data Availability Statement: Not applicable.

Conflicts of Interest: The authors declare no conflict of interest.

References

1. Gibbons, T.B. The performance of Superalloys. *Adv. Mater.* **1990**, *2*, 583–588. [CrossRef]
2. Duhl, D.N. Directionally Solidified Superalloys. In *Superalloys II*; Sims, C.T., Stoloff, N.S., Hagel, W.S., Eds.; John Wiley & Sons, Inc.: New York, NY, USA; Chichester, UK; Brisbane, Australia; Toronto, ON, Canada; Singapore, 1987; pp. 189–214.
3. Doner, M.; Heckler, J.A. Identification of Mechanisms Responsible for Degradation in Thin-Wall Stress-Rupture Properties. *Proc. Int. Symp. Superalloy.* **1988**, 653–662. [CrossRef]
4. Baldan, A. Combined effects of thin-section size, grain size and cavities on the high temperature creep fracture properties of a nickel-base superalloy. *J. Mater. Sci.* **1997**, *32*, 35–45. [CrossRef]
5. Gibbons, T.B. Creep properties of Nimonic 90 in thin section. *Met. Technol.* **1981**, *8*, 472–475. [CrossRef]
6. Doner, M.; Heckler, J.A. Effects of Section Thickness and Orientation on Creep-Rupture Properties of Two Advanced Single Crystal Alloys. *SAE Tech. Pap.* **1985**, *851785*. [CrossRef]
7. Baldan, A. On the thin-section size dependent creep strength of a single crystal nickel-base superalloy. *J. Mater. Sci.* **1995**, *30*, 6288–6298. [CrossRef]
8. Pandey, M.C.; Taplin, D.M.R. Prediction of rupture lifetime in thin sections of a nickel base superalloy. *Scr. Metall. Mater.* **1994**, *31*, 719–722. [CrossRef]
9. Pandey, M.C.; Taplin, D.M.R.; Rao, P.R. An analysis of specimen geometry effect on the creep life of inconel alloy X-750. *Mater. Sci. Eng. A* **1989**, *118*, 33–39. [CrossRef]
10. Seetharaman, V.; Cetel, A.D. Thickness debit in creep properties of PWA 1484. *Proc. Int. Symp. Superalloy.* **2004**, 207–214. [CrossRef]
11. Bensch, M.; Preußner, J.; Hüttner, R.; Obigodi, G.; Virtanen, S.; Gabel, J.; Glatzel, U. Modelling and analysis of the oxidation influence on creep behaviour of thin-walled structures of the single-crystal nickel-base superalloy René N5 at 980 °C. *Acta Mater.* **2010**, *58*, 1607–1617. [CrossRef]
12. Srivastava, A.; Gopagoni, S.; Needleman, A.; Seetharaman, V.; Staroselsky, A.; Banerjee, R. Effect of specimen thickness on the creep response of a Ni-based single-crystal superalloy. *Acta Mater.* **2012**, *60*, 5697–5711. [CrossRef]
13. Bensch, M.; Fleischmann, E.; Konrad, C.H.; Fried, M.; Rae, C.M.F.; Glatzel, U. Secondary Creep of Thin-Walled Specimens Affected by Oxidation. *Proc. Int. Symp. Superalloy.* **2012**, 387–394. [CrossRef]
14. Hüttner, R.; Gabel, J.; Glatzel, U.; Völkl, R. First creep results on thin-walled single-crystal superalloys. *Mater. Sci. Eng. A* **2009**, *510–511*, 307–311. [CrossRef]
15. Hüttner, R.; Völkl, R.; Gabel, J.; Glatzel, U. Creep behavior of thick and thin walled structures of a single crystal nickel-base superalloy at high temperatures—Experimental method and results. *Proc. Int. Symp. Superalloy.* **2008**, 719–724. [CrossRef]
16. Krieg, F.; Mosbacher, M.; Fried, M.; Affeldt, E.; Glatzel, U. Creep and Oxidation Behaviour of Coated and Uncoated Thin Walled Single Crystal Samples of the Alloy PWA1484. *Proc. Int. Symp. Superalloy.* **2016**, 773–779. [CrossRef]
17. Brunner, M.; Bensch, M.; Völkl, R.; Affeldt, E.; Glatzel, U. Thickness influence on creep properties for Ni-based superalloy M247LC SX. *Mater. Sci. Eng. A* **2012**, *550*, 254–262. [CrossRef]
18. Konrad, C.H.; Brunner, M.; Kyrgyzbaev, K.; Völkl, R.; Glatzel, U. Determination of heat transfer coefficient and ceramic mold material parameters for alloy IN738LC investment castings. *J. Mater. Process. Technol.* **2011**, *211*, 181–186. [CrossRef]
19. Körber, S.; Völkl, R.; Glatzel, U. 3D printed polymer positive models for the investment casting of extremely thin-walled single crystals. *J. Mater. Process. Technol.* **2021**, *293*, 117095. [CrossRef]
20. Körber, S.; Fleck, M.; Völkl, R.; Glatzel, U. Anisotropic Growth of the Primary Dendrite Arms in a Single-Crystal Thin-Walled Nickel-Based Superalloy. *Adv. Eng. Mater.* **2022**, *24*, 2101332. [CrossRef]
21. Völkl, R.; Freund, D.; Fischer, B. Economical Creep Testing of Ultrahigh-temperature Alloys. *J. Test. Eval* **2003**, *31*, 35–43.
22. Völkl, R.; Fischer, B. Mechanical Testing of Ultra-high Temperature Alloys. *Exp. Mech.* **2004**, *44*, 121–128. [CrossRef]
23. Kassner, M.E. *Fundamentals of Creep in Metals and Alloys*; Elsevier: Amsterdam, The Netherlands, 2008. [CrossRef]
24. Epishin, A.; Link, T.; Portella, P.D.; Brückner, U. Evolution of the γ/γ' microstructure during high-temperature creep of a nickel-base superalloy. *Acta Mater.* **2000**, *48*, 4169–4177. [CrossRef]
25. Schulze, M.; Seidel, S. *Verdampfungsgleichgewicht und Dampfdruck*; Springer Fachmedien: Wiesbaden, Germany, 2018. [CrossRef]
26. Alcock, C.B.; Itkin, V.P.; Horrigan, M.K. Vapour pressure equations for the metallic elements: 298-2500K. *Can. Metall. Q.* **1984**, *23*, 309–313. [CrossRef]
27. D'Souza, N.; Welton, D.; Kelleher, J.; West, G.D.; Dong, Z.H.; Brewster, G.; Dong, H.B. Microstructure instability of ni-base single crystal superalloys during solution heat treatment. *Proc. Int. Symp. Superalloy.* **2016**, 267–277. [CrossRef]
28. Strößner, J.; Konrad, C.H.; Brunner, M.; Völkl, R.; Glatzel, U. Influence of casting surface on creep behaviour of thin-wall Ni-base superalloy Inconel100. *J. Mater. Process. Technol.* **2013**, *213*, 722–727. [CrossRef]
29. Evans, R.W.; Wilshire, B. *Creep of Metals and Alloys*; U.S. Department of Energy: Oak Ridge, TN, USA, 1985; ISBN 0904357597.
30. Bensch, M.; Sato, A.; Warnken, N.; Affeldt, E.; Reed, R.C.; Glatzel, U. Modelling of High Temperature Oxidation of Alumina-Forming Single-Crystal Nickel-Base Superalloys. *Acta Mater.* **2012**, *60*, 5468–5480. [CrossRef]

Article

The Effect of Temperature and Phase Shift on the Thermomechanical Fatigue of Nickel-Based Superalloy

Ivo Šulák *, Karel Hrbáček and Karel Obrtlík

Institute of Physics of Materials, Czech Academy of Sciences, Žižkova 22, 616 00 Brno, Czech Republic; hrbacek@ipm.cz (K.H.); obrtlik@ipm.cz (K.O.)
* Correspondence: sulak@ipm.cz

Abstract: In this paper, the minimum temperature and phase shift effects on the thermo–mechanical fatigue (TMF) behavior of Inconel 713LC are investigated. TMF tests were performed under 0° (in-phase-IP) and +180° (out-of-phase-OP) phase shifts between mechanical strain and temperature. Cylindrical specimens were cycled at constant mechanical strain amplitude with a strain ratio of $R_\varepsilon = -1$. Tests were performed with temperature ranges of 300–900 °C and 500–900 °C. The heating and cooling rate was 5 °C/s. Fatigue hardening/softening curves and fatigue life data were assessed. Results show that out-of-phase loading was less damaging than in-phase loading. Scanning electron microscopy (SEM) examination of metallographic sections indicated that the life-reducing damage mechanism was intergranular cavitation under in-phase loading. Transmission electron microscopy (TEM) revealed honeycomb structures for IP loading. The plastic strain localization into persistent slip bands was typical for OP loading. For out-of-phase loading, fatigue damage appeared to be dominant. The increase in the temperature range led to a significant decrease in fatigue life. The reduction of fatigue life was far more pronounced for out-of-phase loading. This can be ascribed to the accelerated crack propagation at high tensile stress under out-of-phase loading as well as the amount of accommodated plastic strain deformation. Based on the SEM scrutiny of metallographic sections and TEM observations of dislocation arrangement, the prevailing damage mechanisms were documented and the lifetime behavior was accordingly discussed.

Keywords: nickel-based superalloy; high-temperature fatigue; in-phase; out-of-phase; cyclic stress–strain curves; fatigue life curves

1. Introduction

Nickel-based superalloys are typically employed for gas turbine wheel components in jet propulsion and energy production [1]. These components are often subjected to high variable stresses at elevated temperatures in a harsh environment, where a combination of fatigue, creep, and oxidation can cause premature failure [2]. The harshest loading conditions are those where strain or stress vary simultaneously with temperature, i.e., in the case of thermomechanical fatigue (TMF). Since the inlet temperature of hot gases during service is still well above the melting point of superalloys, turbine wheel components are thus actively cooled [3] and protected with thermal barrier coatings [4–6]. Consequently, noticeable temperature gradients within the component during steady-state operation occur, and together with periodical shut-down and start-up periods of power and propulsion facilities cause severe thermal expansion and contraction that are often reinforced by mechanical stresses or strains associated with centrifugal forces as turbine speed changes. This complex loading history inevitably leads to TMF failure [7]. Several papers were devoted to the TMF behavior of single crystal [8,9] and polycrystal [10–12] nickel-based superalloys. Guth et al. [10] studied the effect of phase shifts on the second-generation superalloy MAR-M247 in the temperature range of 100–850 °C. Results showed that the 0° phase shift between temperature and mechanical strain had the shortest life, followed by the +180° phase shift and the ±90° phase shift. Yamazaki et al. [13] found that in the case of the +180° cycle, the fatigue crack growth rate and thus fatigue life was strongly dependent on the TMF testing temperature.

For the purpose of this study, the first-generation superalloy Inconel 713LC was chosen due to its unique combination of mechanical properties and favorable production costs. It possesses an excellent capability to resist high-temperature creep [14], low cycle fatigue [5,6,15], and high cycle fatigue [16,17]. These properties are balanced by outstanding fracture toughness, which is particularly important at lower temperatures since the power units are periodically shut-down. However, TMF data of Inconel 713 and its low carbon version Inconel 713LC are rare [18]. In the present paper, the TMF behavior of the polycrystalline nickel-based superalloy Inconel 713LC is reported. This study provides a direct comparison of both the TMF load cycle and the effect of the temperature range. It also provides additional information regarding isothermal fatigue at the peak temperature of the TMF cycle. The mechanisms that account for the different lifetimes are described with particular emphasis on the two parameters investigated. The main aim is to better understand the effect of the temperature and phase shift on the lifetime and damage mechanisms. Fatigue hardening/softening curves and maximal and minimal stress evolution curves were plotted. The prevailing damage mechanism explaining the TMF lifetime differences was investigated via scanning electron microscopy (SEM) and transmission electron microscopy (TEM).

2. Materials and Methods

In the current study, rods prepared via an investment-casting technique with a shape close to the TMF test specimens were used. The nickel-based superalloy Inconel 713LC with the nominal chemical composition (wt. %) of 0.06 C, 11.70 Cr, 4.29 Mo, 6.02 Al, 0.64 Ti, 0.16 Ta, 0.23 Fe, 0.008 B, 2.06 Nb, 0.06 Zr, and Ni bal. was supplied by PBS Velká Bíteš a.s., Czech Republic (Velká Bíteš, CZ). Inconel 713LC is typical with coarse dendritic grains (see Figure 1a). The average grain size of the used melt is (2.18 ± 0.78) mm. The microstructure is composed of a face-centered cubic (fcc) γ matrix with a high volume of fcc γ' strengthening precipitates, γ/γ' eutectics, carbides, and casting defects with a size up to 150 µm. The coherent γ' precipitates having a cuboidal shape are shown in Figure 1b.

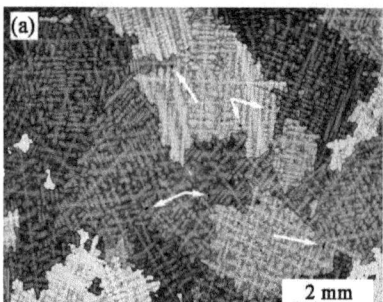

Figure 1. Microstructure of Inconel 713LC (**a**) coarse dendritic grains with small casting defects (white arrows) (**b**) detail of cubic γ' precipitates.

Cylindrical specimens (see Figure 2) with a diameter and gauge length of 7 mm and 16 mm, respectively, were subjected to TMF loading. Two different loading cycles were applied: in-phase (IP-TMF) and out-of-phase (OP-TMF). Figure 3 shows schematically the temperature and mechanical strain courses for these two cycles. TMF experiments were carried out in computer-controlled servo-hydraulic mechanical testing MTS 880 system (MTS Systems Corporation, Eden Prairie, MN, USA) with water-cooled hydraulic grips. Symmetric cyclic straining ($R_\varepsilon = -1$) with triangular shape waveform in total strain amplitude control mode was applied. The temperature was monitored and controlled by a

K-type thermocouple wreathed over the specimen amid the specimen gauge length and covered by heat-resistant fabric. Heating provided a high-frequency inductive heating generator TruHeat HF 3010 (TRUMPF Hüttinger GmbH + Co. KG, Freiburg, Germany). Cooling of the specimens was achieved using the water-cooled grips along with the external airflow system. A capacitive 632.41C-11 extensometer (MTS Systems Corporation, Eden Prairie, MN, USA) equipped with 120-mm-long ceramic tips and a 12 mm base was used to control and measure the strain. The experimental setup is shown in Figure 4. The temperature was varied in two intervals: from 300 °C to 900 °C and from 500 °C to 900 °C. The cycle time was fixed to 240 s and 160 s for the 300–900 °C and 500–900 °C intervals, respectively, which corresponds to the rate of temperature change of 5 °C/s. Since the heating and cooling rate was kept constant, the strain rate differed in dependence on the applied mechanical strain amplitude. The calculated strain rate was in the range of 5×10^{-5} s^{-1} to 1×10^{-4} s^{-1} for the temperature interval of 500–900 °C and 3.3×10^{-5} s^{-1} to 5.8×10^{-5} s^{-1} for the temperature interval of 300–900 °C. This change in the strain rate is small, and it is not expected to make any significant difference in the overall behavior. The mechanical strain was determined and controlled as the total strain minus the thermal strain, which was determined for each specimen by measurement of the coefficient of thermal expansion. A code of practice for TMF tests was followed [19].

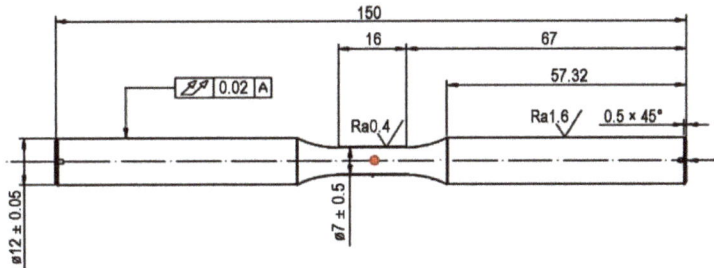

Figure 2. Dimensions (in mm) and shape of the TMF specimen (red dot indicates the position of the thermocouple).

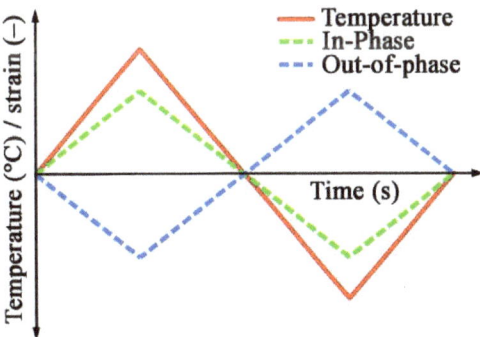

Figure 3. Time courses of temperature and mechanical strain for IP and OP loading.

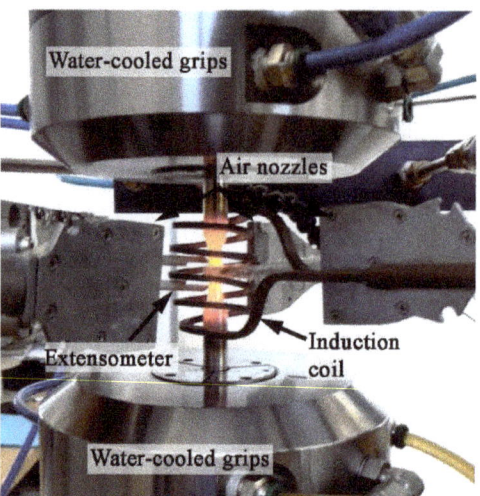

Figure 4. TMF experimental setup showing the specimen heated to 900 °C, water-cooled hydraulic grips, Cu induction coil, air nozzles for specimen cooling, and extensometer.

Experimental data were recorded by the Testsuit program (MTS Systems Corporation, Eden Prairie, MN, USA). A geometric series containing a total of 20 hysteresis loops per decade was chosen for recording. The stress amplitude was evaluated from the recorded hysteresis loops. The plastic strain amplitude was assessed offline from the half of the width of the hysteresis loop.

A Tescan Lyra3 XMU SEM (TESCAN ORSAY HOLDING, Brno, Czech Republic)-equipped Oxford Symmetry electron backscatter diffraction (EBSD) detector (Oxford Instruments, High Wycombe, UK) was adopted to investigate fatigue crack initiation and the propagation mechanism. The internal damage and its relation to fatigue crack propagation were studied on longitudinal metallographic sections of fatigued samples. The dislocation arrangement of Inconel 713LC was also characterized by a high-resolution transmission electron microscope (TEM) JEOL JEM-2100F (JEOL Ltd., Tokyo, Japan) with a field emission gun electron source equipped with a bright field (BF) detector for observation in scanning mode (STEM).

3. Results and Discussion

Figure 5 shows a comparison of the stress response of Inconel 713LC to the applied mechanical strain of 0.35% for IP and OP types of loading and both temperature ranges. For the sake of clarity, only one mechanical strain amplitude (0.35%) was chosen representing the overall behavior of Inconel 713LC under TMF loading. The fatigue hardening/softening curves in the representation of stress amplitude σ_a versus the number of elapsed cycles N are shown in Figure 4a. A stable stress response is typical for both IP and OP cycles in the temperature range of 500–900 °C, suggesting that hardening mechanisms such as the formation of new dislocations, dislocation–dislocation interactions, as well as dislocation–obstacle interactions are in balance with softening mechanisms such as the annihilation of dislocations, the reduction of the effective size of γ' precipitates by shearing, and rafting of γ' precipitates [1,20]. An increase in the temperature range by lowering T_{min} from 500 °C to 300 °C was manifested by a shift of the stress amplitude to higher values, which can be mainly attributed to higher values of Young's modulus at 300 °C (189 GPa) compared to 500 °C (178 GPa). This increased the absolute values of the tensile peak stress (OP) or compressive peak stress (IP) of the TMF cycle (Figure 5b). Considerable fatigue hardening can also be seen while testing in the temperature range of 300–900 °C, indicating the superiority of the above-mentioned hardening mechanisms in this particular temperature range.

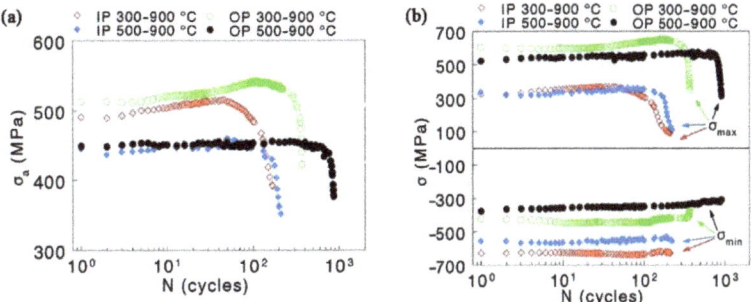

Figure 5. Stress response to applied mechanical strain (**a**) fatigue hardening/softening curves (**b**) evolution of maximal and minimal stresses during the TMF test.

The representative stress–strain hysteresis loops for all TMF loading conditions are shown in Figure 6. The selected hysteresis loops represent the first, $N_f/2$, and N_f loading cycles for a mechanical strain amplitude of 0.35%. In accordance with the hardening/softening curves (Figure 5), a cyclic hardening/softening of the Inconel 713LC superalloy can be observed here as a change in the height and width of the hysteresis loop. Contrary to strain-controlled isothermal fatigue tests [20,21], the hysteresis loops under TMF loading are asymmetrical due to a temperature change. The mean stress then depends on the temperature range as well as on the shift between temperature and applied mechanical strain. When the phase shift is 180° (OP cycle), the hysteresis loop shifts towards tensile stresses and positive mean stress develops. The opposite is true for the IP cycle (0° phase shift) when the compressive mean stress develops.

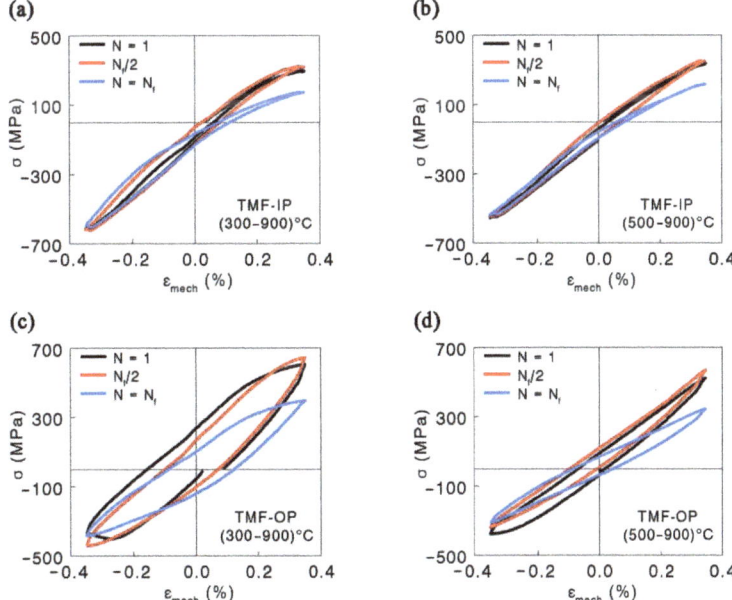

Figure 6. Representative hysteresis loops (**a**) IP 300–900°C (**b**) IP 500–900 °C (**c**) OP 300–900 °C (**d**) OP 500–900 °C.

The diagram of total strain amplitude on the linear ordinate versus the number of cycles to failure on logarithmic abscissa is plotted in Figure 7. Regardless of the temperature range, OP loading revealed a slightly longer fatigue life in the entire range of strain amplitudes compared to IP loading. This behavior was also reported for other fcc metals [10,22,23]. Shifting the minimum temperature from 500 °C to 300 °C resulted in a reduction in fatigue life in both TMF cycles, which can be attributed to the higher values of plastic strain (Figure 6), which is a determining factor in fatigue [20]. Figure 7 also includes the results obtained by Šulák et al. [6] for Inconel 713LC tested under isothermal low cycle fatigue conditions at 900 °C (blue dashed curve). It is clear that TMF straining leads to a significant reduction in fatigue life, showing the importance of these experiments to ensure the safe-life of high-temperature components.

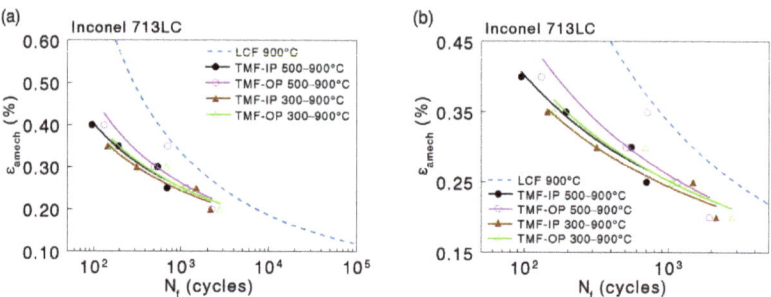

Figure 7. Fatigue life curves in the representation of the mechanical strain amplitude vs. the number of cycles to failure (**a**) overall view (**b**) detailed view focusing on TMF results. LCF data from [6].

Authors have built on the SEM inspection of metallographic sections to reveal that the dominant failure mechanism changes with the phase shift. Figures 8 and 9 depict typical fatigue cracks occurring under TMF loading. Regardless of the TMF conditions, fatigue cracks nucleated from the surface. However, the crack initiation site and fatigue crack propagation mechanisms differ. Fatigue crack initiations at grain boundaries and subsequent propagation of fatigue cracks along grain boundaries are typical for IP loading, as shown in Figure 8a. We can characterize this grain boundary degradation as being rooted in the migration and coalescence of vacancies and a consequential formation of cavities associated with creep damage [14,15,18,24]. Linking of fatigue cracks with internal cavities, as shown in Figure 8b, may accelerate the propagation rate and can explain the lower fatigue life observed for IP loading (Figure 7) although OP loading resulted in considerably higher maximum tensile stress (Figure 5) and higher amplitude of plastic strain (Figure 6) [12]. In addition to surface initiation at grain boundaries, internal degradation with no connection to the surface was observed for IP loading as well (Figure 8c). The inner crack was formed by the coalescence of several cavities and propagated directly along the grain boundary. The Kernel average misorientation (KAM) map showing a local grain misorientation derived from EBSD data is depicted in Figure 8d. In the vicinity of a crack (or a larger cavity), misorientation of adjacent grains occurred due to large plastic deformation ahead of a crack tip. This led to a local increase in stress which may contribute to easier crack propagation along the grain boundary. Figure 9 shows several shallow transgranular fatigue cracks typical for OP loading conditions. No evidence of creep damage under OP TMF loading was found. Zauter et al. [25] stated that when the maximum temperature of the TMF cycle lies within the creep range, creep-induced lifetime reduction occurs only in IP cycling and can be explained by a change in the fracture mode from transgranular to intergranular in OP cycling and IP, respectively. This corresponds well with the findings in this study. Furthermore, surface relief formed in the gauge length of specimens loaded under the OP TMF loading (Figure 8b,c). These structures, called persistent slip markings (PSMs), form along the active slip system, which is most often the {111}-type plane in fcc materials [21,26–28]. PSMs consist of local surface elevations (extrusions) and depressions (intrusions) and form along the intersection of dislocation-rich bands (persistent slip bands—PSBs) with the surface as a consequence of localized cyclic plastic strain in PSBs. When the activity of PSBs is high enough, they arise on free surfaces and form a hill-and-valley relief on the originally flat surface and are often the cause of fatigue crack initiation in the low cycle fatigue domain [29–32]. With an increase in temperature, the PSMs oxidize heavily and their appearance changes [21]. This is associated with a change in the localization of plastic deformation, where at higher temperatures, the plastic deformation is mainly transmitted through γ matrix channels. The change in the temperature range in the case of OP TMF cycling resulted in different PSMs (compare Figure 8b,c). The wavy nature of the PSMs on the specimen cycled in the temperature range of 500–900 °C (Figure 9c) resembled PSMs, where plastic deformation is expected to occur in the matrix channels. An increase in the temperature range (300–900 °C) showed long and narrow PSMs, suggesting that precipitate shearing was dominant [27].

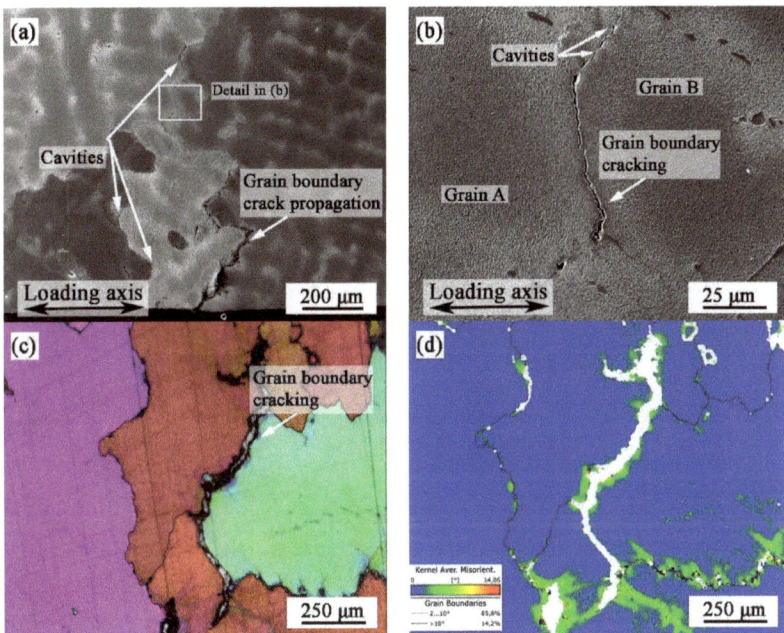

Figure 8. Typical TMF damage under IP loading (**a**) fatigue crack initiation at the grain boundary and intergranular fatigue crack propagation (**b**) detail of internal grain boundary damage (**c**) post-processed EBSD map (band contrast + inverse pole figure coloring) of internal grain boundary damage (**d**) KAM map showing local grain misorientation derived from EBSD data.

The number of secondary fatigue cracks after IP loading was generally lower than that after OP loading, suggesting a higher tendency for plastic strain localization, which corresponds well with the observed hysteresis loops (Figure 6). In contrast, a high number of sharp, straight, and shallow fatigue cracks, as shown in Figure 9, is characteristic of OP loading [10]. These cracks are significant stress concentrators and are extremely dangerous, especially with a decrease in temperature. In the connection with higher maximal tensile stress, sharp fatigue cracks may contribute to a reduction in fatigue life under OP loading in the temperature range of (300–900) °C (see Figure 7).

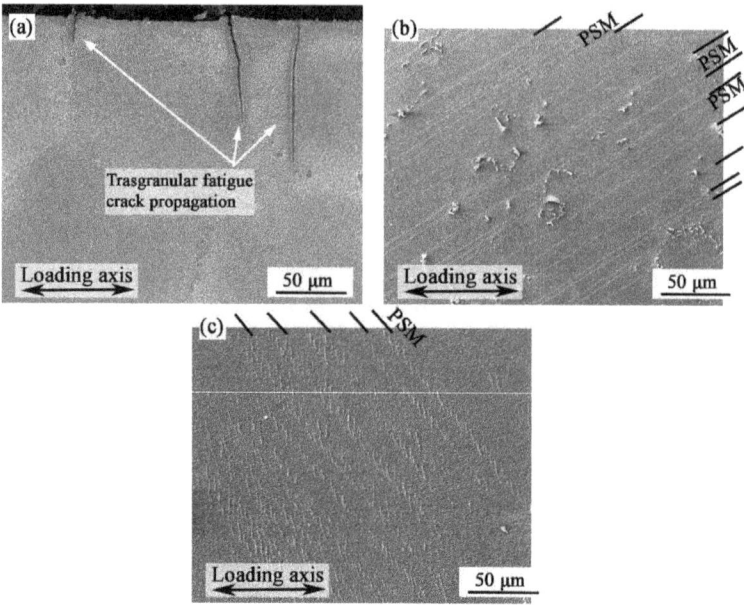

Figure 9. Typical TMF damage under OP loading (**a**) surface fatigue crack initiation and transgranular fatigue crack propagation (**b**) oxidized PSMs manifesting plastic strain localization into PSBs—temperature range 300–900 °C (**c**) wavy PSMs on the specimen surface tested in the temperature range 500–900 °C.

A TEM study was carried out on fractured specimens to investigate the dislocation structure. The STEM bright field images of the dislocation structure after IP TMF loading are presented in Figure 10. A dislocation arrangement and the morphology of γ' precipitates of a specimen cyclically strained with a mechanical strain amplitude of ε_{amech} = 0.30% in the temperature range of 300–900 °C are visible in Figure 10a. The corners of γ' particles are slightly rounded. The dislocations are restricted mainly to the γ phase and γ/γ' interfaces. Figure 10b shows the STEM image from the specimen tested in the temperature range of 500–900 °C with the same mechanical strain amplitude (0.30%). The dislocation density does not seem to be different. However, the dislocation networks with honeycomb morphology characteristic for creep or dwell-fatigue loading [15,33] can be seen (Figure 10c).

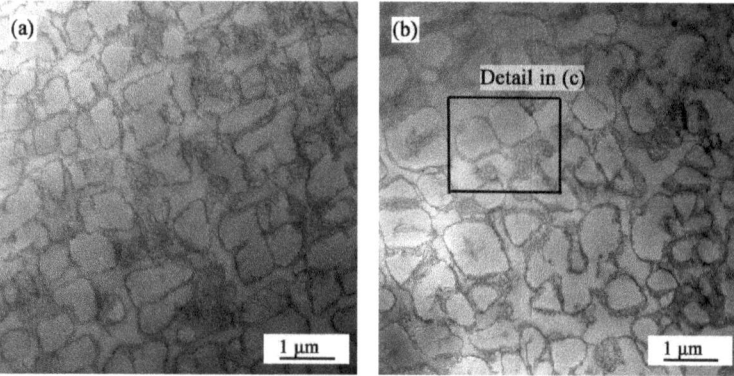

Figure 10. *Cont.*

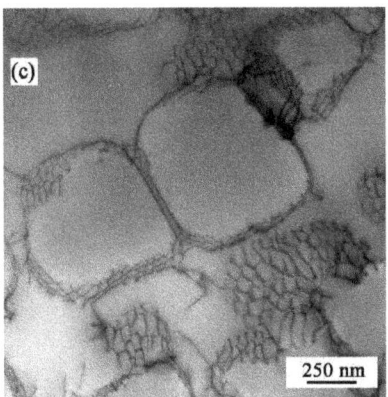

Figure 10. TEM images of dislocation arrangement after TMF tests under IP loading (**a**) 300–900 °C (**b**,**c**) 500–900 °C.

The dislocation arrangement in specimens cyclically loaded with a mechanical strain amplitude of ε_{amech} = 0.30% under OP loading conditions can be found in Figure 10. Contrary to IP TMF loading conditions where interfacial dislocations and clusters of dislocations preferably located in the γ matrix are present, under OP TMF loading, PSBs parallel to {111} slip planes developed [34]. Figure 11a presents PSBs in the form of dislocation-rich slabs for OP TMF loading in the temperature range of 300–900 °C. A detailed view of a dislocation pile-up within PBS parallel to the (111) slip plane is presented in Figure 11b. These slip bands cut both the γ matrix and the γ' precipitates. Observed PSBs are different from the typical ladder-like PSBs observed for pure copper [35]. The narrowing of the temperature range (from 300–900 °C to 500–900 °C) decreased the density of PSBs (as shown in Figure 10c), which corresponds to a lower accumulation of plastic deformation in the material (Figure 5) [20,27]. In another grain (Figure 11d), the arrangement of dislocations was different. The dislocations, similar to the IP stress cycle, are mainly distributed in the channels of the γ matrix. Nonetheless, the dislocation density here is much higher than in the IP cycling (compare with Figure 10). Thus, we can conclude that a higher dislocation density in the γ matrix channels may, in effect, lead to the formation of wavy PSMs (Figure 9c) [15,21].

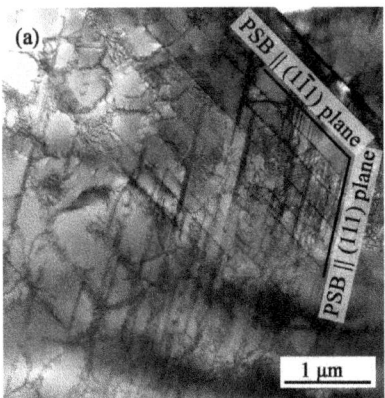 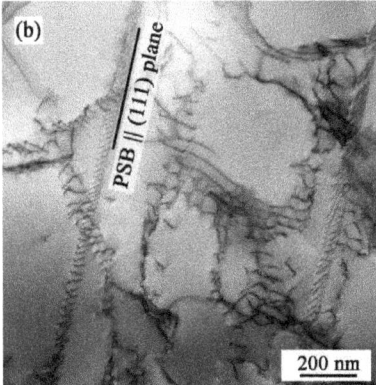

Figure 11. *Cont.*

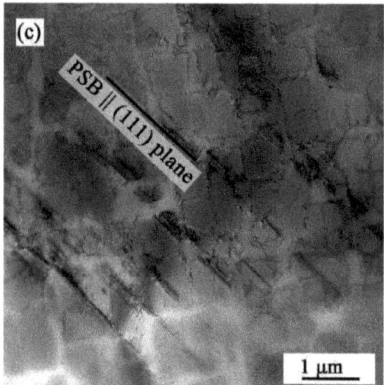
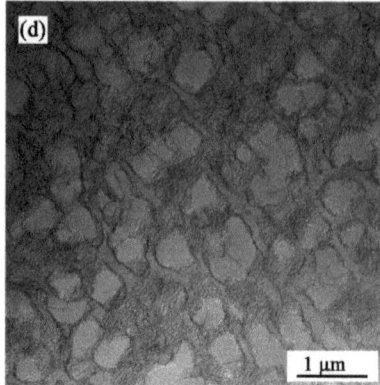

Figure 11. TEM images of dislocation arrangement after TMF tests under OP loading (**a,b**) 300–900 °C, (**c,d**) 500–900 °C.

4. Conclusions

In summary, IP and OP TMF tests within the temperature ranges of 300–900 °C and 500–900 °C were carried out to investigate the cyclic deformation behavior and associated microstructure damage mechanisms, as well as the cracking behavior of the Inconel 713LC superalloy. The following conclusions can be drawn:

(1) A stable stress response is typical for TMF cycling with T_{min} of 500 °C. The shift in T_{min} to 300 °C supported the hardening mechanisms, and cyclic hardening occurred.
(2) The fatigue life decreased with the increase in the TMF temperature range. The decrease is more pronounced for OP loading than for IP loading.
(3) Fatigue cracks initiated from the surface regardless of the TMF conditions.
(4) Transgranular fatigue crack propagation is typical for OP loading whereas cavity formation along grain boundaries under IP loading conditions boosted intergranular fatigue crack propagation.
(5) Dislocation arrangement for IP and OP loading differed. Under OP TMF loading, dislocation-rich bands cutting both γ' precipitates and γ matrix channels formed along {111} slip planes, but under IP TMF loading, honey-comb structures restricted to γ matrix channels typical of creep damage were found.

Author Contributions: Conceptualization, I.Š. and K.O.; methodology, I.Š.; investigation, I.Š.; resources, K.O. and K.H.; data curation, I.Š. and K.O.; writing—original draft preparation, I.Š.; writing—review and editing, K.O. and K.H.; visualization, I.Š.; supervision, K.O. and K.H; funding acquisition, K.H. All authors have read and agreed to the published version of the manuscript.

Funding: The financial support from the Czech Academy of Science is acknowledged.

Data Availability Statement: The raw/processed data required to reproduce these findings are available upon the request.

Acknowledgments: The authors would like to express their gratitude for the support of the Czech Academy of Sciences. The authors would like to express special thanks to the Hans Jürgen Christ for his lifelong work and great scientific contribution in the field of thermomechanical fatigue.

Conflicts of Interest: The authors declare no conflict of interest.

References

1. Reed, R.C. *The Superalloys: Fundamentals and Applications*, 1st ed.; Cambridge University Press: Cambridge, UK, 2008; ISBN 9780521070119.
2. Mallet, O. Influence of Thermal Boundary Conditions on Stress-Strain Distribution Generated in Blade-Shaped Samples. *Int. J. Fatigue* **1995**, *17*, 129–134. [CrossRef]
3. Han, J.-C. Recent Studies in Turbine Blade Cooling. *Int. J. Rotating Mach.* **1900**, *10*, 517231. [CrossRef]

4. Galetz, M.C. Coatings for Superalloys. In *Superalloys*; IntechOpen: London, UK, 2015. [CrossRef]
5. Obrtlík, K.; Čelko, L.; Chráska, T.; Šulák, I.; Gejdoš, P. Effect of Alumina-Silica-Zirconia Eutectic Ceramic Thermal Barrier Coating on the Low Cycle Fatigue Behaviour of Cast Polycrystalline Nickel-Based Superalloy at 900 °C. *Surf. Coat. Technol.* **2017**, *318*, 374–381. [CrossRef]
6. Šulák, I.; Obrtlík, K.; Čelko, L.; Chráska, T.; Jech, D.; Gejdoš, P. Low Cycle Fatigue Performance of Ni-Based Superalloy Coated with Complex Thermal Barrier Coating. *Mater. Charact.* **2018**, *139*, 347–354. [CrossRef]
7. Wang, R.; Jing, F.; Hu, D. In-Phase Thermal–Mechanical Fatigue Investigation on Hollow Single Crystal Turbine Blades. *Chin. J. Aeronaut.* **2013**, *26*, 1409–1414. [CrossRef]
8. Hong, H.U.; Yoon, J.G.; Choi, B.G.; Kim, I.S.; Yoo, Y.S.; Jo, C.Y. Localized Microtwin Formation and Failure during Out-of-Phase Thermomechanical Fatigue of a Single Crystal Nickel-Based Superalloy. *Int. J. Fatigue* **2014**, *69*, 22–27. [CrossRef]
9. Johansson, S.; Moverare, J.; Leidermark, D.; Simonsson, K.; Kanesund, J. Investigation of Localized Damage in Single Crystals Subjected to Thermalmechanical Fatigue (TMF). *Procedia Eng.* **2010**, *2*, 657–666. [CrossRef]
10. Guth, S.; Doll, S.; Lang, K.-H. Influence of Phase Angle on Lifetime, Cyclic Deformation and Damage Behavior of Mar-M247 LC under Thermo-Mechanical Fatigue. *Mater. Sci. Eng. A* **2015**, *642*, 42–48. [CrossRef]
11. Deng, W.; Xu, J.; Hu, Y.; Huang, Z.; Jiang, L. Isothermal and Thermomechanical Fatigue Behavior of Inconel 718 Superalloy. *Mater. Sci. Eng. A* **2019**, *742*, 813–819. [CrossRef]
12. Guth, S.; Lang, K.-H. Influence of Dwell Times on Microstructure, Deformation and Damage Behavior of NiCr22Co12Mo9 under Thermomechanical Fatigue. *Mater. Sci. Eng. A* **2020**, *794*, 139970. [CrossRef]
13. Yamazaki, Y.; Miura, M. Effect of Temperature Condition on Short Crack Propagation in a Single Crystal Ni-Base Superalloy under Thermomechanical Fatigue. *Procedia Struct. Integr.* **2019**, *19*, 538–547. [CrossRef]
14. Azadi, M.; Azadi, M. Evaluation of High-Temperature Creep Behavior in Inconel-713C Nickel-Based Superalloy Considering Effects of Stress Levels. *Mater. Sci. Eng. A* **2017**, *689*, 298–305. [CrossRef]
15. Šulák, I.; Obrtlík, K.; Hutařová, S.; Juliš, M.; Podrábský, T.; Čelko, L. Low Cycle Fatigue and Dwell-Fatigue of Diffusion Coated Superalloy Inconel 713LC at 800 °C. *Mater. Charact.* **2020**, *169*, 110599. [CrossRef]
16. Kunz, L.; Lukáš, P.; Konečná, R. High-Cycle Fatigue of Ni-Base Superalloy Inconel 713LC. *Int. J. Fatigue* **2010**, *32*, 908–913. [CrossRef]
17. Kunz, L.; Lukáš, P.; Konečná, R.; Fintová, S. Casting Defects and High Temperature Fatigue Life of IN 713LC Superalloy. *Int. J. Fatigue* **2012**, *41*, 47–51. [CrossRef]
18. Boutarek, N.; Saïdi, D.; Acheheb, M.A.; Iggui, M.; Bouterfaïa, S. Competition between Three Damaging Mechanisms in the Fractured Surface of an Inconel 713 Superalloy. *Mater. Charact.* **2008**, *59*, 951–956. [CrossRef]
19. Hähner, P. *Validated Code-of-Practice for Strain-Controlled Thermo-Mechanical Fatigue Testing*; Office for Official Publications of the European Communities: Luxembourg, 2006; ISBN 9279022164.
20. Polák, J. *Cyclic Plasticity and Low Cycle Fatigue Life of Metals*, 2nd ed.; Academia: Prague, Czech Republic, 1991.
21. Šulák, I.; Obrtlík, K. AFM, SEM and TEM Study of Damage Mechanisms in Cyclically Strained Mar-M247 at Room Temperature and High Temperatures. *Theor. Appl. Fract. Mech.* **2020**, *108*, 102606. [CrossRef]
22. Petráš, R.; Šulák, I.; Polák, J. The Effect of Dwell on Thermomechanical Fatigue in Superaustenitic Steel Sanicro 25. *Fatigue Fract. Eng. Mater. Struct.* **2021**, *44*, 673–688. [CrossRef]
23. Boismier, D.A.; Sehitoglu, H. Thermo-Mechanical Fatigue of Mar-M247. Part 1. Experiments. *J. Eng. Mater. Technol. Trans. ASME* **1990**, *112*, 68–79. [CrossRef]
24. Šulák, I.; Obrtlík, K. Effect of Tensile Dwell on High-Temperature Low-Cycle Fatigue and Fracture Behaviour of Cast Superalloy MAR-M247. *Eng. Fract. Mech.* **2017**, *185*, 92–100. [CrossRef]
25. Zauter, R.; Christ, H.J.; Mughrabi, H. Some Aspects of Thermomechanical Fatigue of AISI 304L Stainless Steel: Part I. Creep-Fatigue Damage. *Metall. Mater. Trans. A* **1994**, *25*, 401–406. [CrossRef]
26. Antolovich, S.D.; Armstrong, R.W. Plastic Strain Localization in Metals: Origins and Consequences. *Prog. Mater. Sci.* **2014**, *59*, 1–160. [CrossRef]
27. Antolovich, S.D. Microstructural Aspects of Fatigue in Ni-Base Superalloys. *Phil. Trans. R. Soc. A* **2015**, *373*, 20140128. [CrossRef] [PubMed]
28. Babinský, T.; Kuběna, I.; Šulák, I.; Kruml, T.; Tobiáš, J.; Polák, J. Surface Relief Evolution and Fatigue Crack Initiation in René 41 Superalloy Cycled at Room Temperature. *Mater. Sci. Eng. A* **2021**, *819*, 141520. [CrossRef]
29. Mughrabi, H. Cyclic Slip Irreversibilities and the Evolution of Fatigue Damage. *Metall. Mater. Trans. B* **2009**, *40*, 431–453. [CrossRef]
30. Seidametova, G.; Vogt, J.-B.; Proriol Serre, I. The Early Stage of Fatigue Crack Initiation in a 12%Cr Martensitic Steel. *Int. J. Fatigue* **2018**, *106*, 38–48. [CrossRef]
31. Man, J.; Petrenec, M.; Obrtlík, K.; Polák, J. AFM and TEM Study of Cyclic Slip Localization in Fatigued Ferritic X10CrAl24 Stainless Steel. *Acta Mater.* **2004**, *52*, 5551–5561. [CrossRef]
32. Petrenec, M.; Polák, J.; Obrtlík, K.; Man, J. Dislocation Structures in Cyclically Strained X10CrAl24 Ferritic Steel. *Acta Mater.* **2006**, *54*, 3429–3443. [CrossRef]
33. Heep, L.; Schwalbe, C.; Heinze, C.; Dlouhy, A.; Rae, C.M.F.; Eggeler, G. Dislocation Networks in Gamma/Gamma′-Microstructures Formed during Selective Laser Melting of a Ni-Base Superalloy. *Scr. Mater.* **2021**, *190*, 121–125. [CrossRef]

34. Petrenec, M.; Obrtlik, K.; Polak, J. Dislocation Arrangements in Cyclically Strained Inconel 713LC. In *Fracture of Nano and Engineering Materials and Structures*; Springer: Dordrecht, The Netherlands, 2006; pp. 883–884.
35. Wang, R.; Mughrabi, H.; McGovern, S.; Rapp, M. Fatigue of Copper Single Crystals in Vacuum and in Air I: Persistent Slip Bands and Dislocation Microstructures. *Mater. Sci. Eng.* **1984**, *65*, 219–233. [CrossRef]

Communication

Revealing the Role of Cross Slip for Serrated Plastic Deformation in Concentrated Solid Solutions at Cryogenic Temperatures

Aditya Srinivasan Tirunilai [1], Klaus-Peter Weiss [1], Jens Freudenberger [2,3], Martin Heilmaier [1] and Alexander Kauffmann [1,*]

[1] Karlsruhe Institute of Technology (KIT), Engelbert-Arnold-Str. 4, D-76131 Karlsruhe, Germany; aditya.tirunilai@kit.edu (A.S.T.); klaus.weiss@kit.edu (K.-P.W.); martin.heilmaier@kit.edu (M.H.)
[2] Leibniz Institute for Solid State and Materials Research Dresden (IFW Dresden), Helmholtzstr. 20, D-01069 Dresden, Germany; j.freudenberger@ifw-dresden.de
[3] Institute of Materials Science, Technische Universität Bergakademie Freiberg, Gustav-Zeuner-Str. 5, D-09599 Freiberg, Germany
[*] Correspondence: alexander.kauffmann@kit.edu; Tel.: +49-721-608-42346

Abstract: Serrated plastic deformation is an intense phenomenon in CoCrFeMnNi at and below 35 K with stress amplitudes in excess of 100 MPa. While previous publications have linked serrated deformation to dislocation pile ups at Lomer–Cottrell (LC) locks, there exist two alternate models on how the transition from continuous to serrated deformation occurs. One model correlates the transition to an exponential LC lock density–temperature variation. The second model attributes the transition to a decrease in cross-slip propensity based on temperature and dislocation density. In order to evaluate the validity of the models, a unique tensile deformation procedure with multiple temperature changes was carried out, analyzing stress amplitudes subsequent to temperature changes. The analysis provides evidence that the apparent density of LC locks does not massively change with temperature. Instead, the serrated plastic deformation is likely related to cross-slip propensity.

Keywords: cryogenic deformation; serrations; high-entropy alloys

1. Introduction

Metallic materials have shown serrated plastic deformation at different temperatures [1–4]. Recent publications highlight especially intense serrated deformation in high-entropy alloys (HEA) [5–8]. Equiatomic CoCrFeMnNi even exhibits low temperature serrations at 35 K, a temperature higher than previously reported for any other metal or alloy [6]. While a few different hypotheses exist to explain low temperature serrations [9–11], recent results led to a phenomenological model based on dislocation pile ups at LC locks [5,6]. Alternate hypotheses, related to a thermomechanical [9], twinning or martensite-based instability [11] were previously invalidated and are neglected in this communication [5–7].

The model presented in Ref. [6] extends the work of Seeger [10], based on screw dislocation immobility in close-packed crystals at low temperatures [12]. As mobile screw dislocations interact with forest dislocations, they form immobile jogs [12,13]. This restriction is not seen for edge dislocations, for a more detailed view on the differences between edge and screw interaction with forest dislocations refer to Figure 6 in Ref. [6]. The motion of these jogged screw dislocations is associated with vacancy formation [12,13] which becomes increasingly difficult with decreasing temperature. Thus, at temperatures close to 0 K, the motion of jogged screw dislocations is restricted so much that it compromises cross-slip propensity. Correspondingly, as opposed to dislocations cross-slipping out of pile ups at barriers such as LC locks, they would activate dislocation sources on the other side of the pile up as a result of the stress-field at its head [10]. This proliferation and motion of dislocations leads to a macroscopic stress drop, seen as a serration. Ref. [6]

extends the model, accounting for cross-slip propensity as a function of temperature and dislocation density, along with an explanation for transition from continuous to serrated plastic deformation as temperature decreases. In contrast, the original Seeger model [10,14] considers only a critical temperature condition associated with serrations, establishing that at a given temperature, deformation would either be continuous or serrated in nature. It cannot explain initially continuous and subsequently discontinuous deformation, as seen in CoCrFeMnNi for example [6]. This is circumvented by either considering (i) the aforementioned cross-slip propensity variation [6] or, alternatively, (ii) assuming an exponential LC lock density–temperature variation as proposed in Ref. [12]. The present work investigates the validity of these two possibilities by the application of a specific mechanical testing scheme at low temperatures [6,15]. The in situ observation of cross-slip events or the LC lock density as fundamental requisites of the two alternatives exhibits multiple complications when performed at such low temperatures, while microstructural investigations post deformation suffer from partial recovery of dislocation features on heat-up. These issues are presently avoided with a novel interrupted tensile test where serration characteristics are evaluated through stress drop amplitude ($\Delta\sigma_e$). The details of this test are as described in the following section and have been designed specifically for this investigation. Correspondingly, issues that cannot be avoided in other methods are avoided.

2. Materials and Methods

CoCrFeMnNi was synthesized by arc melting of high-purity elements. It was cast, homogenized at 1200 °C for 72 h, rotary swaged, machined and annealed at 800 °C for 1 h. For a detailed overview, please refer to Ref. [16].

Interrupted tensile deformation was carried out between 25 K and 8 K in a sealed chamber with He vapor at ~50 mbar, with multiple interruptions (sequence of events and associated temperatures and strains stated in later sections). The machine used for this was the MTS25 (MTS Systems, Eden Prairie, USA) with a maximum load of 25 kN [17]. The extension was measured by a pair of clip-on extensometers. The specimens had a cylindrical gauge section of 22 mm in length and 4 mm in diameter. Tensile testing was performed at a constant crosshead speed equaling an initial plastic strain rate of $\sim 3 \times 10^{-4}\ s^{-1}$. The strain was measured using two strain gauges within the gauge length of the specimens. Data analysis was carried out using force, time and elongation results through the proprietary software packages Origin 2020b by OriginLab and MATLAB R2018a (MathWorks). For more information, please refer to Ref. [6].

3. Results and Discussion

3.1. Tensile Tests up to Fracture Model Considerations

Figure 1a shows serrated plastic deformation of CoCrFeMnNi at 8, 15 and 25 K, as seen in the engineering stress–strain ($\sigma_e - \varepsilon_e$) curves for tensile tests. Corresponding $\Delta\sigma_e$ were determined from the difference between stress maxima and minima for each serration (Figure 1b). As deformation continues, the intensity of serrations as measured by $\Delta\sigma_e$ increases. Additionally, serrated plastic deformation is initiated at a lower strain at lower temperatures.

According to the model of serrated plastic deformation described above, dislocations pile up at LC locks; at low temperatures, dislocation sources are activated and massive dislocation proliferation events take place at the heads of the pile ups, seen as macroscopic stress drops [10]. This is only noted at low temperatures since mobility of intersected screw dislocations and cross slip is restricted for close-packed crystals at very low temperatures [10,12]. However, this model proved insufficient in explaining results where deformation was continuous at lower strains and discontinuous at larger strains, since the condition was based only on temperature. Different alloys have shown a transition from continuous, to partially serrated and finally fully serrated deformation as temperature decreases in the range T < 50 K [6,14]. Skoczeń et al. [15] proposed that the LC lock density increases exponentially with decreasing temperature to satisfy these results. Correspondingly, the number of pile ups and dislocation proliferation events would increase with decreasing temperature. Alternatively, multiple factors affecting cross slip were instead considered to explain these results in Ref. [6]. These two more recent versions of the model are explained below:

(i) Skoczeń et al. [15] have proposed a rapid increase in LC lock formation rate $\widetilde{F}_{LC}^+$ during deformation with decreasing temperature and assigned it the form:

$$\lg \widetilde{F}_{LC}^+ = A + B \lg T \quad (1)$$

Here, A and B are constants and T is the temperature (note: B is negative due to the inverse variation of $\widetilde{F}_{LC}^+$ and T). $\widetilde{F}_{LC}^+$ expectedly changes by orders of magnitude below 40 K [15]. A minimum strain ε_{min} is required to generate a sufficient LC lock density and dislocation pile ups which result in serrated deformation. At higher temperatures (25 K), $\widetilde{F}_{LC}^+$ is lower, thus, ε_{min} is necessarily larger. At lower temperatures (8 K), the ε_{min} instead decreases significantly due to much higher $\widetilde{F}_{LC}^+$.

(ii) Ref. [6] considers screw dislocation mobility [10], stacking-fault energy (SFE) and dislocation pile up characteristics and their effect on cross-slip propensity. Dislocation density and temperature affect this most significantly. At higher temperatures (i.e., 25 K presently), cross slip is facilitated easily and proceeds sufficiently until a critical strain (ε_{min}), where dislocation density is high enough to compromise screw dislocation mobility, subsequently resulting in serrations. At lower temperatures (8 K), cross slip is so severely restricted that even minor strain results in serrations. Notably, this model considers LC lock density to scale with dislocation density and assumes it to be similar in the temperature range of 8–25 K.

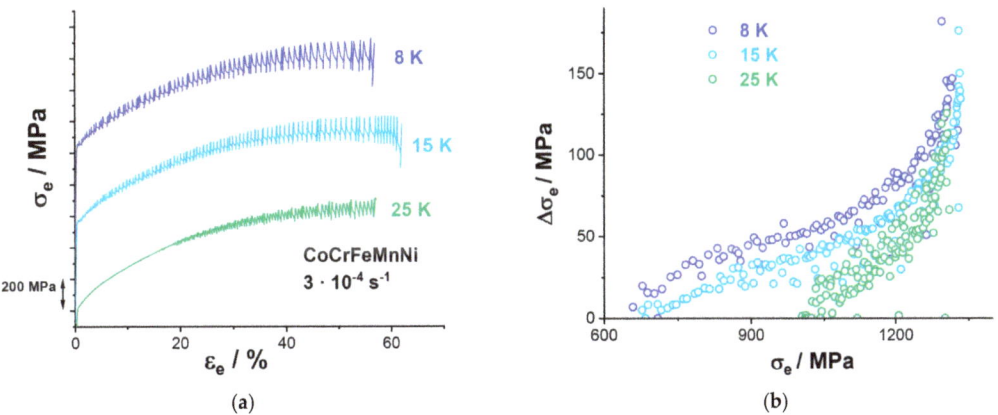

Figure 1. (a) $\sigma_e - \varepsilon_e$ plots of CoCrFeMnNi at 8, 15 and 25 K. (b) $\Delta\sigma_e - \sigma_e$ plots of serrations. Data from [6]. The data in (a) is offset along the vertical axis so that each curve can be clearly distinguished.

3.2. Expectations on Interrupted Tests from the Different Models

To test these models, a temperature change test is carried out. The tensile test is initiated at 8 K and the sample is strained to an engineering strain of approximately $\varepsilon_e \sim 10\%$ and then unloaded. The temperature is then changed to 15 K and the test continued. Stress drop amplitude vs. stress $\Delta\sigma_e - \sigma_e$ is evaluated to confirm the correct model. Regardless of the model, the initial trend at 8 K should have a positive variation, as seen in the experimental data from the uninterrupted test in Figure 1b. Post-interruption, there are two possibilities as illustrated in Figure 2.

(i) Figure 2a shows a condition where instability is met by some physical factor, e.g., minimum dislocation or LC lock density. At 8 K, the necessary LC lock density for serrated plastic deformation is achieved easily. On changing the temperature to 15 K, the LC lock density remains unaffected, hence serrated deformation proceeds onwards. However, based on the lower $\widetilde{F}_{LC}^+$, the $\Delta\sigma_e - \sigma_e$ slope may be lower than at 8 K, possibly even plateauing. This is the generalized interpretation of Ref. [15].

(ii) Alternatively, instability is controlled by cross-slip propensity and, correspondingly, dislocation density and temperature. Thus, after the interruption, $\Delta\sigma_e$ should drop significantly, since cross-slip probability is greater at 15 K than 8 K, evidenced by $\Delta\sigma_e - \sigma_e$ variation (Figure 1b). Since dislocation density increases during further deformation, $\Delta\sigma_e$ should subsequently increase consistently with σ_e (seen in Figure 2b). Thus, the important difference between the expected results is that in one model the $\Delta\sigma_e$ values pre- and post-interruption match [15] and in the other model, there is a distinct change in $\Delta\sigma_e$ [6].

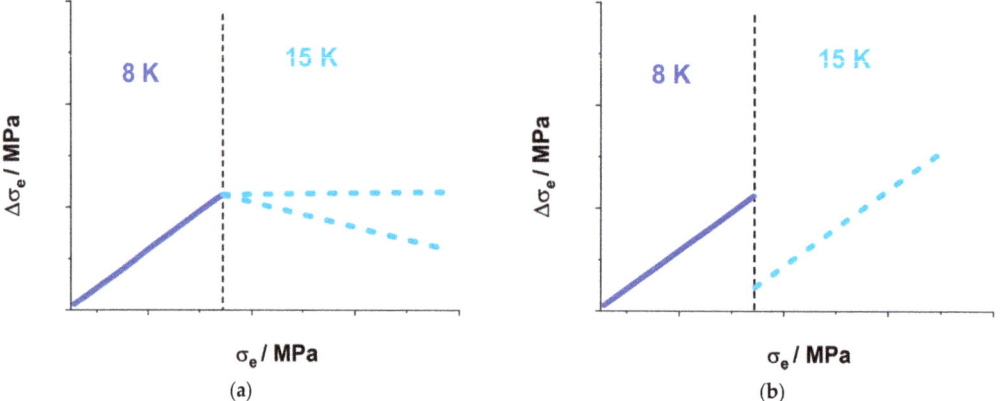

Figure 2. Schematic $\Delta\sigma_e - \sigma_e$ trend for tensile tests with an interrupted temperature change. Both (a) and (b) show possible interpretations of the temperature change based on the models adapted from [6,15], respectively.

3.3. Results of the Interrupted Tests

The temperature change tensile test consists of interruptions at $\varepsilon_e \sim$ 10%, 22% and 40%. The test begins at 8 K and then continues at 15, 25 and back at 8 K after the respective interruptions to verify consistent trends despite pre-deformation. $\sigma_e - \varepsilon_e$ and corresponding $\Delta\sigma_e - \sigma_e$ results are shown in Figure 3.

Figure 3b is clearly indicative of the trend expected for the cross-slip propensity model (compare with Figure 2b). A $\Delta\sigma_e$ drop is noted for temperature changes to both 15 and 25 K. As a final verification, reloading at 8 K would result in a drastically higher $\Delta\sigma_e$ post-interruption when considering cross slip [6] in comparison to a lower $\Delta\sigma_e$ when considering $\widetilde{F}_{LC}^+$ from Equation (1) [15]. The severe increase in $\Delta\sigma_e$ provides further evidence of a cross-slip-based mechanism.

The $\Delta\sigma_e - \sigma_e$ variation of the uninterrupted tests has been included in Figure 3b. The $\Delta\sigma_e - \sigma_e$ values are noticeably lower after the interruptions and temperature changes to 15 and 25 K. This may be explained by the stress τ_{bow} to move the unpinned portion of a jogged screw dislocation using G (shear modulus), b (Burgers vector), l_0 (mobile dislocation length between the jogs) and α (constant ~0.1–0.2) [12]:

$$\tau_{bow} = \frac{\alpha\,G\,b}{l_0} \qquad (2)$$

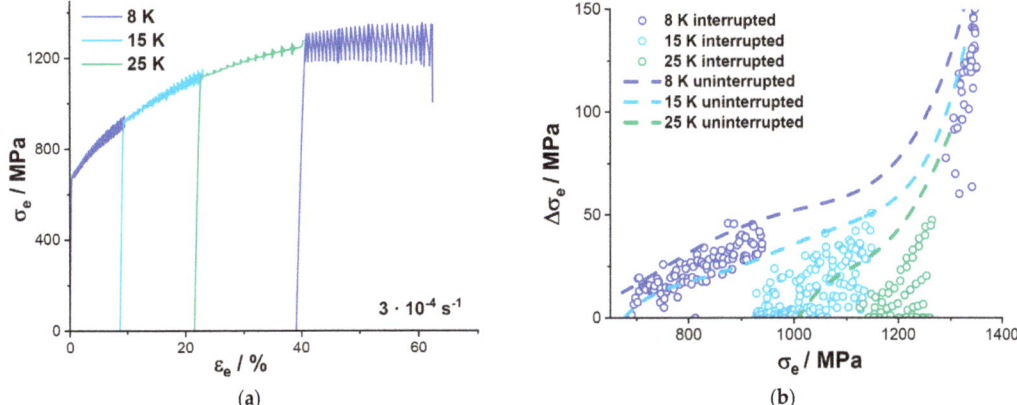

Figure 3. (a) $\sigma_e - \varepsilon_e$ for tensile test at multiple temperatures and (b) the corresponding $\Delta\sigma_e - \sigma_e$ plot. The dashed lines in (b) represent the curves for the uninterrupted test in Figure 1.

As the temperature decreases, the minimum l_0 for a mobile screw dislocation keeps increasing. At a given temperature, screw dislocations of a length less than some l_0 are immobile and at higher temperatures the minimum l_0 is shorter. Accordingly, several dislocations are likely immobile at 8 K but at 15 K the critical l_0 being shorter, the same dislocations become mobile. Combined with the higher stress, cross slip is significantly more viable subsequent to the interruption and reloading at a higher temperature, making serrations less intense. Additionally, since there is an unloading step, unstressed dislocations only partially recede from the stressed state due to dislocation–dislocation interactions. Thus, the uninterrupted and interrupted $\Delta\sigma_e - \sigma_e$ trends should be similar but offset by different states of dislocations.

In the given test, the observed drop in $\Delta\sigma_e$ can only be explained in the absence of an exponential variation of dislocation or LC lock density with temperature. Hence, this experiment reinforces the model given in Ref. [6] where cross slip based on temperature, dislocation density and dislocation mobility govern the serration behavior close to 0 K in face centered cubic alloys.

4. Conclusions

The $\Delta\sigma_e - \sigma_e$ trends observed in tensile tests with deliberate, intermediate temperature steps conducted on CoCrFeMnNi experimentally verify that low temperature serrations in face centered cubic high-entropy alloys are governed by the immobility of screw dislocations. The temperature-dependent cross-slip propensity and dislocation density throughout deformation are the relevant parameters controlling the immobility.

Author Contributions: Conceptualization, A.S.T. and A.K.; methodology, A.S.T., A.K. and K.-P.W.; software, A.S.T. and A.K.; formal analysis, A.S.T. and A.K.; investigation, A.S.T., J.F. and K.-P.W.; resources, A.K., J.F., K.-P.W. and M.H.; data curation, A.S.T. and A.K.; writing—original draft preparation, A.S.T. and A.K.; writing—review and editing, A.S.T., A.K., K.-P.W., J.F. and M.H.; visualization, A.S.T. and A.K.; supervision, A.K.; project administration, A.K. and K.-P.W.; funding acquisition, A.K., J.F. and K.-P.W. All authors have read and agreed to the published version of the manuscript.

Funding: Financial support by the Deutsche Forschungsgemeinschaft within the framework of the Priority Program "Compositionally Complex Alloys High-Entropy Alloys (CCA-HEA)" (SPP 2006) is gratefully acknowledged, under grants no. KA 4631/1-1, FR 1714/7-1 and WE 6279/1-1. We acknowledge support by the KIT-Publication Fund of the Karlsruhe Institute of Technology.

Data Availability Statement: The data presented in this study are available on request from the corresponding author, alexander.kauffmann@kit.edu (A.K.).

Acknowledgments: We gratefully acknowledge the experimental support of V. Tschan. This article is dedicated to our collaborator and friend Hans Jürgen Christ. We appreciate our long-standing and fruitful collaborative work in the development of refractory metal-based alloys for high temperature application.

Conflicts of Interest: The authors declare no conflict of interest.

References

1. Zhang, Y.; Liu, J.P.; Chen, S.Y.; Xie, X.; Liaw, P.K.; Dahmen, K.A.; Qiao, J.W.; Wang, Y.L. Serration and noise behavior in materials. *Prog. Mater. Sci.* **2017**, *90*, 358–460. [CrossRef]
2. Pustovalov, V.V. Serrated deformation of metals and alloys at low temperatures (Review). *Low Temp. Phys.* **2008**, *34*, 683–723. [CrossRef]
3. Cottrell, A.H. LXXXVI. A note on the Portevin-Le Chatelier effect. *Lond. Edinb. Dublin Philos. Mag. J. Sci.* **1953**, *44*, 829–832. [CrossRef]
4. Lebyodkin, M.A.; Lebedkina, T.A.; Brechtl, J.; Liaw, P.K. Serrated Flow in Alloy Systems. In *High-Entropy Materials: Theory, Experiments, and Applications*; Brechtl, J., Liaw, P.K., Eds.; Springer: Cham, Switzerland, 2021; pp. 523–644. [CrossRef]
5. Tirunilai, A.S.; Sas, J.; Weiss, K.-P.; Chen, H.; Szabó, D.V.; Schlabach, S.; Haas, S.; Geissler, D.; Freudenberger, J.; Heilmaier, M.; et al. Peculiarities of deformation of CoCrFeMnNi at cryogenic temperatures. *J. Mater. Res.* **2018**, *33*, 3287–3300. [CrossRef]
6. Tirunilai, A.S.; Hanemann, T.; Weiss, K.-P.; Freudenberger, J.; Heilmaier, M.; Kauffmann, A. Dislocation-based serrated plastic flow of high entropy alloys at cryogenic temperatures. *Acta Mater.* **2020**, *200*, 980–991. [CrossRef]
7. Naeem, M.; He, H.; Harjo, S.; Kawasaki, T.; Lin, W.; Kai, J.-J.; Wu, Z.; Lan, S.; Wang, X.-L. Temperature-dependent hardening contributions in CrFeCoNi high-entropy alloy. *Acta Mater.* **2021**, *221*, 117371. [CrossRef]
8. Naeem, M.; He, H.; Zhang, F.; Huang, H.; Harjo, S.; Kawasaki, T.; Wang, B.; Lan, S.; Wu, Z.; Wang, F.; et al. Cooperative deformation in high-entropy alloys at ultralow temperatures. *Sci. Adv.* **2020**, *6*, eaax4002. [CrossRef] [PubMed]
9. Basinski, Z.S. The instability of plastic flow of metals at very low temperatures. *Proc. R. Soc. A* **1957**, *240*, 354–358. [CrossRef]
10. Seeger, A. The mechanism of glide and work hardening in Face-Centered Cubic and Hexagonal-Close Packed metals. In *Dislocations and Mechanical Properties of Crystals*; Fisher, J.C., Ed.; John Wiley & Sons, Inc.: New York, NY, USA, 1958; pp. 243–330.
11. Han, W.; Liu, Y.; Wan, F.; Liu, P.; Yi, X.; Zhan, Q.; Morrall, D.; Ohnuki, S. Deformation behavior of austenitic stainless steel at deep cryogenic temperatures. *J. Nuclear Mater.* **2018**, *504*, 29–32. [CrossRef]
12. Seeger, A. CXXXII. The generation of lattice defects by moving dislocations, and its application to the temperature dependence of the flow-stress of F.C.C. crystals. *Lond. Edinb. Dublin Philos. Mag. J. Sci.* **1955**, *46*, 1194–1217. [CrossRef]
13. Hirth, J.P.; Lothe, J. Glide of jogged dislocations. In *Theory of Dislocations*; McGraw-Hill Book Company: New York, NY, USA; St. Louis, MO, USA; San Francisco, CA, USA; Toronto, ON, Canada; London, UK; Sydney, Australia, 1968; pp. 535–556.
14. Obst, B.; Nyilas, A. Experimental evidence on the dislocation mechanism of serrated yielding in f.c.c. metals and alloys at low temperatures. *Mater. Sci. Eng. A* **1991**, *137*, 141–151. [CrossRef]
15. Skoczeń, B.; Bielski, J.; Sgobba, S.; Marcinek, D. Constitutive model of discontinuous plastic flow at cryogenic temperatures. *Int. J. Plast.* **2010**, *26*, 1659–1679. [CrossRef]
16. Tirunilai, A.S.; Hanemann, T.; Reinhart, C.; Tschan, V.; Weiss, K.-P.; Laplanche, G.; Freudenberger, J.; Heilmaier, M.; Kauffmann, A. Comparison of cryogenic deformation of the concentrated solid solutions CoCrFeMnNi, CoCrNi and CoNi. *Mater. Sci. Eng. A* **2020**, *783*, 139290. [CrossRef]
17. Sas, J.; Weiss, K.-P.; Bagrets, N. Cryomak—The overview of cryogenic testing facilities in Karlsruhe. *Acta Metall. Slovaca* **2015**, *21*, 330–338. [CrossRef]

Article

Crashworthiness Analysis of Square Aluminum Tubes Subjected to Oblique Impact: Experimental and Numerical Study on the Initial Contact Effect

Konstantina D. Karantza * and Dimitrios E. Manolakos

Laboratory of Manufacturing Technology, School of Mechanical Engineering, National Technical University of Athens, Heroon Polytechniou 9, 15780 Athens, Greece
* Correspondence: konstantinakarantza@mail.ntua.gr; Tel.: +30-210-772-3688

Abstract: This study investigates the crashworthiness behavior of square aluminum thin-walled tubes subjected to both axial and oblique impact loading, emphasizing the effects of crushing angle and initial contact between impactor and tube on the plastic collapse initiation and energy absorption capacity. A parametric study in crushing angle is conducted until 15°, while the two examined types of initial contact between impactor and tube consist of a contact-in-edge case and a contact-in-corner one, aiming to capture the effect of initial contact on both plastic collapse and energy absorption. Both experimental quasi-static tests and numerical simulation via finite element modeling in LS-DYNA software are carried out for the evaluation of the crushing response of the tested tubes. The 5° oblique cornered crushing revealed the greatest energy absorption, reflecting the most efficient loading case as significant tearing failure occurred around the tube corners in axial crushing due to a higher peak crushing force, while the increase in crushing angle caused a drop in energy absorption and peak force regarding the oblique loading. Finally, an initial contact-in-corner case revealed higher energy absorption compared to both axial and edged oblique loading, while peak force showed a slight decrease with crushing angle in that case.

Keywords: crashworthiness; square tubes; energy absorption; oblique impact; LS-DYNA; initial contact

Citation: Karantza, K.D.; Manolakos, D.E. Crashworthiness Analysis of Square Aluminum Tubes Subjected to Oblique Impact: Experimental and Numerical Study on the Initial Contact Effect. *Metals* **2022**, *12*, 1862. https://doi.org/10.3390/met12111862

Academic Editor: Ezio Cadoni

Received: 13 October 2022
Accepted: 28 October 2022
Published: 1 November 2022

Publisher's Note: MDPI stays neutral with regard to jurisdictional claims in published maps and institutional affiliations.

Copyright: © 2022 by the authors. Licensee MDPI, Basel, Switzerland. This article is an open access article distributed under the terms and conditions of the Creative Commons Attribution (CC BY) license (https://creativecommons.org/licenses/by/4.0/).

1. Introduction

Crashworthiness is a design philosophy that aims to control the extent of impact damage, thus increasing the safety levels of structures subjected to impact loading. Thin-walled structures have been widely proved as the most efficient energy-absorbing components, providing high crashworthy performance with low weight, which highlights them as the most preferable devices for crashworthy structures. In general, designing for crashworthiness aims to control the extent of impact damage by dissipating the crushing kinetic energy under a progressive collapse, converting it to plastic deformation energy for the crushed structure. Tan et al. [1] highlighted that axial and oblique impact loading are the two main crushing conditions based on the statistical data of an accident probability analysis, which revealed that axial and angular crushing modes represent almost about 35% and 36% of car crashes, respectively. In this direction, the research community has turned its interest into studying the crashworthiness response of axially or obliquely crushed structures, aiming to capture the effect of loading angle on energy absorption and failure-mode stability, while significant attention is paid to identifying a critical crushing angle, which reacts to a sharp decrease in energy-absorption capacity due to the occurrence of unstable global bending-deformation mode during plastic collapse.

Kim and Wierzbicki [2] identified two different types of oblique impact consisting of angled loading and off-axis loading, where the tube and impactor are moving vertically towards an angled crushing surface in the first case, while in the case of off-axis loading the tube and the bottom holder are obliquely positioned to the impactor at the proper crushing angle. A non-linear finite element analysis (FEA) was conducted in PAM-CRASH software, aiming to reveal the most preferable loading case with respect to crushing angle, examining square and rectangular tubes. Additionally, Reyes et al. [3] examined the behavior of aluminum square tubes subjected to off-axis oblique impact loading until a 30° crushing angle by conducting numerical simulations in LS-DYNA and utilizing experimental data for their validation procedure. The results revealed the wall thickness and the initial length as the key geometrical parameters, which mainly affected the crashworthiness performance, while a 5° critical crushing angle was captured reacting to a significant drop in energy-absorption capability due to the unstable global bending mode of collapse. Moreover, Han and Park [4] investigated numerically the crushing response of mild steel square tubes subjected to angled impact loading. Their work derived an analytical expression for mean crushing force under oblique impact, while a critical crushing angle was identified, focusing on the transition from a progressive to an unstable mode of collapse, such as Euler-type buckling, reacting thus to a significant drop in energy absorption due to the progressive decrease in crushing force during collapse.

Additionally, the crashworthiness response of more novel designs has been further investigated in order to assess the effects of geometrical characteristics and cross-section shape on energy absorption and on critical crushing angle considering the case of oblique impact. Tran et al. [5] developed a theoretical approach for energy-absorption assessment of multi-cell square tubes under oblique impact, validating their results via FEA simulations. Pirmohammad et al. [6] studied multi-cell tubes subjected to oblique quasi-static loading until an angle of 27° by carrying out FEA simulation in LS-DYNA. Different geometries were examined and the complex proportional assessment (COPRAS) method was utilized to identify the optimum geometry with respect to energy-absorption maximization, revealing that circularly multi-cell tubes proved to be the most efficient design. Further, Azimi et al. [7] studied homo-polygonal multi-cell aluminum tubes subjected to axial and oblique loading under FEA simulation in LS-DYNA, validated against experimental tests. The COPRAS method was utilized for capturing the optimum cross-section geometry and cell dimensions. The superiority of multi-cell tubes against conventional ones was highlighted, as their crashworthiness performance gains seemed to be significantly higher, especially at low crushing angles.

Additionally, the crushing response of tapered thin-walled tubes subjected to axial and oblique impact has been also investigated. Liu et al. [8] studied tapered star-shaped aluminum tubes by conducting a numerical multi-objective optimization to identify the optimum cross-section topology with respect to peak-crushing-force (PCF) minimization and specific-energy-absorption (SEA) maximization. The results showed that an almost 10% increase in SEA can be achieved by the optimal design of a star-shaped tube, while greater wall thickness and taper angle react to an increase in critical crushing angle from 10° to 15°. Further, Qi et al. [9] investigated the response of tapered square tubes by studying numerically several multi-cell configurations, indicating that multi-cell tubes revealed greater energy absorption compared to single-cell ones. Song et al. [10] conducted a numerical study on windowed tubes under oblique loading, highlighting their greater energy-absorption capacity compared to conventional tubes, while an optimum window design seemed to be capable of increasing the critical crushing angle. Moreover, the crashworthiness behavior of functionally graded thickness (FGT) tubes under oblique crushing has also been examined. Mohammadiha et al. [11] indicated that the optimal thickness distribution alongside tube length is affected by the crushing angle in the case of oblique loading, while Baykasoglou et al. [12] revealed that the FGT effect on energy absorption seems to be stronger in high crushing angles where a 93% increase in SEA can be achieved. Finally, Crutzen et al. [13] studied a beneficial wall-thickness distribution for obliquely crushed square tubes in order to avoid the occurrence of an unstable global bending mode during plastic collapse, which would reduce significantly energy-absorption capacity.

Furthermore, Bai et al. [14] studied numerically the crushing behavior of obliquely loaded novel octagonal sandwiched tubes reinforced with an internal plate. The internal plate thickness revealed a significant impact on optimum design, which was further affected by the loading angle. Additionally, the crashworthiness response of obliquely loaded polymer-fiber-reinforced tubes [15] and hybrid tubes with composite external layers [16] has been also examined for identifying the optimum design and ply orientation in order to maximize energy absorption. In addition, other novel designs such as honeycomb structures [17], lateral corrugated tubes [18], and double conical tubes [19] have been also investigated and proved to be of significant crashworthy efficiency with honeycomb structures, reaching the greatest energy-absorption capacity among the others in the case of oblique impact loading.

The effect of foam-filling has been widely studied regarding the crushing response of thin-walled structures under axial and oblique impact loading. Qi et al. [20] examined foam-filled circular and conical aluminum tubes subjected to oblique crushing by conducting FEA simulations. Foam-filled tubes revealed an improvement in PCF and SEA compared to empty tubes, while foam-filled conical tubes reached a maximum increase in SEA of 106%. Gao et al. [21] conducted a multi-objective optimization for foam-filled ellipse tubes via FEA modeling, showing a 3% drop in PCF and 27% increase in SEA compared to empty circular and square tubes. Finally, Borvik et al. [22] examined empty and foam-filled aluminum circular tubes under oblique loading until 30° via both quasi-static tests and numerical simulations in LS-DYNA. The provided results revealed that foam-filling increases crashworthiness efficiency, while the energy-absorption drop in high crushing angles seemed to be more strongly affected in the case of foam-filled tubes rather than empty ones.

The current work studies the crashworthiness behavior of thin-walled aluminum square tubes subjected to axial and oblique impact loading. Quasi-static experimental tests and FEA numerical simulations in LS-DYNA are carried out for both the validating procedure of the developed models and the evaluation of crashworthiness performance. A parametric analysis of crushing angle varying until 15° is conducted in order to investigate the effect of crushing angle on energy absorption, plastic collapse initiation, and on the stability of collapse mode. Finally, two different types of initial contact between impactor and tube are examined, containing a contact-in-edge case and a contact-in-corner one, in order to capture the effect of initial contact on plastic collapse by assessing the change in PCF and energy-absorption capability.

2. Materials and Methods

2.1. Examined Test Cases and Specimens

The current work studied the crashworthiness behavior of thin-walled square tubes subjected to axial and oblique impact loading. The investigation was carried out by conducting experimental tests in INSTRON 4482 testing machine (Instron, Norwood MA, USA) (Figure 1) and finite element numerical simulation in LS-DYNA (Livermore Software Technology Corporation, Livermore, CA, USA). The examined specimens consisted of thin-walled aluminum square tubes produced via an extrusion process. All specimens were cut at a length of 100 mm, while the dimensions of their square cross-sections were 50 mm in width and 1.5 mm in wall thickness, revealing a specimen mass of 78.57 g.

The examined cases of the tested tubes contained axial and oblique loading of 5°–10°–15° crushing angles representing the loading cases 1–4 respectively, according to which the specimens were numbered properly. All oblique loading cases were carried out representing off-axis crushing conditions with the tube and the bottom holder being positioned at the proper angle to the impactor. Further, the bottom tube end was fixedly supported to the bottom holder via an external configuration of the last one in order to avoid any sliding during collapse in the case of oblique crushing. The examination of the selected oblique loading cases aims for a parametric analysis of crushing angle in order to assess its effect on energy-absorption capacity, plastic collapse initiation, and the stability of the occurring failure mechanism during plastic deformation. Finally, in each oblique loading case, two different types of initial contact between impactor and tubular specimen are examined, containing a contact-in-edge case and a contact-in-corner one regarding the top end of tube as depicted in Figure 2 where edged and cornered initial contacts are labelled as "a" and "b" respectively. The above investigation selection aims to estimate the effect of initial contact under oblique impact loading on plastic collapse initiation and in consequence of the collapse progress and energy-absorption capability of the crushed structure.

Figure 1. Experimental test configuration.

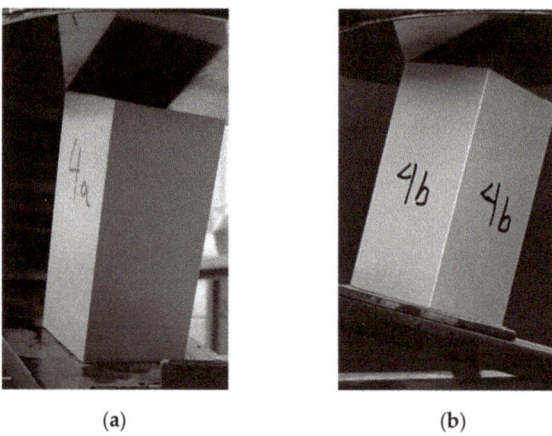

Figure 2. Type of examined initial contact between tube and impactor. (**a**) contact-in-edge case; (**b**) contact-in-corner case.

2.2. Material Characterization

The tested aluminum tubes are made of AA6060-T6 subjected to an extrusion process. In order to develop effective numerical models, the mechanical properties and the hardening behavior during plastic deformation of AA6060-T6 tube need to be identified for the material-modeling procedure. For this reason, an experimental tension test is conducted according to ASTM E8M-2004 standards in an INSTRON 4482 testing machine (Instron, Norwood, MA, USA) at room temperature under a loading rate of 10 mm/min. Although numerical simulations are examined under a significantly higher loading rate, representing the crushing conditions, the selected loading rate of the tension test is considered reliable due to the fact that AA6060-T6 has been proved as lightly sensitive to strain rate [23], and thus it can be modelled as rate-insensitive with good accuracy.

In addition, the tension test is conducted twice in order to avoid any possible material defects or data recording mismatches securing thus the validity of the provided results. Figure 3 depicts the stress–strain curve of AA6060-T6 as revealed from the processing of the measured force and recorded displacement data. Finally, the material mechanical properties such as density, Poisson ratio, and Young's modulus are estimated according to open literature data, while yield stress, ultimate stress, plastic strain failure, and the stress-plastic strain hardening curve are extracted from the provided tension test curves as summarized at Tables 1 and 2.

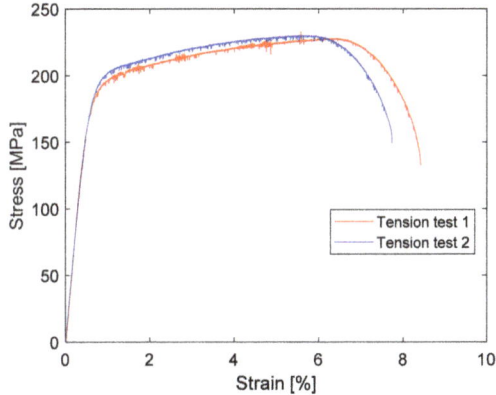

Figure 3. AA6060-T6 material stress–strain curve from tension tests.

Table 1. AA6060-T6 material properties.

Description	Variable	Value
Density (kg/m^3)	ρ	2700
Young modulus (GPa)	E	70
Poisson ratio (-)	ν	0.33
Yield stress (MPa)	σ_Y	180
Ultimate tensile strength (MPa)	UTS	229
Failure plastic strain (%)	ε_{pf}	7.93

Table 2. AA6060-T6 stress-plastic strain hardening curve.

Stress, σ (MPa)	Plastic Strain, ε_p (%)
180	0.00
200	0.40
208	1.05
216	2.25
220	2.84
225	3.92
228	4.96
229	5.77

2.3. Quasi-Static Compression Tests

All experimental compression tests were conducted in an INSTRON 4482 dual-column testing machine (Instron, Norwood, MA, USA) under quasi-static conditions of 10 mm/min loading rate. In all test cases, the maximum specimen shortening reached about 60 mm, while in the case of oblique loading, the tube and the bottom holder were rotated to the proper angle, representing off-axis oblique loading conditions. Furthermore, in the case of oblique loading, the bottom tube end was fixedly supported to the bottom holder in order to avoid any sliding during plastic collapse. For each experimental test, a proper data recording system was utilized for force (F) and displacement (x) measurements in order to provide the respective experimental F-x curve, which is taken into consideration for both crashworthiness performance evaluation of the examined tubes and the validating procedure of the developed finite element models. Moreover, different states of plastic collapse are captured during the specimen-deformation process in order to identify the failure mechanism occurring and its formation characteristics. Finally, the experimental results regarding crashworthiness response parameters and collapse mode are set into comparison with the numerical results for validating the accuracy of the developed models, although the examined strain rate of quasi-static tests and FEA simulations differ from each other, as AA6060-T6 has been described as strain-rate insensitive with sufficient accuracy [23].

2.4. Crashworthiness Response Parameters

This study utilizes several widely used crashworthiness response parameters regarding energy-absorption capacity, plastic collapse initiation, and crushing efficiency in order to evaluate the crashworthiness performance of the examined tubes. The energy-absorption mechanism of thin-walled tubes contains the bending energy, which is dissipated by the bending of rotated folds and the membrane energy, which in turn is dissipated by the extension of formulated plastic folds [24]. In more specificity, the utilized crashworthiness indicators include peak crushing force (PCF), mean crushing force (MCF), energy absorption (EA), specific energy absorption (SEA), and crushing force efficiency (CFE). The assessment of the above crashworthiness response parameters is carried out via the provided force-displacement curves.

EA refers to total energy absorption, which is dissipated during plastic collapse as the crushing kinetic energy is transformed into plastic deformation energy. Considering F(x) as the instantaneous crushing force and d as the maximum impactor displacement, EA is computed as the total area below the force-displacement curve as depicted in the following:

$$EA = \int_0^d F(x)dx \qquad (1)$$

However, a more reliable indicator for assessing the energy-absorption capacity of structures is SEA, which expresses the absorbed energy per unit mass of the crushed structure (m). Thus, SEA reflects a more indicative parameter than EA for comparing the crashworthiness performance for structures of different material, dimensions, and cross-section geometry. Thus, for a crushed structure of ρ material density, A cross-sectional area, and d maximum crushing shortening, SEA is expressed as follows:

$$SEA = \frac{EA}{m} = \frac{EA}{\rho \cdot A \cdot d} \qquad (2)$$

Regarding the crushing force indicators, PCF and MCF contain the two metrics, which reflect the plastic collapse initiation and the energy-absorption capacity, respectively. In more detail, PCF refers to maximum crushing force required for plastic collapse initiation and is responsible for the initial formulation of the first plastic convolution, while MCF is defined as the ratio of energy absorption to the maximum impactor displacement. In fact, MCF represents a constant sustained force during post-buckling region of force-displacement curve in which the plastic collapse progresses, formulating local force peaks and lows reflecting the formulation of external and internal folds, respectively. Thus, PCF and MCF can be expressed, respectively, as:

$$PCF = \max \{F(x)\} \qquad (3)$$

$$\text{MCF} = \frac{\text{EA}}{\text{d}} \qquad (4)$$

Finally, CFE is defined as the ratio of MCF to PCF where MCF reflects the mean sustained force under which the plastic collapse progresses by folding deformation. Thus, considering PCF as the maximum crushing force, the CFE parameter reflects the uniformity of crushing force fluctuation and is expressed as:

$$\text{CFE} = \frac{\text{MCF}}{\text{PCF}} \qquad (5)$$

Thus, the desirable characteristics of an efficient energy absorber include high enough EA and SEA revealing a sufficiently high energy-absorption capability under low weight. Similarly, high MCF will allow for high EA levels, while PCF must be sufficiently high to allow for high crushing-force levels, restricted however by a reasonable upper limit, which must be reached by the crushing force in order to deform plastically the crushed structure, as in the opposite case the level of absorbed energy will be negligible due to inelastic deformation.

2.5. Finite Element Modeling

For the crashworthiness response evaluation of the examined tubes, numerical simulations are also carried out utilizing non-linear explicit dynamic LS-DYNA code [25] as the modeling tool in this study. Finite element (FE) models are developed for the examined cases of axial and oblique impact loading in order to investigate the crashworthiness behavior and capture the effect of crushing angle and initial type of contact between tube and impactor. Therefore, for each simulated oblique impact loading case, two different FE models are created, representing each examined initial contact type between impactor and tube, as Figure 4 depicts regarding edged and cornered oblique crushing conditions. For the development of the FE models, the geometry and the dimensions are initially defined regarding the parts from which the tested configuration consists of, including the tube specimen, bottom holder, and impactor. Next, an element mesh is generated, selecting the type of finite elements and the mesh density, while following this the material properties are adjusted properly. After this, the boundary conditions are defined regarding the tube-end support, constraining properly its nodal degrees of freedom (DOFs), while also boundary conditions for avoiding any penetration phenomena are considered for the interface contacts. The modeling procedure is completed by determining the loading characteristics and the time termination.

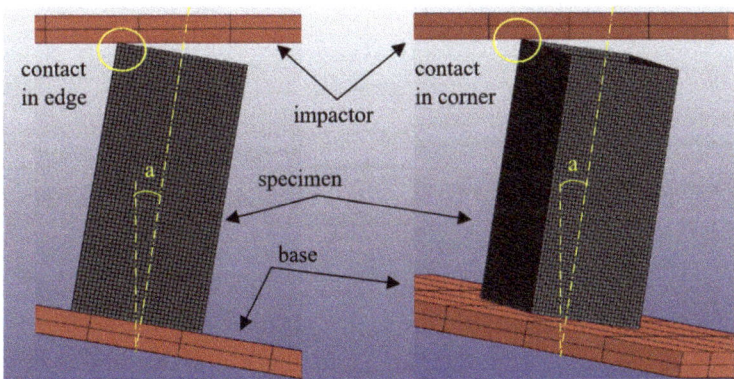

Figure 4. FE models for oblique impact loading under an edged (**left**) and a cornered (**right**) initial contact.

More specifically, thin-walled tested tubes are modelled via four-node shell elements as their accuracy in predicting the failure mechanism regarding the collapse mode and the number of formulated folds has been proved more reliable than others in the case of thin-walled structures [26]. In contrast, impactor and bottom-holder base are modelled via eight-node solid elements, as they are treated as compact rigid bodies. In addition, the shell element tubular square cross-section is dimensioned according to its mean circumference, with a width of 48.5 mm, while the finite element mesh density is generated properly, such that the shell element dimensions are adjusted at 1.5×1.5 mm, as a dimensioning just about equal to the tube-wall thickness provides reliable results regarding the fold formulation of thin-walled structures [27]. Further, for the shell element formulation, the Belytschko–Lin–Tsay formula with five integration points through shell thickness is considered. The proposed element formulation model is based on the Reissner–Mindlin kinematic assumption [28], which takes into account the superposition of mid-surface displacements and rotations to express plate deformation, considered as thus more suitable for shell elements. Additionally, a viscous and stiffness hourglass control is adjusted according to the Flanagan–Belytschko stiffness formula under an hourglass coefficient of 0.1, thus avoiding an hourglass formation of elements, which results in zero energy-deformation modes and volumetric blocking [7], which could bring instabilities during numerical solution.

For the material modeling, the AA6060-T6 tube material is approached via an isotropic elastic-plastic model utilizing the 'Mat024 piecewise linear plasticity' keyword of LS-DYNA, which is capable of capturing AA6060-T6 linearly hardening behavior sufficiently. In specific, the 'Mat024' keyword utilizes the material properties such as density, Poisson ratio, and Young's modulus, which are introduced according to the data listed in Table 1, and considers further AA6060-T6's linear hardening behavior, which is identified by a number of stress-plastic strain points, according to the results from the AA6060-T6 tension test in Table 2. In addition, the 'Mat024' model is capable of accounting for failure criteria and the strain-rate effect [24], which however is not implemented in this study, as AA6060-T6 has been described as strain-rate insensitive [23] with sufficient accuracy. In contrast, a failure plastic strain of 7.93% according to Table 1's results is implemented during FE material modeling in order to capture the tearing failure around tube corners during their plastic collapse, as revealed in some of the experimental tests. Thus, this study considers aluminum-tube hardening behavior with only the effect of plastic strain implementing in addition a failure plastic strain penalty. Regarding the steel plates of the impactor and bottom holder base, the 'Mat020 rigid' keyword is utilized as both impactor and bottom base are of significantly higher mass and stiffness and are considered thus as undeformable and rigid bodies. For 'Mat020' model of impactor and bottom holder, steel material properties are considered, such as 7830 kg/m^3 density, 200 GPa Young's modulus and a Poisson ration of 0.3, while for each body the kinematic DOFs are also adjusted properly, allowing only a vertical displacement for impactor and constraining each kinematic DOF for the bottom holder base, which is stationary during the test.

Following this, the boundary conditions for the interface contacts are implemented in order to prevent from any penetration between the interacting structural members. Thus, at first a 'nodes-to-surface' contact algorithm between tube and rigid bodies of impactor and holder base is adjusted in order to avoid any penetration of the nodes of the tube shell elements with the surface of the impactor and bottom base. The 'nodes-to-surface' contact algorithm implements a penalty formulation that allows the separate definition of the tube slave nodes and the master contacting surfaces of rigid impactor and base, thus preventing any penetration in the interfaces. Additionally, Coulomb friction conditions are considered by applying static and dynamic friction coefficients of 0.61 and 0.47, respectively, according to open literature data for aluminum–steel interface contacts. In addition, an 'automatic single surface' contact algorithm is further implemented to detect self-interaction of tube shell elements contacting each other during the formulation of folds, as the plastic collapse progresses. For this reason, shell elements' nodal normal projections are used by the 'automatic single surface' algorithm in order to prevent tube elements from penetrating the specimen surface. Coulomb friction static and dynamic coefficients are adjusted to equal 1.2 and 1.4, respectively, according to open literature data for aluminum–aluminum interface contacts. More, boundary conditions regarding the fixed bottom tube end are also adjusted, constraining the bottom tube nodal DOFs against any displacement and rotation.

Completing the FE-modeling procedure, the loading characteristics of each examined test case are simulated by applying a constant loading rate of 1 m/s until 60 mm of maximum impactor vertical displacement, while the examined crushing angle is implemented by rotating tube and bottom holder base to the proper angle. The significantly higher loading rate of FEA simulations compared to that of quasi-static tests is not only due to the strain-rate insensitivity of AA6060-T6, but also mainly due to the fact that the explicit time integration method, which is utilized during numerical solution, requires a reasonable time step, revealing thus simultaneously reliable results and avoiding extremely large calculation times. Finally, a time step of 1 ms is adjusted for recording the force-displacement data and the collapse states of the FEA simulations.

3. Results and Discussion
3.1. Modeling Verification

At first, a comparison between experiments and FE models was carried out in order to secure the validity of the developed models for accounting for both numerical and experimental results (Table 3) in the evaluation of the crashworthiness performance of the square tubes in terms of energy-absorption capacity and plastic collapse initiation. Thus, both tests and simulations showed a sufficient agreement in crushing force fluctuation during collapse, as the respective force-displacement curves depict in all examined cases shown in Figures 5–8. As a result, a sufficient agreement in the predicted PCF and EA was revealed too, providing deviations between tests and simulations below 7.6% and 6.8%, respectively, as Table 4 depicts.

Table 3. Experimental and numerical results for crashworthiness parameters.

Loading Case	Method	PCF (kN)	MCF (kN)	EA (kJ)	SEA (kJ/kg)	CFE (-)
0°	Experiment	44.02	15.28	0.913	19.36	0.347
	Simulation	43.96	14.74	0.885	18.76	0.335
5°—edge	Experiment	25.21	14.90	0.894	18.97	0.591
	Simulation	23.30	14.63	0.878	18.62	0.628
5°—corner	Experiment	26.14	17.06	1.022	21.68	0.653
	Simulation	24.80	16.73	1.004	21.90	0.675
10°—edge	Experiment	23.66	14.15	0.848	18.01	0.598
	Simulation	23.20	13.85	0.831	17.63	0.597
10°—corner	Experiment	24.62	16.29	0.977	20.73	0.661
	Simulation	24.98	15.74	0.945	20.04	0.630
15°—edge	Experiment	21.07	13.58	0.815	17.28	0.644
	Simulation	21.57	13.79	0.827	17.55	0.639
15°—corner	Experiment	24.17	15.93	0.956	20.27	0.659
	Simulation	23.49	14.84	0.890	18.89	0.632

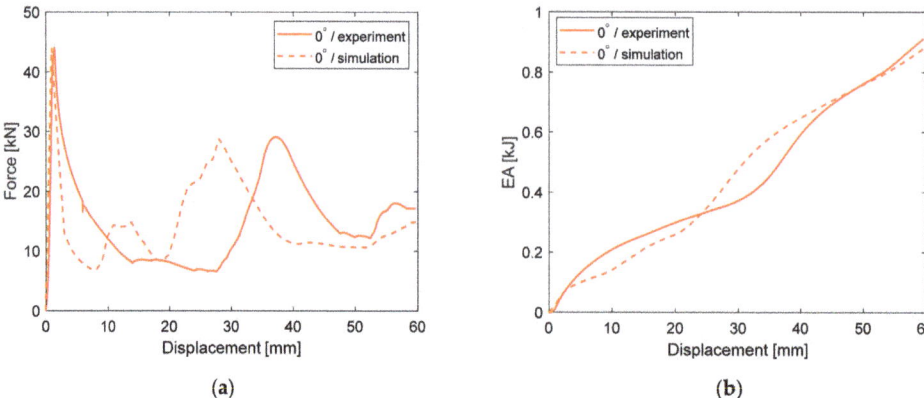

Figure 5. Experimental and numerical results of 0° crushing angle for axial impact. (**a**) force-displacement curve; (**b**) EA-displacement curve.

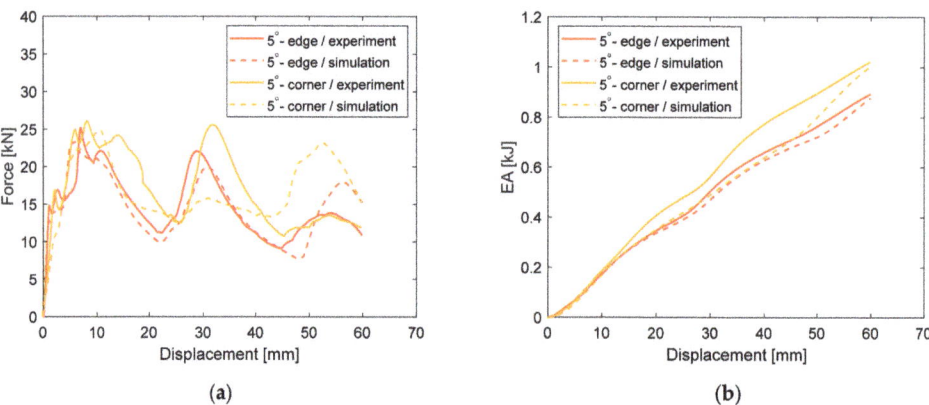

Figure 6. Experimental and numerical results of 5° crushing angle for cornered and edged oblique impact. (**a**) force-displacement curve; (**b**) EA-displacement curve.

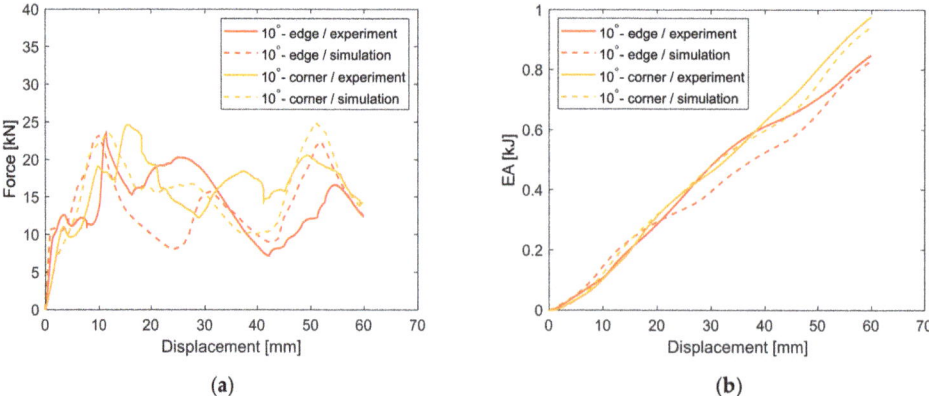

Figure 7. Experimental and numerical results of 10° crushing angle for cornered and edged oblique impact. (**a**) force-displacement curve; (**b**) EA-displacement curve.

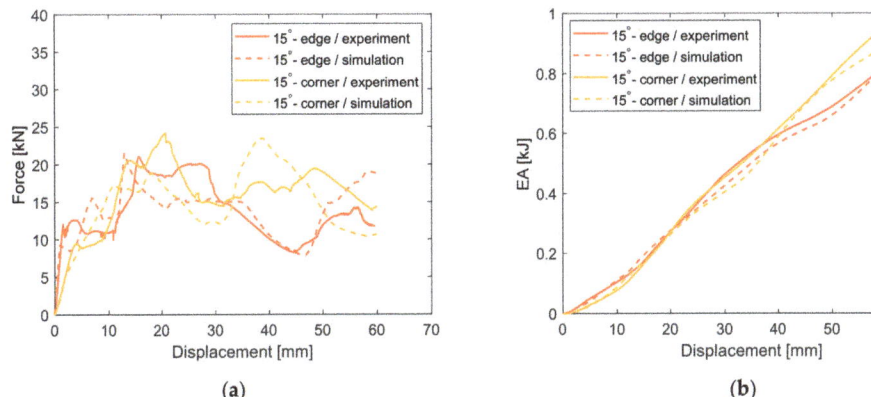

Figure 8. Experimental and numerical results of 15° crushing angle for cornered and edged oblique impact. (**a**) force-displacement curve; (**b**) EA-displacement curve.

Table 4. Deviation in PCF and EA between experiments and simulations.

Loading Case	Deviation in PCF (%)	Deviation in EA (%)
0°	0.13	3.07
5°—edge	7.57	1.82
5°—corner	5.15	1.79
10°—edge	1.97	2.12
10°—corner	1.44	3.35
15°—edge	2.38	1.56
15°—corner	2.81	6.82

Further, simulations seemed to capture sufficiently the plastic collapse mechanisms that occurred in tests, as Figures 9–15 depict, predicting both the collapse mode and the number of formulated folds in most cases, which is additionally reflected by the number of local peaks and lows in crushing-force fluctuation during impactor displacement. More specifically, for an axially crushed tube, the test showed an inextensional deformation mode with three formulated folds during collapse, while tearing failure was also observed around tube corners, due to the high bending-moment concentration. In similar direction, the simulation revealed a mixed collapse mode, showing two initial extensional folds and one inextensional fold at following, predicting in addition the occurrence of tearing around tube corners. Moreover, for 5° and 10° obliquely crushed tubes either under an edged or a cornered initial contact with the impactor, both tests and simulations revealed three inextensional folds during plastic deformation; however, the slight tearing around tube corners in tests was not captured by simulations due to a lower PCF, which was not proved to be great enough to react to material failure, due to stress concentration in the tube corners. Finally, regarding 15° obliquely crushed tubes under either an initial contact with the impactor on an edge or in a corner, both simulations and tests agreed on an inextensional deformation mode during collapse, formulating two folds, while the slight tearing failure observed in tests was not captured by simulations as was achieved in the axial impact loading case. Therefore, all simulations showed a sufficient agreement with the experiments regarding both the collapse mechanism and the number of formulated folds, while tearing failure was only captured in axial impact loading, as the in the case of obliquely crushed tubes the tearing extent was significantly lower without thus affecting the energy-absorption capacity, which was predicted with sufficient accuracy by FE models.

Figure 9. Plastic deformation progress of 0° crushing angle for axial impact. (**a**) experiment; (**b**) simulation.

Figure 10. Plastic deformation progress of 5° crushing angle for edged oblique impact. (**a**) experiment; (**b**) simulation.

Figure 11. Plastic deformation progress of 5° crushing angle for cornered oblique impact. (**a**) experiment; (**b**) simulation.

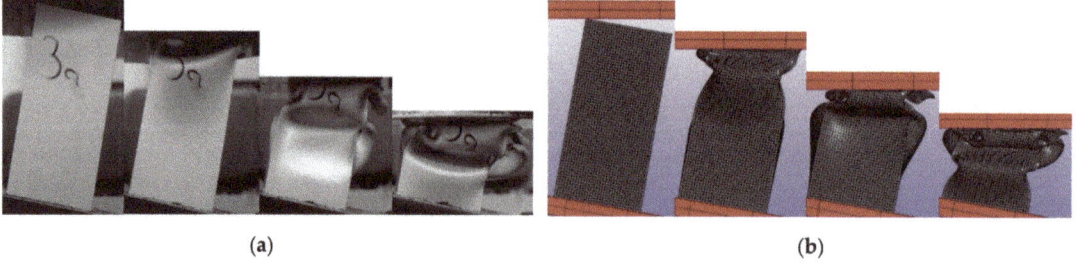

Figure 12. Plastic deformation progress of 10° crushing angle for edged oblique impact. (**a**) experiment; (**b**) simulation.

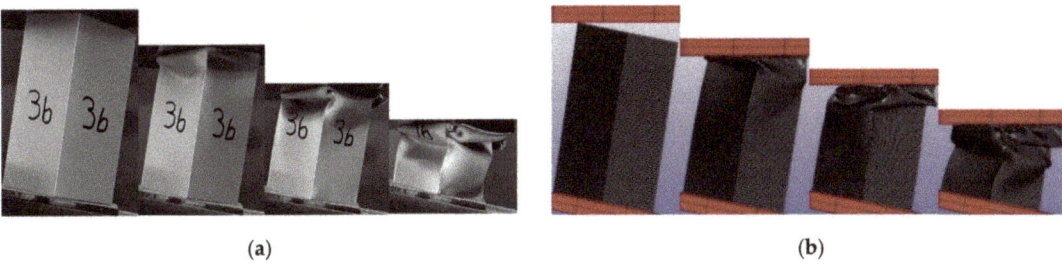

Figure 13. Plastic deformation progress of 10° crushing angle for cornered oblique impact. (**a**) experiment; (**b**) simulation.

Figure 14. Plastic deformation progress of 15° crushing angle for edged oblique impact. (**a**) experiment; (**b**) simulation.

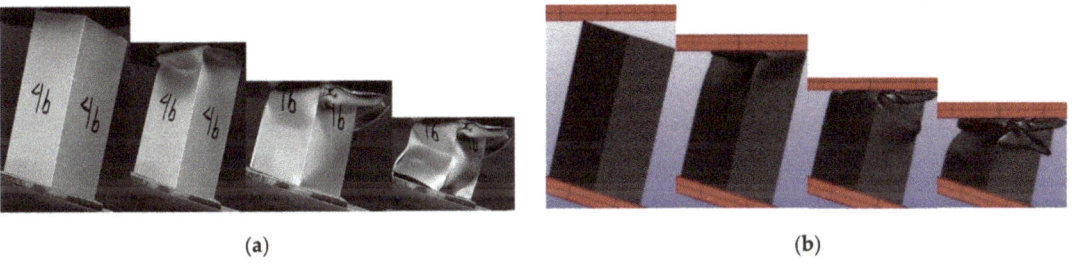

Figure 15. Plastic deformation progress of 15° crushing angle for cornered oblique impact. (**a**) experiment; (**b**) simulation.

3.2. Force-Displacement Characteristics

For the evaluation of crashworthiness performance, force-displacement and EA-displacement curves are provided regarding both experimental and numerical results as Figures 5–8 depict. Thus, the proper crashworthiness indicators are estimated in order to assess the energy-absorption capacity and the plastic collapse initiation regarding the examined loading cases. In more specificity, the plastic collapse initiation of axially crushed tubes occurred around 44 kN, formulating the first plastic fold, while at the next, two more folds were deformed as reflected by the local peaks and lows in crushing force, the fluctuation of which was sustained around an MCF of 15 kN during collapse. In fact, tearing failure was also observed around tube corners due to high stress concentration, which is captured by the drop in the rate of EA increase during impactor displacement range from 20 mm to 40 mm as shown in Figure 5b. That decrease in the rate of EA increase is caused by the unstable behavior of tearing failure, which was predicted slightly earlier during collapse progress by the FE model compared to the experiment. For this reason, a deviation between experiment and simulation is provided regarding the displacement region in which the EA slope reduction occurred in Figure 5b, where the tearing failure is predicted at a lower displacement by the FE model.

For the 5° obliquely crushed tube, regarding the initial contact in the edge case between impactor and tube, the plastic collapse initiation was revealed around 24 kN, while the force fluctuation during collapse was sustained around 14.8 kN, formulating three local peaks and lows of crushing force, reflecting the deformation of three inextensional plastic folds. Further, the slight tube tearing that occurred did not seem enough to react to any significant EA decrease, as its increase rate during collapse seems to be almost constant, as Figure 6b depicts. However, the 5° crushing angle reacted to both PCF and EA decrease, due to the additional bending moment introduced by the lateral crushing-force component, thus facilitating both plastic collapse initiation and progress, as less bending energy was required for the formulation of plastic folds. Regarding the 5° obliquely crushed tube under an initial contact with the impactor on a corner, a PCF of 25 kN revealed the plastic collapse initiation, and also revealed a progressive behavior, with three inextensional folds reflected by the crushing-force distribution around 17 kN MCF. Despite the angled loading, the 5° cornered oblique crushing revealed greater energy-absorption capacity compared to both 5° edged loading and axial impact, as the tearing failure was of a significantly lower extent, which in the case of the axially collapsed tube reacted to an important decrease in EA.

The 10° obliquely crushed tube under an initial edge contact with the impactor, revealed a PCF of 23.5 kN while MCF was sustained about 14 kN reflecting the formulation of three plastic folds during collapse, as depicted by the local force peaks in force-displacement curve in Figure 7a. The greater crushing angle of 10° compared to previous cases reacted to lower PCF, while further energy-absorption capability revealed a decrease too as the angled loading introduced an additional bending moment, which facilitated the plastic fold deformation and thus plastic collapse initiation and progress. In the case of the 10° obliquely crushed tube with an initial corner contact with impactor, PCF and MCF were captured at about 24.8 kN and 16 kN, respectively, revealing a PCF drop due to the increased crushing angle compared to axial and 5° oblique impact. However, the EA seemed to be greater compared to the 10° oblique edged loading due to the cornered initial contact and slighter tearing extent, which is reflected in Figure 7b, where EA increases linearly during collapse under a more constant slope compared to that of the edged oblique loading. However, both PCF and EA revealed a slight decrease compared to 5° oblique loading for both edged and cornered initial contact types, due to the higher crushing angle.

Finally, the 15° obliquely crushed tube under an edged initial contact with the impactor revealed a PCF of 21.2 kN, reflecting a plastic collapse initiation, while MCF was captured about 13.7 kN reflecting the lowest energy-absorption capacity among all examined cases due to the crushing angle effect, which reduced EA with loading angle increase. The occurrence of slight tearing failure resulted in only a slight drop in the rate of EA increase during the final stages of collapse, as captured in Figure 8b by the reduced EA curve slope. However, regarding the 15° cornered oblique impact case, PCF and EA were at greater levels due to the cornered contact between tube and impactor, which introduced a lower bending moment due to the angled loading compared to the edged oblique crushing. As a result, EA increased linearly at a greater rate compared to the edged oblique loading.

3.3. Deformation Modes

The plastic collapse mode is also investigated by evaluating the observations of both experiments and simulations by capturing different states of collapse shown in Figures 9–15 for all examined cases. The importance of analyzing the failure mechanism that occurred is its effect on the crushing force fluctuation during plastic deformation, as the local peaks and lows in crushing force reflect the formulation of external and internal plastic folds, respectively. Thus, the mode of collapse mechanism affects further the energy-absorption capacity of a crushed structure and for this reason it is considered an additional indicator of crashworthiness performance in terms of a progressive and stable collapse, which will allow for high EA levels. As Figure 9 depicts, the axially crushed tube revealed three plastic folds during collapse, while tearing failure also occurred around the tube corners due to the high stress concentration. In fact, the experiment showed three inextensional folds in contrast with the FE simulation, which in turn revealed a mixed collapse mode by initially formulating two extensional folds and one inextensional fold. The tearing failure reacted to the EA decrease as its rate of increase dropped slightly due to the unstable behavior of the tearing mechanism.

Both 5° and 10° obliquely crushed tubes under either an edged or a cornered initial contact with impactor revealed three inextensional folds during their collapse, in which both experiments and simulation showed an absolute agreement. Further, slight tearing failure was captured in the experimental observations, the low extent of which did not seem to affect significantly the energy-absorption capability, as EA showed a constant increase rate during the collapse affected by the stable and progressive behavior of inextensional folding deformation.

Moreover, 15° obliquely crushed tube formulated two inextensional folds during plastic collapse under both edged and cornered initial contact with impactor. Further, the experiments revealed slight tearing around tube corners, which however was restricted in low extent without so affecting significantly both failure stability and energy-absorption capacity in consequence. Therefore, all examined crushed tubes revealed an inextensional collapse mode formulating three plastic folds except the 15° crushed tubes, which deformed under two folds, while further slight tearing failure was observed around tube corners, which however was captured from the FE model only in the case of axial impact.

The inextensional folding deformation was caused by the non-uniform circumferential distribution of the bending moment, which reacted to a stretching and a compression of the cross-section in different directions, thus causing the formulation of rectangular convolutions during plastic deformation. More specifically, as Figure 16 depicts, for the case of 10° edged oblique crushing, the formulation of inextensional folds is caused by the circumferentially non-uniform resultant bending moments M_x and M_y, which lead to either external or internal buckling on the cross-section sides depending on their sign. Finally, as Figure 17 shows, the inextensional folding mode was revealed in all crushed tubes, showing in fact a significant superiority against the slight tearing failure occurring around tube corners, thus providing high EA capacity due to its stable and progressive collapse.

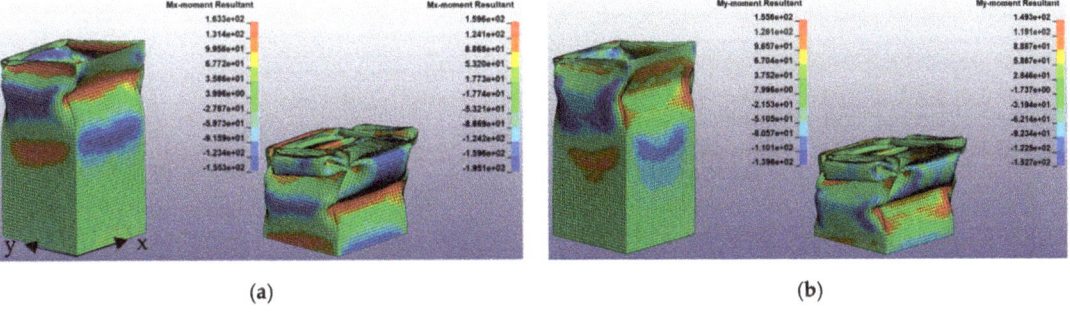

Figure 16. Bending moment circumferential distribution during collapse for 10° edged oblique impact. (**a**) M_x moment; (**b**) M_y moment.

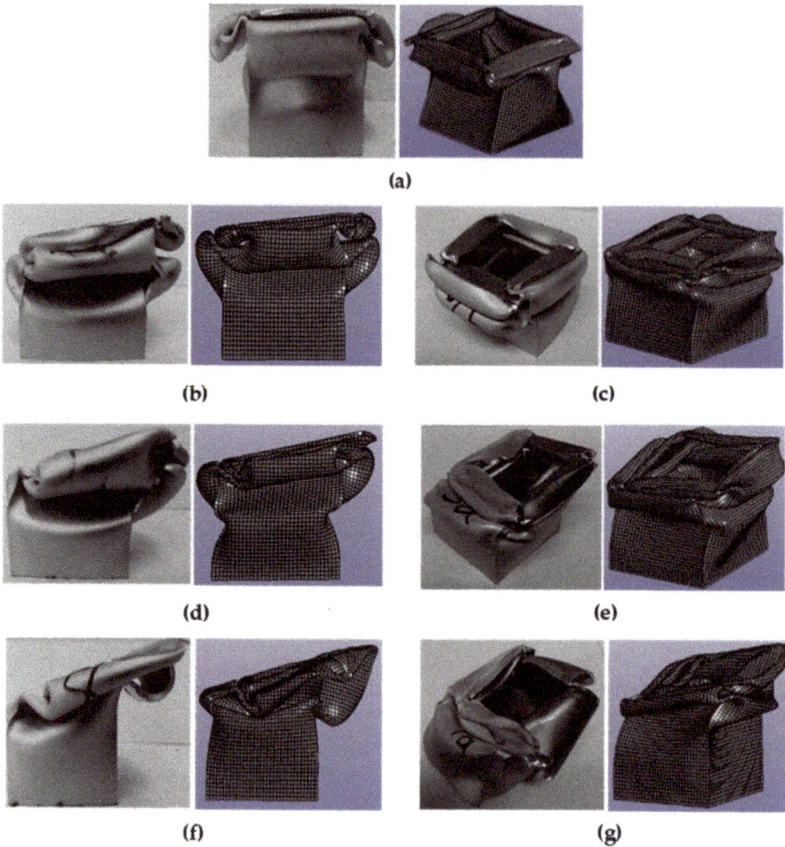

Figure 17. Final views of fully crushed tubes. (**a**) axial; (**b**) 5°—edge; (**c**) 5°—corner; (**d**) 10°—edge; (**e**) 10°—corner; (**f**) 15°—edge; (**g**) 15°—corner.

3.4. Crushing Angle Effect

Both numerical and experimental results agreed on the effect of crushing angle on crashworthiness performance in terms of plastic collapse initiation and energy-absorption capacity. In more specificity, the increase in crushing angle resulted in a PCF decrease due to the additional bending moment introduced by the lateral force component, which facilitated plastic collapse initiation. Figure 18 depicts the above tendency showing a PCF drop with the crushing angle either for an edge initial contact or a cornered one between tube and impactor. In fact, PCF reveals a significant decrease in oblique impact loading, reducing by about 43% from axial to 5° oblique impact, while at higher angles the PCF drop seems to flatten out under corner contact between impactor and tube in contrast to edged oblique loading in which PCF seems to linearly decrease with crushing angle even at higher angles.

Further, Figure 19 shows that the increase in crushing angle reacted to lower EA, as additional introduced bending moment due to angled loading facilitated the plastic collapse progress by reducing the necessary plastic bending moment required for folding deformation. However, the 5°obliquely crushed tube under an initial corner contact with the impactor revealed the greatest energy-absorption capacity, lying about 1.022 kJ despite the 5° angled loading, as in the case of axial impact loading the significant tearing failure around tube corners resulted in an EA decrease. For this reason, the EA of the axially collapsed tube proved lower compared to the 5° obliquely crushed tube under corner contact in which the tearing occurred to a significantly lesser extent without thus affecting EA, which was maintained sufficiently. However, the case of 5° edged oblique impact revealed slightly lower EA compared to axial impact, as expected due to the greater crushing angle. Therefore, increased crushing angles resulted in EA drop without accounting for any tearing effect as shown more accurately for all examined oblique loading cases, where a tearing effect was observed to a lower extent.

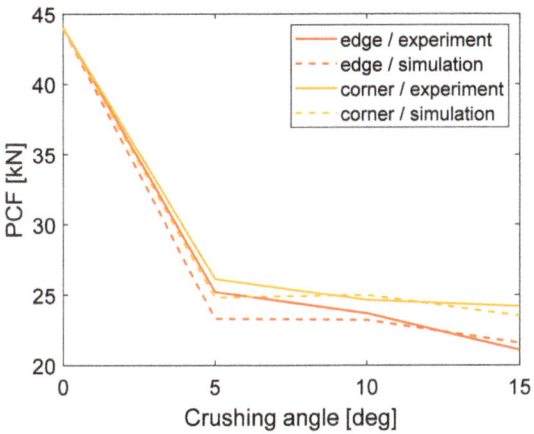

Figure 18. PCF variation with crushing angle.

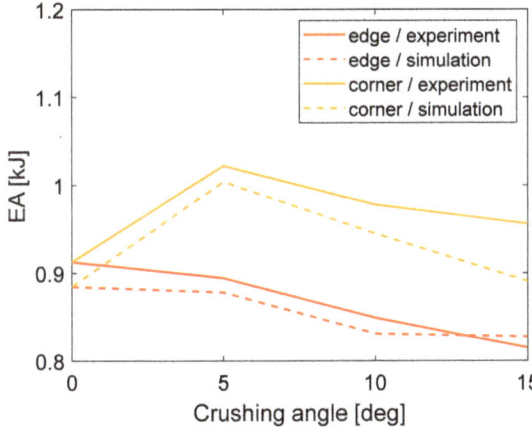

Figure 19. EA variation with crushing angle.

Finally, as Figure 20 illustrates, CFE revealed an increase at low crushing angles for either an edged or a cornered oblique impact compared to axial crushing, as in the case of cornered oblique loading PCF showed a reduction with angle while EA increased due to the absence of significant tearing. Thus, considering MCF as proportional to EA, CFE seemed to be increased at low angles compared to axial impact. Additionally, in the case of edged oblique loading, although both PCF and EA showed a decrease with angle, the drop in PCF seemed stronger than the one of EA and in consequence in MCF, thus revealing a CFE increase also at low crushing angles. Moreover, higher CFE levels were captured for cornered oblique impact compared to the edged one, while the difference between them seemed to decrease at higher angles. In specific, CFE was flattened out with a crushing angle increase according to experiments for cornered oblique impact, while simulations showed a linear drop of CFE with crushing angle. In contrast, a CFE increase was captured at high angles for edged oblique crushing conditions. Therefore, the cornered oblique impact condition under low crushing angles can be evaluated as the most beneficial loading case providing sufficiently low PCF, resulting in lesser tearing failure and achieving high enough EA and CFE levels.

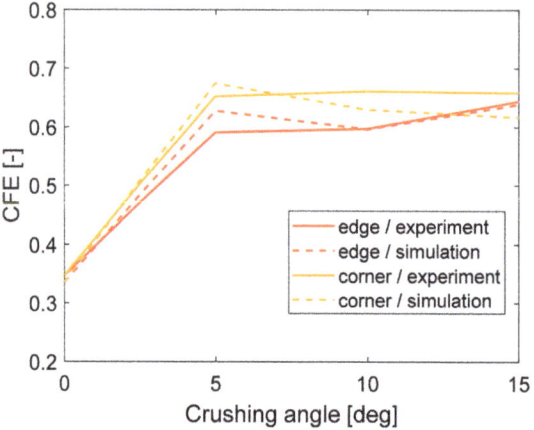

Figure 20. CFE variation with crushing angle.

3.5. Initial Contact Effect

The type of initial contact between impactor and tube did not seem to cause differences in the plastic collapse mechanism in terms of deformation mode and number of formulated folds, as all examined cases revealed the same collapse characteristics for either an edged or a cornered oblique impact under a certain crushing angle. Regarding the effect of initial contact on PCF, as Figure 18 depicts for all examined crushing angle range, cornered oblique impact revealed greater PCF compared to edged one, while their difference was captured slightly higher as the crushing angle increased. In fact, PCF seems to flatten out at high crushing angles under an initial corner contact between tube and impactor, while in contrast PCF decreased linearly under edged oblique impact conditions, showing more a sharper drop at angles above 10°. Thus, edged oblique crushing seems to better facilitate plastic collapse initiation, providing lower PCF levels compared to cornered oblique loading.

Further, EA was revealed greater in the case of cornered oblique impact at all examined crushing angle range compared to edged oblique loading providing higher energy capacity for the crushed tubes as shown in Figure 19. In fact, 5° obliquely crushed tube under an initial contact in corner with impactor revealed the greatest EA among all examined cases, even compared to an axially collapsed tube in which the tearing failure reacted to an EA drop. Moreover, the increased EA in the case of cornered oblique impact compared to the edged one is premised on the fact that the additional bending moment (M_{add}) due to angled loading introduced by the lateral force component (F_l) as Figure 21 shows, is slightly lower compared to the one in the case of edged oblique crushing. Therefore, the required deformation energy for plastic collapse progress is revealed at a slightly greater magnitude for cornered oblique impact, thus resulting in a greater EA for a certain crushing angle. That is because for short tubes and low crushing angles, the deformation energy is mainly reflected by the bending moment (M_{Fc}), which is provided by the compressive crushing force component (F_c) and which, combined with M_{add}, should result in the plastic bending moment M_p required for plastic fold formulation. Therefore, considering M_{Fc} as proportional to EA and in consequence considering MCF according to analytical expressions of various past studies [29,30], the EA of a cornered obliquely crushed structure is revealed to be greater compared to that of edged oblique impact, as h_e is greater than h_c according to Equation (8).

$$M_{add} = F_l \cdot h = F \cdot \sin\alpha \cdot h \tag{6}$$

$$M_p = M_{add} + M_{F_c} \tag{7}$$

$$\tan\alpha = \frac{b}{2(L-h_e)} = \frac{\sqrt{2}\,b}{2(L-h_c)} \tag{8}$$

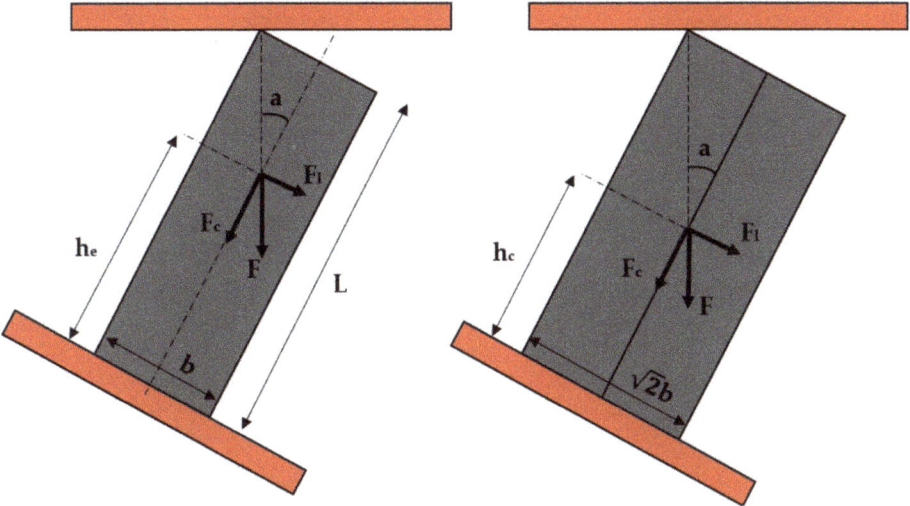

Figure 21. Bending moment due to angled loading for edged (**left**) and cornered (**right**) oblique impact.

Finally, CFE is also captured at higher levels for cornered oblique crushing conditions compared to edged ones, as in the first case the benefit in EA and in consequence in MCF overcomes the greater observed PCF revealing thus higher CFE. In fact, the maximum CFE is obtained at about 0.675 for the 5° obliquely crushed tube under an initial contact in corner with impactor. Additionally, CFE seems to flatten out at higher angles regarding cornered oblique impact, while edged oblique crushing revealed a CFE increase at higher loading angles, thus reducing the difference in CFE between the two examined types of initial contact at high angles. Therefore, cornered oblique impact under low crushing angles is considered the most beneficial loading scenario resulting in the greatest EA capacity under the highest CFE.

4. Conclusions

This study investigated the crashworthiness behavior of thin-walled aluminum square tubes subjected to both axial and oblique impact, emphasizing the effect of crushing angle and initial contact between tube and impactor on plastic collapse initiation and energy-absorption capability. Two types of initial contact were examined, consisting of an edged and a cornered contact. Both quasi-static tests and FE simulation in LS-DYNA were performed, while the provided experimental and numerical results were compared for the modeling validation and the evaluation of crashworthiness performance.

Both experiments and simulations showed a sufficient agreement on PCF and EA, while the observed collapse mechanism was also captured by the FE models, revealing inextensional folds and predicting their number accurately. However, tearing failure around tube corners was only captured for the axially crushed tube as in the case of obliquely crushed tubes the low extent of the tearing that occurred was not shown by the simulations, as the significantly lower PCF did not seem enough to cause material failure.

In more specificity, in all examined cases three inextensional plastic folds were revealed during collapse, except the 15° obliquely crushed tubes, where two inextensional folds were formulated. Additionally, all conducted experiments revealed slight tearing failure around tube corners, which however was of a significantly lower extent in the case of obliquely crushed tubes. The 5° obliquely crushed tube under an initial corner contact with the impactor revealed the greatest EA and CFE reflecting, and thus the most beneficial loading condition for the examined square tubes. Compared to the axially crushed tube, the greater energy capacity of the 5° cornered collapsed tube was revealed mainly due to the lower tearing extent, which in the case of axial impact reacted to the EA drop due to its greater magnitude.

Observing the results regarding oblique impact loading, the increase in crushing angle resulted in PCF drop as the lateral force component reacted to additional bending moment, thus facilitating the plastic collapse initiation. Further, the bending moment due to angled loading resulted in a lower deformation energy required for folding formulation, thus revealing an EA decrease as the crushing angle increased. In specific, however, the effect of tearing failure on EA was proved stronger than that of the crushing angle in low loading angles, thus resulting in a lower EA in the case of axial impact compared to 5° cornered crushing. As a result, CFE was maximized in low crushing angles, while as the angle became higher CFE seemed to flatten out or slightly increase.

Finally, regarding the effect of the initial contact type between impactor and tube, cornered oblique impact revealed the greatest PCF and EA at all examined angle ranges. More specifically, the difference in PCF between edged and cornered oblique collapse was increased at higher crushing angles, as PCF almost flattened out at high angles, while regarding edged oblique impact PCF seemed to linearly decrease with the loading angle. Further, cornered oblique impact reacted to lower bending moment due to the angled loading compared to the edged one, thus revealing significantly greater EA as higher deformation energy was then required for folding formulation. However, the initial contact type between impactor and tube did not seem to affect mechanism of the mode of plastic collapse and the number of formulated folds. CFE was also higher for an initial corner contact between the impactor and tube, while the difference in CFE compared to edged oblique loading was decreased at higher angles. Therefore, cornered oblique impact under a low crushing angle was proved to be the most beneficial loading case, providing the greatest EA and CFE between all examined cases.

Author Contributions: Conceptualization, K.D.K. and D.E.M.; methodology, K.D.K.; software, K.D.K.; validation, K.D.K.; formal analysis, K.D.K.; investigation, K.D.K.; resources, K.D.K. and D.E.M.; data curation, K.D.K.; writing—original draft preparation, K.D.K.; writing—review and editing, K.D.K.; visualization, K.D.K.; supervision, D.E.M.; project administration, D.E.M.; funding acquisition, D.E.M. All authors have read and agreed to the published version of the manuscript.

Funding: This research received no external funding.

Institutional Review Board Statement: Not applicable.

Informed Consent Statement: Not applicable.

Data Availability Statement: Not applicable.

Conflicts of Interest: The authors declare no conflict of interest.

References

1. Pei, T.S.; Saffe, S.N.M.; Rusdan, S.A.; Hamran, N.N.N. Oblique impact on crashworthiness: Review. *Int. J. Eng. Technol. Sci.* **2017**, *4*, 32–48. [CrossRef]
2. Kim, H.S.; Wierzbicki, T. Numerical and analytical study on deep biaxial bending collapse of thin-walled beams. *Int. J. Mech. Sci.* **2000**, *42*, 1947–1970. [CrossRef]
3. Reyes, A.; Langseth, M.; Hopperstad, O.S. Crashworthiness of aluminum extrusions subjected to oblique loading: Experiments and numerical analyses. *Int. J. Mech. Sci.* **2002**, *44*, 1965–1984. [CrossRef]

4. Han, D.C.; Park, S.H. Collapse behavior of square thin-walled columns subjected to oblique loads. *Thin-Walled Struct.* **1999**, *35*, 167–184. [CrossRef]
5. Tran, T.; Hou, S.; Han, X.; Nguyen, N.; Chau, M. Theoretical prediction and crashworthiness optimization of multi-cell square tubes under oblique impact loading. *Int. J. Mech. Sci.* **2014**, *89*, 177–193. [CrossRef]
6. Pirmohammad, S.; Marzdashti, S.E. Crushing behavior of new designed multi-cell members subjected to axial and oblique quasi-static loads. *Thin-Walled Struct.* **2016**, *108*, 291–304. [CrossRef]
7. Azimi, M.B.; Asgari, M.; Salaripoor, H. A new homo-polygonal multi-cell structures under axial and oblique impacts; considering the effect of cell growth in crashworthiness. *Int. J. Crashworthiness* **2020**, *25*, 628–647. [CrossRef]
8. Liu, W.; Jin, L.; Luo, Y.; Deng, X. Multi-objective crashworthiness optimization of tapered star-shaped tubed under oblique impact. *Int. J. Crashworthiness* **2021**, *26*, 328–342. [CrossRef]
9. Qi, C.; Yang, S.; Dong, F. Crushing analysis and multiobjective crashworthiness optimization of tapered square tubes under oblique impact loading. *Thin-Walled Struct.* **2012**, *59*, 103–119. [CrossRef]
10. Song, J. Numerical simulation on windowed tubes subjected to oblique impact and a new method for the design of obliquely loaded tubes. *Int. J. Impact Eng.* **2013**, *54*, 192–205. [CrossRef]
11. Mohammadiha, O.; Ghariblu, H. Crush behavior optimization of multi-tubes filled by functionally graded foam. *Thin-Walled Struct. Part B* **2016**, *98*, 627–639. [CrossRef]
12. Baykasoglu, C.; Baykasoglu, A.; Cetin, M.T. A comparative study on crashworthiness of thin-walled tubed with functionally graded thickness under oblique impact loadings. *Int. J. Crashworthiness* **2019**, *24*, 453–471. [CrossRef]
13. Crutzen, Y.; Inzaghi, A.; Mogilevsky, M.; Albertini, C. *Computer Modelling of the Energy Absorption Process in Box-Type Structures under Oblique Impact*; Automotive Automation Ltd.: Shropshire, UK, 1996; pp. 1293–1298.
14. Bai, Z.; Liu, J.; Zhu, F.; Wang, F.; Jiang, B. Optimal design of a crashworthy octagonal thin-walled sandwich tube under oblique loading. *Int. J. Crashworthiness* **2015**, *20*, 401–411. [CrossRef]
15. Patel, S.; Vusa, V.R.; Soares, C.G. Crashworthiness analysis of polymer composites under axial and oblique impact loading. *Int. J. Mech. Sci.* **2019**, *156*, 221–234. [CrossRef]
16. Zarei, H.R. Experimental and numerical crashworthiness investigation of hybrid composite aluminium tubes under dynamic axial and oblique loading. *Int. J. Automot. Eng.* **2015**, *5*, 1084–1093.
17. Ma, F.; Liang, H.; Pu, Y.; Wang, Q.; Zhao, Y. Crashworthiness analysis and multi-objective optimization for honeycomb structures under oblique impact loading. *Int. J. Crashworthiness* **2022**, *27*, 1128–1139. [CrossRef]
18. Jamal-Omidi, M.; Benis, A.C. A numerical study on energy absorption capability of lateral corrugated composite tube under axial crushing. *Int. J. Crashworthiness* **2021**, *26*, 147–158. [CrossRef]
19. Zhang, Y.; Wang, J.; Chen, T.; Lu, M.; Jiang, F. On crashworthiness design of double conical structures under oblique impact. *Int. J. Vehicle Design* **2018**, *7*, 20–45. [CrossRef]
20. Qi, C.; Yang, S. Crashworthiness and lightweight optimisation of thin-walled conical tubes subjected to an oblique impact. *Int. J. Crashworthiness* **2014**, *19*, 334–351. [CrossRef]
21. Gao, Q.; Wang, L.; Wang, Y.; Wang, C. Crushing analysis and multiobjective crashworthiness optimization of foam-filled ellipse tubes under oblique impact loading. *Thin-Walled Struct.* **2016**, *100*, 105–112. [CrossRef]
22. Borvik, T.; Hopperstad, O.S.; Reyes, A.; Langseth, M.; Solomos, G.; Dyngeland, T. Empty and foam-filled circular aluminium tubes subjected to axial and oblique quasistatic loading. *Int. J. Crashworthiness* **2003**, *8*, 481–494. [CrossRef]
23. Chen, Y.; Clausen, A.H.; Hopperstad, O.S.; Langseth, M. Stress-strain behaviour of aluminium alloys at a wide range of strain rates. *Int. J. Solids Struct.* **2009**, *46*, 3825–3835. [CrossRef]
24. Baroutaji, A.; Sajjia, M.; Olabi, A.G. On the crashworthiness performance of thin-walled energy absorbers: Recent advances and future developments. *Thin-Walled Struct.* **2017**, *108*, 137–163. [CrossRef]
25. Hallquist, J.O. *LS-DYNA Theory Manual*; Livermore Software Technology Corporation: Livermore, CA, USA, 2006.
26. Pled, F.; Yan, W.; Wen, C. Crushing modes of aluminium tubes under axial compression. In Proceedings of the 5th Australasian Congress on Applied Mechanics, Brisbane, Australia, 10–12 December 2007; pp. 178–180. Available online: https://hal.archives-ouvertes.fr/hal-01056929 (accessed on 25 May 2022).
27. Kilicaslan, C. Numerical crushing analysis of aluminum foam-filled corrugated single- and double- circular tubes subjected to axial impact loading. *Thin-Walled Struct.* **2015**, *96*, 82–94. [CrossRef]
28. Haufe, A.; Schweizerhof, K.; Dubois, P. *Properties and Limits: Review of Shell Element Formulations*; Developer Forum DYNA More: Filderstadt, Germany, 2013.
29. Abramowicz, W.; Jones, N. Dynamic progressive buckling of circular and square tubes. *Int. J. Impact Eng.* **1986**, *4*, 243–270. [CrossRef]
30. Wierzbicki, T.; Abramowicz, W. On the crushing mechanics of thin-walled structures. *J. Appl. Mech.* **1983**, *50*, 727–734. [CrossRef]

Article

Low Cycle Fatigue Performance of Additively Processed and Heat-Treated Ti-6Al-7Nb Alloy for Biomedical Applications

Maxwell Hein [1,2,*], David Kokalj [3], Nelson Filipe Lopes Dias [3], Dominic Stangier [3], Hilke Oltmanns [4], Sudipta Pramanik [1], Manfred Kietzmann [4], Kay-Peter Hoyer [1,2], Jessica Meißner [4], Wolfgang Tillmann [3] and Mirko Schaper [1,2]

[1] Chair of Materials Science (LWK), Paderborn University, Warburger Str. 100, 33098 Paderborn, Germany; pramanik@lwk.upb.de (S.P.); hoyer@lwk.upb.de (K.-P.H.); schaper@lwk.upb.de (M.S.)
[2] DMRC-Direct Manufacturing Research Center, University of Paderborn, Mersinweg 3, 33100 Paderborn, Germany
[3] Institute of Materials Engineering (LWT), TU Dortmund University, Leonhard-Euler-Str. 2, 44227 Dortmund, Germany; david.kokalj@tu-dortmund.de (D.K.); filipe.dias@tu-dortmund.de (N.F.L.D.); dominic.stangier@tu-dortmund.de (D.S.); wolfgang.tillmann@udo.edu (W.T.)
[4] Department of Pharmacology, Toxicology and Pharmacy, University of Veterinary Medicine, Bünteweg 17, 30559 Hannover, Germany; hilke.oltmanns@tiho-hannover.de (H.O.); manfred.kietzmann@tiho-hannover.de (M.K.); jessica.meissner@tiho-hannover.de (J.M.)
* Correspondence: hein@lwk.upb.de; Tel.: +49-5251-60-5447

Abstract: In biomedical engineering, laser powder bed fusion is an advanced manufacturing technology, which enables, for example, the production of patient-customized implants with complex geometries. Ti-6Al-7Nb shows promising improvements, especially regarding biocompatibility, compared with other titanium alloys. The biocompatible features are investigated employing cytocompatibility and antibacterial examinations on Al_2O_3-blasted and untreated surfaces. The mechanical properties of additively manufactured Ti-6Al-7Nb are evaluated in as-built and heat-treated conditions. Recrystallization annealing (925 °C for 4 h), β annealing (1050 °C for 2 h), as well as stress relieving (600 °C for 4 h) are applied. For microstructural investigation, scanning and transmission electron microscopy are performed. The different microstructures and the mechanical properties are compared. Mechanical behavior is determined based on quasi-static tensile tests and strain-controlled low cycle fatigue tests with total strain amplitudes ε_A of 0.35%, 0.5%, and 0.8%. The as-built and stress-relieved conditions meet the mechanical demands for the tensile properties of the international standard ISO 5832-11. Based on the Coffin–Manson–Basquin relation, fatigue strength and ductility coefficients, as well as exponents, are determined to examine fatigue life for the different conditions. The stress-relieved condition exhibits, overall, the best properties regarding monotonic tensile and cyclic fatigue behavior.

Keywords: laser powder bed fusion; Ti-6Al-7Nb; titanium alloy; biomedical engineering; low cycle fatigue; microstructure; nanostructure

---

1. Introduction

Biomedical materials often are used for replacing lost or diseased biological structures [1–3]. Implants require adequate properties, including mechanical properties, such as high wear resistance, corrosion resistance, excellent biocompatibility, osseointegration, and non-cytotoxicity, to avoid revision surgeries [4].

Titanium and its alloys are generally used in biomedical applications due to their excellent biocompatibility, high corrosion resistance, and superb mechanical properties, such as low elastic modulus and high strength. Titanium alloys are the most widely used metallic materials for load-bearing biomedical applications [5–9]. Ti-6Al-7Nb is an (α + β) titanium alloy with high specific strength and corrosion resistance, accompanied by excellent biocompatibility, and is used as an orthopedic and dental alloy [10–12]. Ti-6Al-4V, which already is of high interest in the biomedical industry, is commonly used but has slight disadvantages towards Ti-6Al-7Nb, regarding corrosion resistance and biocompatibility [13]. Previous studies showed that elements such as titanium, niobium, zirconium, gold, and tin are biocompatible, whereas aluminum, vanadium, chromium, and nickel are probably hazardous elements for the human body [14,15]. Vanadium was found to be cytotoxic and there are assumptions regarding vanadium ion release in service [16,17]. Various efforts have been carried out to address the issue of cytotoxicity of the alloying elements of Ti-6Al-4V, for example, by replacing all the alloying elements, as in the alloy Ti-35Nb-7Zr-5Ta [18]. Due to the hazardous vanadium in Ti-6Al-4V, the development of Ti-6Al-7Nb, through the substitution of vanadium by niobium, offered an alternative for load-bearing implant materials [16,19,20].

Laser powder bed fusion (LPBF), also called laser beam melting, is, regarding additive manufacturing (AM) of metals, one of the most established techniques [21–23]. In LPBF processes, the fast heating rates and cooling rates, respectively, result in a characteristic microstructure and therefore characteristic mechanical properties [24–26]. Conventional fabrication methods are limited due to manufacturing constraints producing patient-specific implants. Through AM, or rather LPBF, extraordinary biomedical implant topologies, such as porous and foam structures, are feasible [27–29].

Many available research results focus on the mechanical properties of different additively manufactured alloys, such as 316L, 17-4 PH, or Ti-6Al-4V [30,31]. However, there remains a significant gap in terms of fatigue properties analysis, since most available research on dynamic behavior is related to high cycle fatigue (HCF) testing and the crack growth analysis of additively manufactured parts of steel alloys [32–36] and titanium alloys [37–42]. Only few studies deal with the low cycle fatigue (LCF) properties of additively manufactured alloys in general [43–47]. Even fewer studies were performed with a focus on post-treatments for titanium alloys. Previous studies addressed the fatigue behavior of additively processed Ti-6Al-4V for different conditions and loading situations [38,42,48,49]. Examinations of the quasi-static behavior of Ti-6Al-7Nb are at hand but none of them include LCF behavior [17,50–53]. In previous studies, extremely high loads were reported during stumbling. Such load peaks during uncontrolled movements are difficult to investigate systematically. Implants are usually examined for high cycle fatigue, although the rare extreme loads could endanger the implant. Therefore, in addition to monotonic material characterization and high cycle fatigue tests, implants should also be investigated at low cycles with higher load levels [54,55]. Given that the physiological—sometimes extreme—loading during service life as an implant is cyclic, the fatigue performance of laser beam-melted Ti-6Al-7Nb requires attention. To combine LPBF with the promising quasi-static properties of Ti-6Al-7Nb, the microstructural and mechanical properties of additively manufactured Ti-6Al-7Nb in different conditions are investigated in this work. Heat treatments are an additional part of this examination, to initiate microstructural changes for the relaxation of the tensed crystal lattice and the beneficial crack growth behavior, as well as to achieve a homogenous, decreased residual stress state [56–58]. Generally, titanium alloys are heat-treated due to high residual stresses as well as due to the brittle α'-phase, occurring after LPBF. Moreover, thermal post-treatments induce an improvement of the quasi-static mechanical properties of different titanium alloys [52,59]. To sum up, since there is a lack of studies addressing fatigue behavior, it is necessary to focus on an investigation of the microstructure under different conditions, as-built and heat-treated, and their effect on quasi-static and LCF behaviors.

2. Materials and Methods

2.1. Manufacturing Procedure, Materials, and Mechanical Characterization

The specimens were fabricated using an LT12 SLM machine (DMG MORI AG, Bielefeld, Germany) with a beam spot size of 35 µm. For data preparation, the software Materialise Magics (Version 21.1, Materialise GmbH, Munich, Germany) was applied. To obtain dense material with LPBF (relative density $\varphi > 99.9\%$), the following parameters were used: laser power P = 227 W, laser scanning speed v = 1.675 mm s^{-1}, hatch distance h = 0.077 mm. A constant scanning strategy of 5 mm stripes by layer-wise rotation of the scanning vectors of 67° alongside a defined layer thickness of 50 µm was applied. As contour parameters, a laser power P_c = 123 W and a scanning speed v_c = 0.512 mm s^{-1} were applied. The Ti-6Al-7Nb powder is supplied by ECKART TLS GmbH (Bitterfeld, Germany). The Ti-6Al-7Nb powder particles were examined concerning particle size distribution (PSD) with a Mastersizer 2000 (Malvern Panalytical GmbH, Kassel, Germany) using laser diffraction. The powder batch had a nominal PSD comprising 26.9 µm (D_{10}) and 52.1 µm (D_{90}) with a log-transformed normal distribution centered at 37.6 µm (D_{50}), see Figure 1a. Different heat treatments were performed under a vacuum atmosphere. Recrystallization annealing (HT1 = 925 °C for 4 h) was conducted to obtain equiaxed α with β at grain boundary triple points, β annealing (HT2 = 1050 °C for 2 h) to receive Widmanstätten α + β colony structures, and stress relieving (HT3 = 600 °C for 4 h) to decrease the undesirable residual stresses due to the LPBF process, as well as a decomposition of the martensitic α′ to α-phase [59,60]. The heat treatment process routes are presented in Figure 1c. The powder morphology was investigated employing the scanning electron microscope (SEM) Zeiss Ultra Plus (Carl Zeiss AG, Oberkochen, Germany). The powder consists of mainly spherical particles with a few agglomerations on bigger particles, see Figure 1b. Microstructural investigations were accomplished with the SEM equipped with an electron backscatter diffraction (EBSD) unit to detect phases and corresponding grain orientations. EBSD data was post-processed using the MATLAB-based (Version R2019a 9.6, The MathWorks, Inc., Natick, MA, USA) toolbox MTEX (Version 5.6.0) [61]. MTEX is a free available toolbox for analyzing and modelling crystallographic textures. Microscopic observations and classification of the fractured surfaces for quasi-static and fatigue tests were also carried out using SEM. Microstructural study of samples on nanoscale was undertaken by transmission electron microscopy (TEM). For TEM, thin-slice samples ($\approx$ 400 µm) were cut by Struers Sectom-5 (Struers GmbH, Willich, Germany) and further polished to a thickness of $\approx$ 100 µm. At last, 3 mm diameter circular disc samples were punched from the thin samples. Twin jet electropolishing of the thin foils was performed with the Struers Tenupol-5 (Struers GmbH, Willich, Germany) using an electrolyte containing perchloric acid (60 mL), butanol (340 mL), and methanol (600 mL). Electropolishing was executed at a voltage of 21 V, a current of 35 mA, and a temperature of −22 °C. TEM investigations were executed using a cold field emission gun equipped with JEOL JEM-ARM 200F (JOEL Ltd., Tokyo, Japan). TEM, high-resolution TEM (HRTEM), and high-angle annular dark-field scanning transmission electron microscopy (HAADF-STEM) images were taken. Energy dispersive spectroscopy (EDS) was performed during HAADF-STEM imaging. EDS maps were collected with a 30 nm step size with 10 s dwell time per step. The crystalline phase composition of Ti-6Al-7Nb was characterized through X-ray diffraction (XRD; D8 Advance, Bruker, Madison, WI, USA) using a Cu-Kα radiation source (λ = 0.154187 nm) operating at a current of 40 mA and a voltage of 35 kV. The as-built and heat-treated Ti-6Al-7Nb parts were measured in θ-θ geometry within a scanning range from $2\theta = 30°$ to $2\theta = 90°$ applying a step width of 0.035° and an exposure time of 1.5 s per step. All tests for mechanical characterization were performed at an ambient temperature. Tensile specimens were built according to Figure 1d. The loading direction was parallel to the building direction (BD) of the samples.

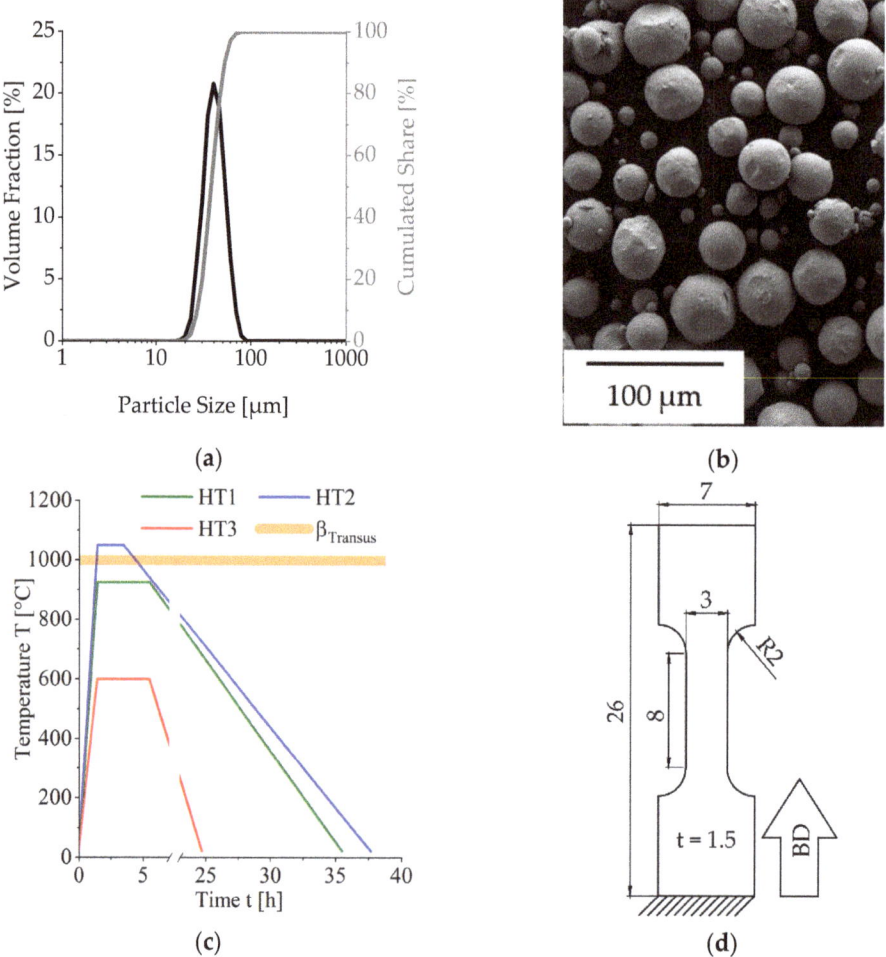

Figure 1. (a) Particle size distribution of Ti-6Al-7Nb; (b) powder morphology (SEM image); (c) schematic overview of the different heat treatments conducted—recrystallization annealing (HT1), β annealing (HT2), and stress relieving (HT3), as well as the $β_{transus}$ temperature for the alloy; (d) geometry and building direction (BD) of tensile and fatigue specimens.

All tensile specimens were blasted with high-grade Al_2O_3. The blasting material corresponded to a particle size of 70 μm–250 μm, used at 4 bar air pressure with a SMG 25 DUO (MHG Strahlanlagen GmbH, Düsseldorf, Germany). The tensile and LCF tests were performed utilizing a servo-hydraulic test-rig MTS 858 table-top system (MTS Systems Corporation, Eden Prairie, MN, USA) equipped with a 20 kN load cell and an extensometer 632.29F-30 (MTS Systems Corporation, Eden Prairie, MN, USA). The tensile test procedure corresponded to a displacement-controlled execution with a crosshead speed of 1.5 mm min^{-1}, according to DIN EN ISO 6892-1. The LCF tests were strain-controlled, at total strain amplitudes $ε_A$ of 0.35%, 0.5%, and 0.8%, with a R-ratio of −1 (compression–tensile fatigue), and a strain rate of $6 × 10^{-3}$ s^{-1}. At least three specimens per condition were tested with both monotonic and cyclic tests for each strain amplitude. The monotonic material properties, such as Young's moduli E, tensile yield strengths $R_{p0,2}$, ultimate tensile strengths R_m, and plastic elongations A, were obtained from the static tensile test. The determination of the fatigue material constants requires the performance of several fatigue tests under cyclic loading and a R-ratio of −1. Cyclically loaded materials often have unstable and changing stress amplitude during the test, due to cyclic hardening or softening. Therefore, the stress amplitude for the stabilized state must be used, which occurs at the half number of cycles to fracture. The plastic strain amplitude can either be calculated with the stress amplitude, the Young's modulus and the total strain amplitude (see Equation (4)), or by measuring the thickness of the recorded stable hysteresis loop recorded during the fatigue tests [62–65]. For evaluation of the total stress amplitude, σ_a, from the S–N curves, the Basquin relation is suitable, as follows:

$$\frac{\Delta \sigma}{2} = \sigma_a = \sigma'_f \cdot (2N_f)^b. \tag{1}$$

In Equation (1), σ'_f delineates the fatigue strength coefficient and b is the fatigue strength exponent. This equation fits the high-stress and low-stress fatigues [66]. For a better description of the high-stress fatigue region, the dependence between the number of reversals to failure $2N_f$ and the plastic strain amplitude $\Delta \varepsilon_p/2$, also called the Coffin–Manson relation, is used to describe the total fatigue life, as follows:

$$\frac{\Delta \varepsilon_p}{2} = \varepsilon'_f \cdot (2N_f)^c, \tag{2}$$

with the fatigue ductility coefficient, ε'_f, and the fatigue ductility exponent, c [66–68]. The total strain amplitude, $\Delta \varepsilon_A/2$, can be divided into two components, the plastic strain amplitude, $\Delta \varepsilon_p/2$, and the elastic strain amplitude, $\Delta \varepsilon_e/2$. The total strain amplitude, $\Delta \varepsilon/2$, of fatigue life curves is described as the sum of elastic and plastic strain amplitude, as follows, by Suresh [69]:

$$\frac{\Delta \varepsilon_A}{2} = \frac{\Delta \varepsilon_e}{2} + \frac{\Delta \varepsilon_p}{2}. \tag{3}$$

$\Delta \varepsilon_e/2$ is described by Hooke's law as the quotient of σ_a to the Young's modulus E as follows:

$$\frac{\Delta \varepsilon_e}{2} = \frac{\Delta \sigma}{2E} = \cdot \frac{\sigma_a}{E}. \tag{4}$$

Together with the modified Basquin equation and the Coffin–Manson relation, combining Equations (1)–(4), one obtains the following:

$$\varepsilon_A = \frac{\sigma'_f}{E} \cdot (2N_f)^b + \varepsilon'_f \cdot (2N_f)^c. \tag{5}$$

The first and second terms on the right-hand side of Equation (5) are the elastic ε_e and plastic ε_p components, respectively, of the total strain amplitude ε_A. Equation (5) can be used as the basis for the strain life approach to fatigue design. For determination of the fatigue life, a schematic illustration of the as-built condition is presented in Figure 2. The intersection of the curves for ε_e and ε_p describes the transition point, where the plastic and elastic strains are identical. From this point, the LCF life is governed more by elastic than plastic strain. For the as-built condition this point is around 110 cycles, see Figure 2. According to the Coffin–Manson–Basquin approximation, one is able to predict the fatigue life depending on the applied strain amplitude for the different conditions [66–68,70].

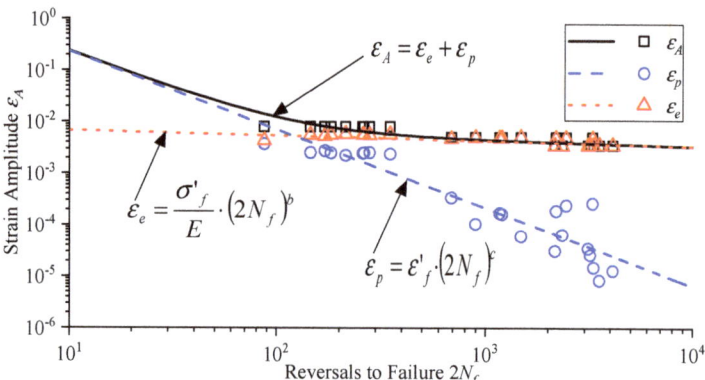

Figure 2. Example of the Coffin–Manson–Basquin curve according to Equation (5) for the as-built condition.

2.2. Biocompatibility Investigations

Two different surface conditions were analyzed to determine the optimum surface roughness for biocompatibility. On the one hand, the as-received surface was investigated without any post-treatment (NT). On the other hand, the surface was post-treated utilizing blasting with high-grade Al_2O_3 (ab). Blasting lead to rougher surfaces and to an increased surface area, resulting in faster osseointegration and higher survival rates for moderately rough implant surfaces: this was demonstrated by clinical studies [71–75]. The surface roughness was measured with an optical 3D macroscope VR-3100 (KEYENCE GmbH, Neu-Isenburg, Germany). The arithmetical mean roughness, Ra, and average roughness, Rz, were determined on the surfaces of the untreated and blasted specimens. A total of 5 specimens per condition were measured with 5 line measurements (length $\approx$ 6.8 mm) per side.

2.2.1. Cytocompatibility Testing

Biocompatibility studies were performed with different cell types. Therefore, murine fibroblasts (L-929; CLS Cell Lines Service, Eppelheim, Germany), human osteosarcoma cells (HOS 87070202; European Collection of Authenticated Cell Cultures, Merck, Germany), and human umbilical vein endothelial cells (HUVEC; Promocell, Heidelberg, Germany) were used. L-929 cells were grown and passaged in RPMI-1640 medium (Biochrom GmbH, Berlin, Germany), while HOS cells were grown and passaged in Eagle's MEM (EMEM)/Hanks' (Carl Roth GmbH + Co. KG, Karlsruhe, Germany). Both media were supplemented with fetal calf serum (FCS) superior (10%, Biochrom GmbH, Berlin, Germany) and penicillin/streptomycin (1%, Biochrom GmbH, Berlin, Germany). The EMEM/Hanks' medium was also supplemented with non-essential amino acids (1%, Biochrom GmbH, Berlin, Germany) and L-glutamine (2 mmol, Biochrom GmbH, Berlin, Germany). HUVEC cells were grown and passaged in an endothelial cell grown medium kit (Promocell, Heidelberg, Germany) and were supplemented with penicillin/streptomycin (1%). All cell lines were grown and passaged in cell culture flasks or multi-well plates (Greiner Bio-One, Frickenhausen, Germany). For passaging, a trypsin (0.05%)/ethylene-diamine-tetraacetic (0.02%) acid solution (Biochrom GmbH, Berlin, Germany) was used. The cells were plated with a density of 50,000 cells per well in 24-well culture plates. After hot air sterilization of the test samples, the cells were seeded on the top of the samples ($\approx 5 \times 5$ mm^2), which were embedded in agarose (2%, Agarose NEEO, Carl Roth GmbH + Co. KG, Karlsruhe, Germany) in respective culture mediums. After 72 h incubation (in a humidified atmosphere at 37 °C and 5% CO_2), cell proliferation was determined. Therefore, a crystal violet staining (CV) assay was performed [76]. After 72 h, the cells were fixed with glutaraldehyde (2%, Sigma-Aldrich, Merck KGaA, Darmstadt, Germany) in phosphate-buffered saline (PBS) for 20 min. Subsequently, the supernatant was removed, and the cells were stained with CV (0.1%, Carl Roth GmbH + Co. KG, Karlsruhe, Germany) in deionized water. After 30 min, the pigment was removed, and the samples were removed from the agarose. After washing with deionized water, the test samples were air-dried. For 1 h, Triton X-100 (2%, Sigma-Aldrich, Merck KGaA, Darmstadt, Germany) in deionized water was added, so that CV was dissolved from the cells. Finally, the supernatant was transferred in a 96-well microtiter plate and the absorbance was determined at 570 nm using a 96-well microplate reader (MRX microplate reader, Dynatech Laboratories, Denkendorf, Germany). The experiments were carried out six times. Furthermore, the supernatant of L-929 and HOS was analyzed of the cytokine Il-6 expression (R&D Systems DuoSet, R&D Systems Inc., Minneapolis, MN, USA).

2.2.2. Antibacterial Examinations

To analyze the bacterial behavior on test samples, two different samples were used—*Escherichia coli* (ATCC 25922) and *Staphylococcus aureus* (ATCC 25923). The bacteria were cultivated on Columbia agar plates with sheep blood (7%, Fisher Scientific GmbH, Schwerte, Germany) for 24 h. The test samples were also embedded in agarose (2%, Agarose LM, Gerbu Biotechnik GmbH, Gailberg, Germany). Bacteria (10^{-6} CFU ml^{-1}) were seeded on the test samples. After 48 h of incubation at 37 °C, cell proliferation was analyzed, similarly to the cell proliferation, with a CV assay. The experiment was performed in four biological replicates.

3. Results and Discussion

3.1. Microstructure and Nanostructure of as-Built and Heat-Treated Ti-6Al-7Nb

The microstructure of ($\alpha + \beta$) titanium strongly depends on the cooling rates and the quenching parameters, from the β-phase field at higher temperatures and the following heat treatment, respectively [77,78]. Figure 3 shows EBSD maps of the Ti-6Al-7Nb alloy after additive manufacturing and subsequent heat treatments. The EBSD map of the additively processed Ti-6Al-7Nb shows a very fine, acicular microstructure, see Figure 3a,c. The fine-lamellar α' grain structures are strongly oriented inside the prior β grains. The size of these fine needles decreased with an increased cooling rate during solidification [37,79,80]. Figure 3b shows the reconstructed parent β grains using MTEX software. During solidification, the bcc β-phase preferentially grows in the <100> direction; therefore, the elongated, columnar primary β grains evolve parallel to the BD [81–83]. The resulting anisotropy and primary β grains affect the mechanical properties of the specimens [84]. Due to the fast cooling and passing the $\beta_{Transus}$ temperature, these β grains transform to α'-phase, according to the Burgers relation, in 12 possible transformation variants [81,85–87]. Due to the high cooling rate, the probability of α' formation is very high [37,88,89]. The rapid cooling leads to a martensitic transformation and a limitation of diffusional transformation [90]. Figure 3c shows the cross-section perpendicular to the BD. Areas with similar crystallographic orientations are observable inside the grain boundaries of the probable parent β grains. Figure 3d demonstrates the microstructure for post-treatment HT1. An unexpected, coarse lamella-like microstructures evolved instead of equiaxed grains [60]. The microstructure is dependent on the initial microstructure and dislocations before the heat treatment. HT2 results in a coarse microstructure and huge grains due to temperatures above $\beta_{Transus}$ and grain growth. The grain orientation seems random, see Figure 3e. Figure 3f exhibits the microstructure after stress relief treatment (HT3). Only minor changes in the microstructure occurred compared with the as-built condition. Areas with similar grain orientation are present and the prior β grain boundaries are noticeable. The detected phase distribution of α, α', and β titanium, respectively, is summarized in Table 1.

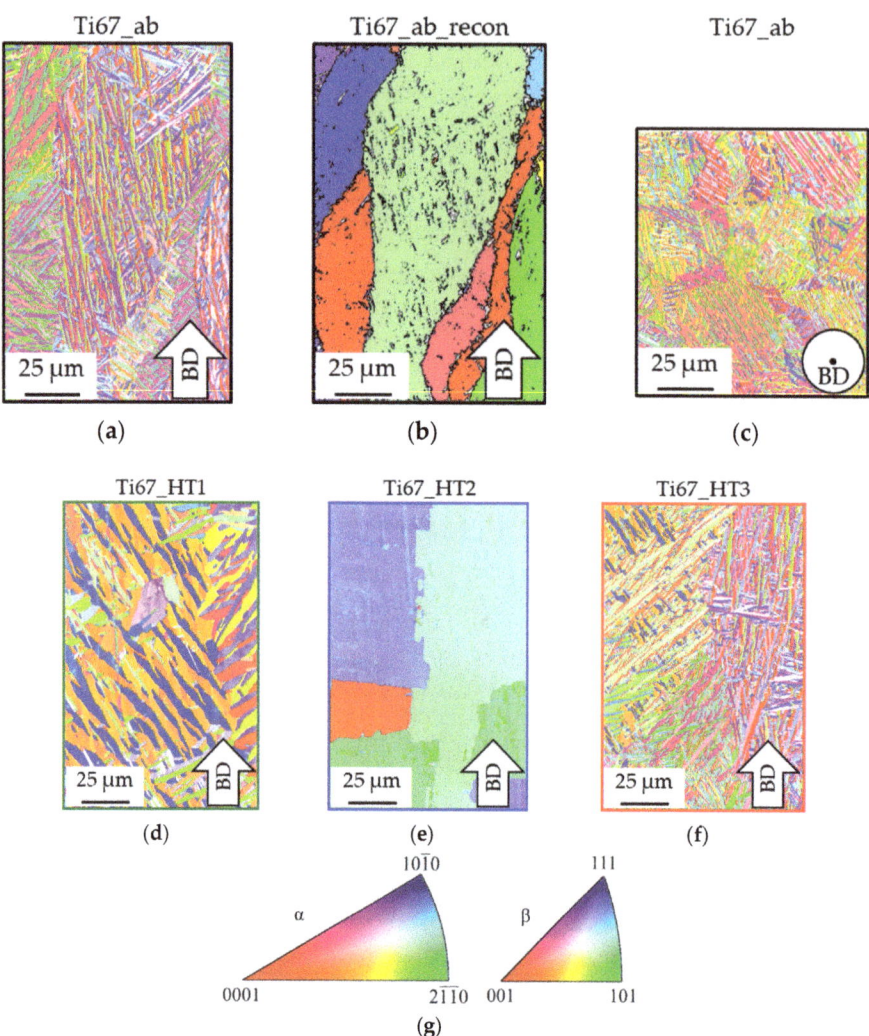

Figure 3. (**a**) Inverse pole figure (IPF) of as-built Ti-6Al-7Nb; (**b**) β parent grain reconstruction of (**a**); IPF of as-built (top view) (**c**); HT1 (**d**); HT2 (**e**); and HT3 (**f**) Ti-6Al-7Nb; and (**g**) color legend for inverse pole figure.

Figure 4 summarizes the TEM images, the HRTEM images, and the fast Fourier transformation (FFT) pattern of the as-built sample. The α' martensite laths are visible in Figure 4a. The width of the lath (dark contrast in Figure 4a) is 280 nm. Figure 4b is the FFT pattern from the white circular area in Figure 4a. Figure 4c is the FFT pattern from the yellow circular area in Figure 4b. The diffraction pattern confirms the hexagonal crystal structure of α' martensite in both laths. In addition, an HRTEM image of the interface between two laths (black square region in Figure 4a) is presented in Figure 4d. The red line highlights the interface between the two α' martensite laths. Figure 4e is the magnified HRTEM image of the black square region in Figure 4d. The $(0\bar{1}11)$ plane of one lath intersects the $(\bar{1}010)$ plane of another lath at 45°. Furthermore, Figure 4f shows the HAADF-STEM image and Figure 4g the EDS maps of the as-built sample. Here, laths of α′ martensite are apparent. The EDS maps of the black square region in Figure 4g show that no segregation of alloying elements is observable, and the titanium, aluminum, and niobium distribution is homogenous in the laths.

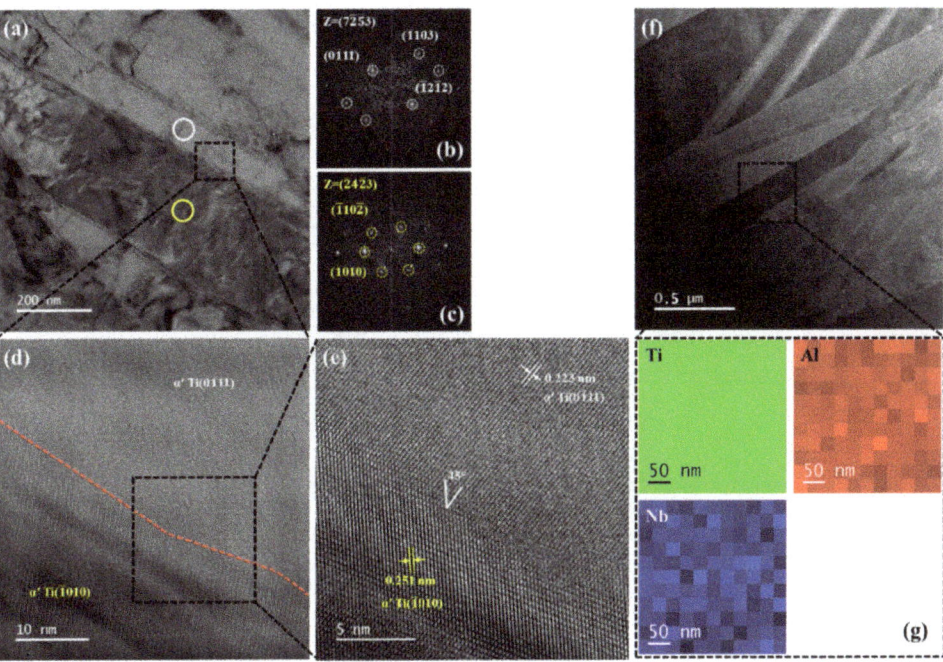

Figure 4. (**a**) TEM image of the as-built sample; (**b**,**c**) FFT pattern taken from the white and yellow circular areas in (**a**); (**d**) HRTEM image of the black square region in (**a**), the red line shows the boundary between two α′-martensite laths; (**e**) HRTEM image of the black square region in (**d**); (**f**) HAADF-STEM image of the as-built sample; (**g**) EDS maps of the black square region in (**f**).

Figure 5 shows the TEM images, HRTEM images, and diffraction pattern of the HT3 sample. The α laths are apparent in Figure 5a. In Figure 5a, some α laths are relatively coarse (≈170 nm). However, the α laths are relatively fine (≈ 40 nm). In this context, Figure 5b shows an HRTEM image taken from the region within the black square in Figure 5a. The lattice plane of two α laths is identified as ($\bar{1}010$). The red lines point out the boundary between two α laths. For a deeper insight, Figure 5d is the magnified view of the black square region in Figure 5b. The interplanar spacing of the ($\bar{1}010$) planes are measured as ≈ 0.252 nm. However, the two ($\bar{1}010$) planes are misoriented by 18°. The diffraction pattern in Figure 5c confirms the zone axis to be ($\bar{1}2\bar{1}1$). Again, the diffraction pattern shows the ($\bar{1}010$) planes to be misoriented by 18°. Figure 5e shows the HAADF-STEM image and EDS maps of the stress relief treated sample, where α laths are visible. Finally, Figure 5f is the EDS maps of a triple junction between three α laths from the black square region in Figure 5e. An enrichment of niobium and a depletion of aluminum is observed at the triple junction. The α lath boundary regions are enriched in niobium and depleted in aluminum. This leads to the white contrast in the HAADF-STEM image in Figure 5e.

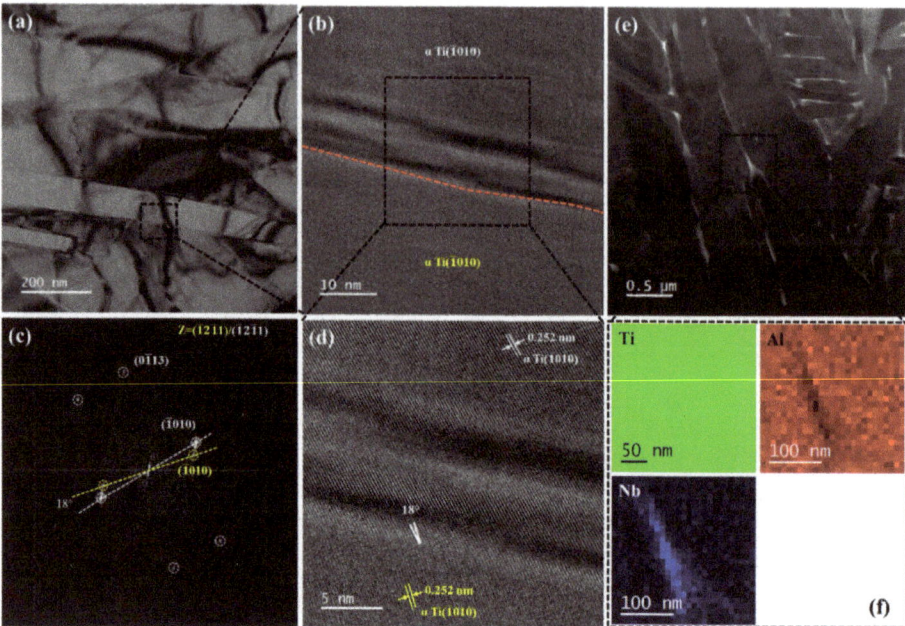

Figure 5. (a) TEM image of the HT3 sample; (b) HRTEM image from the black square area in (a), the red line indicates the boundary between two α-laths; (c) FFT pattern from the black square area in (b); (d) HRTEM image from the black square in (b); (e) HAADF-STEM image of the HT3 sample; (f) EDS maps of black square region in (e).

Figure 6 shows XRD patterns for the as-built and heat-treated Ti-6Al-7Nb samples. The XRD pattern of the sample in the as-built condition is characteristic for the hexagonal close-packed (hcp) α'/α structure of titanium alloys. Due to the hexagonal structure of both the α' and α phases with similar lattice parameters, a separation or distinction between both phases based on XRD is not possible [59]. Since additively manufactured alloys are characterized by high local cooling rates, the α' martensitic phase is assumed as dominating component, based on the findings of Xu et al. [52]. Stress relief heat treatment at 600 °C does not lead to any significant microstructural changes in the sample HT3, since the $\alpha' \rightarrow \alpha + \beta$ transformation is known to start at higher temperatures (≈760 °C) [59]. As shown for sample HT1, recrystallization annealing at 925 °C leads to the formation of the bcc β-structure of Ti and probably to a decomposition of α' to α. HT2 treatment was carried out above the $\beta_{Transus}$ temperature and did result in decomposition of the as-fabricated α'/α-phase [59]. Therefore, the β-phase is formed in the HT2 treated specimens. This is in good agreement with reported transformations and temperatures between 735 °C and 1050 °C for the $\alpha \rightarrow \beta$ reaction [59]. Post-treatment HT1 and HT2 lead to a decrease in the α'/α-Ti peak width, indicating grain growth. For the samples HT1 and HT2 additional peaks were observed between 2θ = 42–45°. Due to the large width of the peaks, several similar phases with small differences in stoichiometry are most likely. As reported by Bolzoni et al., titanium and aluminum can form binary phases, e.g., titanium aluminates, above 660 °C due to diffusion processes [91]. Therefore, the formation of small fractions of Ti_3Al, $TiAl$, and/or $TiAl_3$ is concluded for the heat treatments HT1 and HT2.

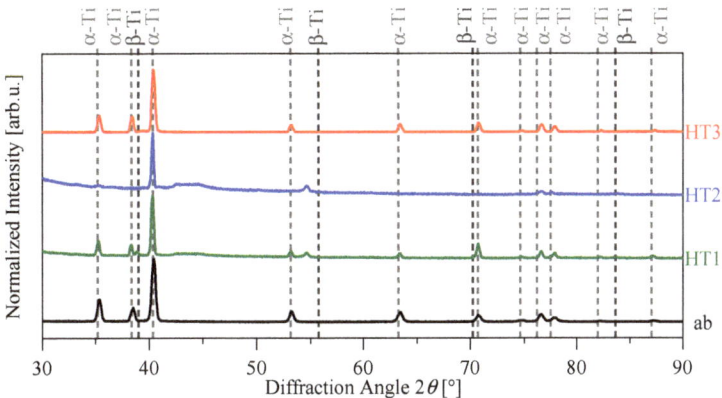

Figure 6. XRD patterns of as-built and heat-treated Ti-6Al-7Nb samples.

Table 1. Phase fraction detected by means of EBSD and results of quasi-static and fatigue tests of the LPBF fabricated, as-built, and heat-treated Ti-6Al-7Nb, including the minimum values regarding standard ISO 5832-11 [92].

State	Condition	α Phase	β Phase	$R_{p0.2}$ [MPa]	R_m [MPa]	E [GPa]	A [%]	σ'_f	b	ε'_f	c
ab	as-built	97.0%	3.0%	940 ± 14	1109 ± 3	105 ± 3	14.4 ± 0.9	89.7	−0.102	765.04	−1.518
HT1	925 °C/4 h	98.3%	1.7%	870 ± 11	934 ± 7	108 ± 6	8.8 ± 2.3	49.9	−0.027	191.97	−1.166
HT2	1050 °C/2 h	97.8%	2.2%	718 ± 4	791 ± 6	115 ± 3	12.0 ± 1.1	52.2	−0.036	10.103	−0.719
HT3	600 °C/4 h	97.7%	2.3%	1045 ± 9	1110 ± 10	116 ± 2	12.5 ± 0.9	100.7	−0.097	139.84	−1.167
ISO	–	–	–	800	900	–	10	–	–	–	–

3.2. Mechanical Properties of Ti-6Al-7Nb

Figure 7 displays the results for different mechanical properties of the monotonic tensile tests with images of fracture surfaces. The stress–strain curves for the monotonic test are presented in Figure 7a. For the different conditions, the mean values of Young´s moduli, E, tensile yield strengths, $R_{p0,2}$, ultimate tensile strengths, R_m, and plastic elongations, A, are compared to the international standard for implants for surgery ISO 5832-11 in Figure 7c and Table 1 [92]. The martensitic structure of the additively processed (α + β) titanium leads to high values for the mechanical properties regarding ultimate tensile and yield strength compared with wrought and conventionally processed materials, especially with an α-dominant equiaxed microstructure [7]. The different material behavior might be explained by the different microstructures and phase composition, which have a strong impact on the mechanical properties. Typically, in conventional material, the size and morphology of the α grains determine the mechanical properties, while in LPBF the colony size and the size of the α or rather α′ laths control the properties. Plastic deformation tends to be the movement of dislocations. Conventional material with larger α grains enables the deformation with less dislocation pileups, whereas the smaller grains in the additively processed material increase the dislocation pileups [93]. The tensile properties are, inter alia, caused by the higher residual stresses, the martensitic transformation during fast cooling to the α′ instead of the α phase, and grain refinement as a strengthening mechanism described by the Hall–Petch equation [94]. The Young´s moduli vary within a certain range between 105 GPa–116 GPa. Due to heat treatments and microstructural changes, the average grain size increases. Sliding effects are mainly detected between grains; therefore, the breaking elongation should increase [95]. Another reason for changes in breaking elongation is the presence of β titanium. An increased amount of bcc β titanium should improve the ductile behavior [57,95]. EBSD phase analysis showed the highest fraction of β titanium for as-built conditions (3.0%), the lowest fraction for HT1 (1.7%), and between for HT2 (2.2%)

and HT3 (2.3%), in accordance to the breaking elongation, see Figure 7c. According to the XRD analysis only for HT1 and HT2 β titanium is verified and may, therefore, explain the lower tensile strength for these conditions but not the reduced breaking elongations.

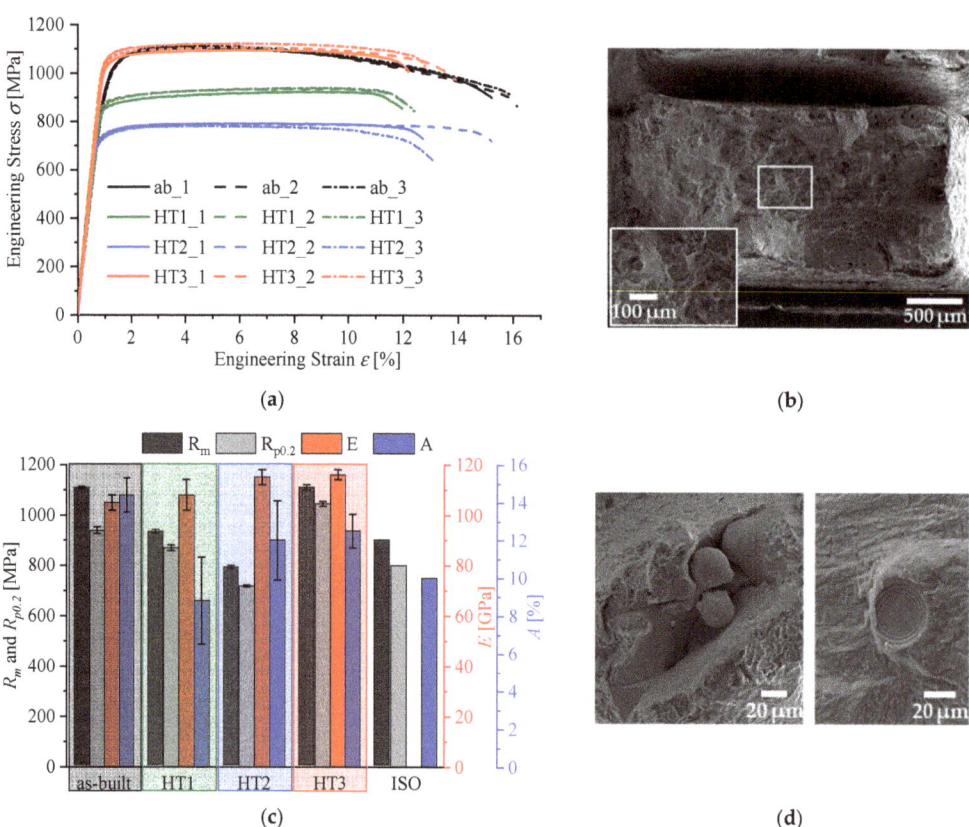

Figure 7. (**a**) Stress–strain curves of the as-built and heat-treated samples; (**b**) SEM images of the tensile fracture surface of an as-built Ti-6Al-7Nb specimen; (**c**) mechanical properties of the different conditions including minimum values for Ti-6Al-7Nb according to ISO 5832-11 (ISO) [92]; (**d**) SEM images of a characteristic LPBF defects, lack of fusion with unmelted powder particles (left) and gas pore (right), on a tensile fracture surface of an as-built Ti-6Al-7Nb specimen.

The ε_A-N_f plots for the different conditions are approximated with the Coffin–Manson–Basquin relation based on the elastic and plastic strain superposition explained in Figure 2. The results are depicted in Figure 8a. The approximations represent the relation between the total strain ε_A and the fatigue life N_f of Ti-6Al-7Nb alloy in different conditions. The Coffin–Manson–Basquin fatigue life approximation are determined as follows:

$$\varepsilon_{A,\,\text{as-built}} = 765.04 \cdot (2N_f)^{-1.518} + 0.8545 \cdot (2N_f)^{-0.102}; \tag{6}$$

$$\varepsilon_{A,\text{HT1}} = 191.97 \cdot (2N_f)^{-1.166} + 0.4616 \cdot (2N_f)^{-0.027}; \tag{7}$$

$$\varepsilon_{A,\,\text{HT2}} = 10.103 \cdot (2N_f)^{-0.719} + 0.4535 \cdot (2N_f)^{-0.036}; \tag{8}$$

$$\varepsilon_{A,\,\text{HT3}} = 139.84 \cdot (2N_f)^{-1.167} + 0.8677 \cdot (2N_f)^{-0.097}. \tag{9}$$

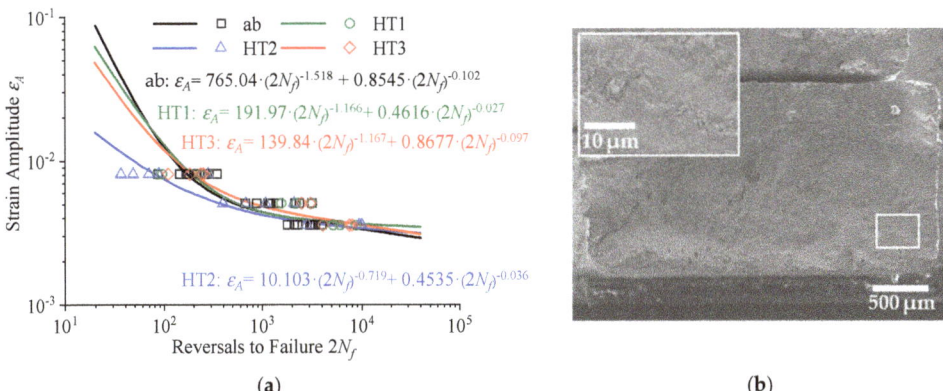

Figure 8. (**a**) Coffin–Manson–Basquin plots for Ti-6Al-7Nb samples in as-built, HT1, HT2, and HT3 conditions; (**b**) SEM images of a fracture surface of an as-built Ti-6Al-7Nb fatigue specimens, with fatigue striations, probably oriented perpendicular to the crack growth direction.

Figure 8a depicts the fatigue life of the different conditions. The fatigue performance for higher strain amplitudes ($2N_f < 10^2$ reversals) of the as-built condition is better, compared with the HT1, HT2, and HT3 conditions. The fatigue performance in LCF range is connected to the monotonic tensile performance in terms of tensile strength. Higher strengths tend to lead to higher, tolerable strain amplitudes and higher fatigue life. Fatigue strength is generally high as higher monotonic strength hinders microplasticity and eventually local damage [42]. The curves of the as-built HT1 and HT3 conditions intersect around 10^2 cycles and are comparable up to 10^4 cycles. While the performance for higher strain amplitudes ($2N_f < 10^4$ reversals) for the HT2 condition is the worst, the performance gets better for lower strain amplitudes. As the process-induced pores and defects are not affected by these post-treatments, other causes are likely to be decisive, such as the microstructures or residual stresses. For post treatment HT2 the reduced strength and the microstructural notches could lead to accumulation of local damage and finally results in early crack initiation [42]. The fatigue crack growth behavior of post-treated Ti-6Al-7Nb is affected by the microstructure. Depending on the crack growth direction and grain long axis different crack growth rates are probable [96–98]. For lower strain amplitudes, the three heat-treated conditions show superior fatigue behavior compared with the as-built conditions, attributed to the reduced residual stresses and microstructural features [52,56–59,94]. Corresponding to the fatigue ductility exponent c, the heat-treated specimens show smaller gradients, while the as-built condition has the lowest value resulting in the shortest fatigue life. HT1 and HT3 exhibit close fatigue ductility exponent c and, therefore, are probably favorable for HCF applications.

In general, due to miniaturization of samples the fatigue and monotonic tensile behavior could be affected [42]. The monotonic material properties, such as Young´s moduli E, tensile yield strengths $R_{p0,2}$, ultimate tensile strengths Rm, plastic elongations A, the fatigue parameters—such as fatigue strength coefficients σ'_f and exponents b—and fatigue ductility coefficients ε'_f and exponents c, of the Coffin–Manson–Basquin equation, are given in Table 1.

3.3. Fracture Behavior

The fracture surface of the as-built Ti-6Al-7Nb alloy demonstrates mainly ductile fracture behavior, see Figure 7b. The propensity of cleavage fracture decreases with decreasing grain size. Therefore, the fracture surface of the additively processed materials shows cleavage facets with high amounts of dimples at the grain boundaries [58]. These fractured surfaces show small, shallow dimples on quasi-cleavage fracture surfaces, and transgranular facets, confirming the minor brittle fracture behavior of the additively processed specimens. Features, as a result of the LPBF fabrication, such as pores, unmelted powder particles, and defects occur on fracture surfaces in as-built and heat-treated conditions, see Figure 7d. These build defects typically are perpendicular to the build direction and pulled apart by the tensile load during monotonic and fatigue tests [99,100]. The fracture surfaces of fatigue tests tend to be smoother for lower strain amplitudes. Distinct fatigue striations, perpendicular to the crack growth direction, can be detected for all conditions and strain amplitudes, see Figure 8b. The striations are close to the probable crack initiation spot. Typical forced rupture areas are difficult to spot.

3.4. Characterization of Roughness, Cytocompatibility, and Antibacterial Effects

The roughness of the specimens without surface treatment (NT) exhibits values for Ra of 7.5 ± 0.3 µm and Rz of 49.6 ± 2.2 µm. Due to unmelted powder particles and the surface of the layer-wise fabrication, the roughness of NT specimens is higher than for the blasted specimens (ab) with Ra of 4.8 ± 0.2 µm and Rz of 34.3 ± 1.2 µm, which have smoother surfaces, on account of the mechanically post-processed treatment.

There was no effect of cell proliferation of the two examined samples on murine L-929 cells and human HUVEC cells. A slight tendency of diminished cell proliferation was observed in human HOS cells on blasted Ti-6Al-7Nb. Thus, the examined samples show a proper biocompatibility behavior for the different cell lines, used in the present study, see Figure 9. Furthermore, no increase in cytokine release of Il-6 was detected (data not shown). Concerning antibacterial effects, *E. coli* proliferation was not influenced by the different samples, while the proliferation of *S. aureus* was reduced by the blasted surface of Ti-6Al-7Nb, see Figure 9.

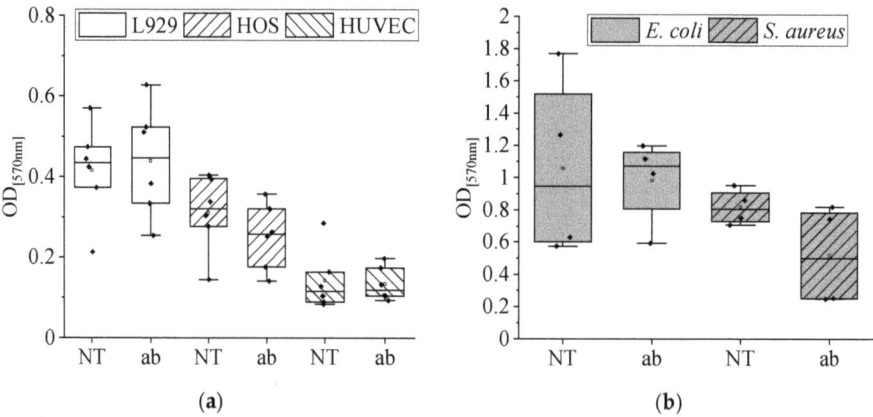

Figure 9. (a) Effects of different surfaces on the proliferation of L929, HOS and HUVEC cells after 72 h incubation, and (b) bacterial proliferation of *E. coli* and *S. aureus* after 24 h incubation; OD = optical density at 570 nm; $n = 4$–6.

As described by Schweikl et al., blasting with Al_2O_3 particles led to incorporation of Al into the outer surface layers [101]. Therefore, the increased concentration of Al on the specimens' surfaces could result in decreasing the proliferation behavior of HOS cells and *S. aureus* [102–104]. Rough sandblasted surfaces with sharp ridges and edges could influence the proliferation and it seems to appear that the quality of cell contact on rough surfaces is related to the minimum width of the cavity [101]. As reported in various studies, blasting leads to the presence of severely plastic deformed layers and, therefore, to strain-hardening, as well as to compressive residual stresses [105–107]. Due to the antibacterial effect regarding the proliferation of *S. aureus* and because of probably better fatigue properties of blasted specimens, the specimens for the mechanical tests were blasted with Al_2O_3 particles.

4. Conclusions

The microstructure and the monotonic and fatigue behaviors of additively processed Ti-6Al-7Nb were investigated. Different surface conditions were examined regarding biocompatible properties, such as cytocompatibility and antibacterial effects. The following conclusions can be drawn from these investigations:

- Ti-6Al-7Nb shows significantly different microstructures in the as-built, stress-relieved, recrystallized, and β-annealed conditions. While the α'/α structure is dominant within the as-built state, the formation and precipitation of the β-phase are manageable by a vacuum heat treatment above 925 °C as analyzed using XRD and EBSD.
- There are significant differences in the monotonic tensile properties of the various conditions. Concerning the ISO values, the as-built and stress-relieved conditions (HT3) are favorable. Tensile and yield strength, as well as breaking elongation, are higher than the demanded values, but the specimens' dimensions have to be considered. HT1 and HT2 do not fulfill the requirements, probably due to the present β-phase. Heat treatments, such as HT1 and HT2, can significantly affect the microstructure and may tend to soften the lattice structures and decrease the residual stresses and, therefore, lead to significantly reduced tensile strength.
- LCF life is higher for lower strain amplitudes in the heat-treated specimens than in as-built conditions, which can be attributed to decreased residual stresses as well as to microstructural differences. HT2 shows the highest fatigue life for lower strain amplitudes, while the as-built condition has higher service life for higher strain amplitudes. Regarding overall performance, quasi-static results, and LCF performance, stress relief treatment (HT3) is favorable. The material behavior for HCF loading still has to be determined.
- Transgranular facets characterize the fracture surface of the additively processed Ti-6Al-7Nb, confirming the ductile behavior. Small, shallow dimples on quasi-cleavage fracture surfaces are visible. Fatigue fracture surfaces are characterized by fatigue striations and remaining forced rupture surfaces. Defects may have less impact on quasi-static but a high impact on fatigue behavior.
- Both surfaces, the untreated and the blasted, show good biocompatibility in different cell types (fibroblasts, osteosarcoma cells, and endothelial cells). Only a slight antiproliferative effect was observed for blasted Ti-6Al-7Nb in osteoblasts. An increase in cytokine release of Il-6 was not observed.
- Blasting with high-grade Al_2O_3 is preferable regarding biocompatibility and antibacterial effects. Blasted Ti-6Al-7Nb exhibits an antibacterial effect against *S. aureus* in comparison with not post-treated Ti-6Al-7Nb. *E. coli* was able to grow on both surfaces of Ti-6Al-7Nb similarly. On account of remaining aluminum on the blasted surface, glass bead blasting could be taken into consideration.

Author Contributions: Conceptualization, M.H., K.-P.H. and M.S.; data curation, M.H.; formal analysis, M.H., K.-P.H. and M.S.; funding acquisition, K.-P.H., J.M., W.T. and M.S.; investigation, M.H., D.K., N.F.L.D., D.S., H.O., S.P. and J.M.; methodology, M.H., K.-P.H. and M.S.; project administration, M.H.; resources, J.M., W.T. and M.S.; software, M.H. and D.K.; supervision, K.-P.H., J.M., W.T. and M.S.; validation, M.H., K.-P.H. and M.S.; visualization, M.H.; writing—original draft preparation, M.H.; writing—review and editing, M.H., D.K., N.F.L.D., D.S., H.O., S.P., M.K., K.-P.H., J.M., W.T. and M.S. All authors have read and agreed to the published version of the manuscript.

Funding: This research was funded by the Deutsche Forschungsgemeinschaft (DFG), grant numbers SCHA 1484/45-1, ME 4991/2-1 and TI 343/167-1.

Data Availability Statement: The data that support the findings of this study are available from the corresponding author upon reasonable request.

Acknowledgments: The research was performed with the equipment and base of the LWK and DMRC research infrastructure. The authors are grateful to the staff members of the LWK and DMRC.

Conflicts of Interest: The authors declare no conflict of interest. The funders had no role in the design of the study; in the collection, analyses, or interpretation of data; in the writing of the manuscript, or in the decision to publish the results.

References

1. Abdel-Hady Gepreel, M.; Niinomi, M. Biocompatibility of Ti-alloys for long-term implantation. *J. Mech. Behav. Biomed. Mater.* **2013**, *20*, 407–415. [CrossRef] [PubMed]
2. Geetha, M.; Singh, A.K.; Asokamani, R.; Gogia, A.K. Ti based biomaterials, the ultimate choice for orthopaedic implants—A review. *Prog. Mater. Sci.* **2009**, *54*, 397–425. [CrossRef]
3. Long, M.; Rack, H. Titanium alloys in total joint replacement—A materials science perspective. *Biomaterials* **1998**, *19*, 1621–1639. [CrossRef]
4. Kurtz, S.M.; Ong, K.L.; Schmier, J.; Mowat, F.; Saleh, K.; Dybvik, E.; Kärrholm, J.; Garellick, G.; Havelin, L.I.; Furnes, O.; et al. Future clinical and economic impact of revision total hip and knee arthroplasty. *J. Bone Jt. Surg. Am.* **2007**, *89* (Suppl. S3), 144–151. [CrossRef]
5. Vandenbroucke, B.; Kruth, J.-P. Selective laser melting of biocompatible metals for rapid manufacturing of medical parts. *Rapid Prototyp. J.* **2007**, *13*, 196–203. [CrossRef]
6. Iijima, D. Wear properties of Ti and Ti–6Al–7Nb castings for dental prostheses. *Biomaterials* **2003**, *24*, 1519–1524. [CrossRef]
7. Niinomi, M. Mechanical biocompatibilies of titanium alloys for biomedical applications. *J. Mech. Behav. Biomed. Mater.* **2008**, *1*, 30–42. [CrossRef]
8. Hollander, D.A.; von Walter, M.; Wirtz, T.; Sellei, R.; Schmidt-Rohlfing, B.; Paar, O.; Erli, H.-J. Structural, mechanical and in vitro characterization of individually structured Ti-6Al-4V produced by direct laser forming. *Biomaterials* **2006**, *27*, 955–963. [CrossRef] [PubMed]
9. Murr, L.E.; Quinones, S.A.; Gaytan, S.M.; Lopez, M.I.; Rodela, A.; Martinez, E.Y.; Hernandez, D.H.; Martinez, E.; Medina, F.; Wicker, R.B. Microstructure and mechanical behavior of Ti-6Al-4V produced by rapid-layer manufacturing, for biomedical applications. *J. Mech. Behav. Biomed. Mater.* **2009**, *2*, 20–32. [CrossRef]
10. Kobayashi, E.; Wang, T.J.; Doi, H.; Yoneyama, T.; Hamanaka, H. Mechanical properties and corrosion resistance of Ti-6Al-7Nb alloy dental castings. *J. Mater. Sci. Mater. Med.* **1998**, *9*, 567–574. [CrossRef]
11. Kobayashi, E.; Mochizuki, H.; Doi, H.; Yoneyama, T.; Hanawa, T. Fatigue Life Prediction of Biomedical Titanium Alloys under Tensile/Torsional Stress. *Mater. Trans.* **2006**, *47*, 1826–1831. [CrossRef]
12. Srimaneepong, V.; Yoneyama, T.; Kobayashi, E.; Doi, H.; Hanawa, T. Comparative study on torsional strength, ductility and fracture characteristics of laser-welded alpha+beta Ti-6Al-7Nb alloy, CP Titanium and Co-Cr alloy dental castings. *Dent. Mater.* **2008**, *24*, 839–845. [CrossRef]
13. Liu, X.; Chu, P.; Ding, C. Surface modification of titanium, titanium alloys, and related materials for biomedical applications. *Mater. Sci. Eng. R Rep.* **2004**, *47*, 49–121. [CrossRef]
14. Kuroda, D.; Niinomi, M.; Morinaga, M.; Kato, Y.; Yashiro, T. Design and mechanical properties of new β type titanium alloys for implant materials. *Mater. Sci. Eng. A Struct. Mater.* **1998**, *243*, 244–249. [CrossRef]
15. Li, Y.; Yang, C.; Zhao, H.; Qu, S.; Li, X.; Li, Y. New Developments of Ti-Based Alloys for Biomedical Applications. *Materials* **2014**, *7*, 1709–1800. [CrossRef]
16. López, M.; Gutiérrez, A.; Jiménez, J. In vitro corrosion behaviour of titanium alloys without vanadium. *Electrochim. Acta* **2002**, *47*, 1359–1364. [CrossRef]
17. Chlebus, E.; Kuźnicka, B.; Kurzynowski, T.; Dybała, B. Microstructure and mechanical behaviour of Ti—6Al—7Nb alloy produced by selective laser melting. *Mater. Charact.* **2011**, *62*, 488–495. [CrossRef]
18. Surmeneva, M.; Grubova, I.; Glukhova, N.; Khrapov, D.; Koptyug, A.; Volkova, A.; Ivanov, Y.; Cotrut, C.M.; Vladescu, A.; Teresov, A.; et al. New Ti–35Nb–7Zr–5Ta Alloy Manufacturing by Electron Beam Melting for Medical Application Followed by High Current Pulsed Electron Beam Treatment. *Metals* **2021**, *11*, 1066. [CrossRef]
19. Tamilselvi, S.; Raman, V.; Rajendran, N. Corrosion behaviour of Ti–6Al–7Nb and Ti–6Al–4V ELI alloys in the simulated body fluid solution by electrochemical impedance spectroscopy. *Electrochim. Acta* **2006**, *52*, 839–846. [CrossRef]
20. Metikoš-Huković, M.; Kwokal, A.; Piljac, J. The influence of niobium and vanadium on passivity of titanium-based implants in physiological solution. *Biomaterials* **2003**, *24*, 3765–3775. [CrossRef]
21. Schmidt, M.; Merklein, M.; Bourell, D.; Dimitrov, D.; Hausotte, T.; Wegener, K.; Overmeyer, L.; Vollertsen, F.; Levy, G.N. Laser based additive manufacturing in industry and academia. *CIRP Ann.* **2017**, *66*, 561–583. [CrossRef]
22. Bourell, D.; Kruth, J.P.; Leu, M.; Levy, G.; Rosen, D.; Beese, A.M.; Clare, A. Materials for additive manufacturing. *CIRP Ann.* **2017**, *66*, 659–681. [CrossRef]
23. Wohlers, T.T.; Campbell, I.; Diegel, O.; Kowen, J. *Wohlers Report 2018. 3D Printing and Additive Manufacturing State of the Industry: Annual Worldwide Progress Report*; Wohlers Associates, Inc.: Fort Collins, CO, USA, 2018; ISBN 0991333241.

24. Attar, H.; Prashanth, K.G.; Chaubey, A.K.; Calin, M.; Zhang, L.C.; Scudino, S.; Eckert, J. Comparison of wear properties of commercially pure titanium prepared by selective laser melting and casting processes. *Mater. Lett.* 2015, *142*, 38–41. [CrossRef]
25. Dai, N.; Zhang, L.-C.; Zhang, J.; Chen, Q.; Wu, M. Corrosion behavior of selective laser melted Ti-6Al-4 V alloy in NaCl solution. *Corros. Sci.* 2016, *102*, 484–489. [CrossRef]
26. Yang, Y.; Chen, Y.; Zhang, J.; Gu, X.; Qin, P.; Dai, N.; Li, X.; Kruth, J.-P.; Zhang, L.-C. Improved corrosion behavior of ultrafine-grained eutectic Al-12Si alloy produced by selective laser melting. *Mater. Des.* 2018, *146*, 239–248. [CrossRef]
27. Melchels, F.P.; Domingos, M.A.; Klein, T.J.; Malda, J.; Bartolo, P.J.; Hutmacher, D.W. Additive manufacturing of tissues and organs. *Prog. Polym. Sci.* 2012, *37*, 1079–1104. [CrossRef]
28. Herzog, D.; Seyda, V.; Wycisk, E.; Emmelmann, C. Additive manufacturing of metals. *Acta Mater.* 2016, *117*, 371–392. [CrossRef]
29. Guo, N.; Leu, M.C. Additive manufacturing: Technology, applications and research needs. *Front. Mech. Eng.* 2013, *8*, 215–243. [CrossRef]
30. Lewandowski, J.J.; Seifi, M. Metal Additive Manufacturing: A Review of Mechanical Properties. *Annu. Rev. Mater. Res.* 2016, *46*, 151–186. [CrossRef]
31. Yap, C.Y.; Chua, C.K.; Dong, Z.L.; Liu, Z.H.; Zhang, D.Q.; Loh, L.E.; Sing, S.L. Review of selective laser melting: Materials and applications. *Appl. Phys. Rev.* 2015, *2*, 41101. [CrossRef]
32. Tolosa, I.; Garciandía, F.; Zubiri, F.; Zapirain, F.; Esnaola, A. Study of mechanical properties of AISI 316 stainless steel processed by "selective laser melting", following different manufacturing strategies. *Int. J. Adv. Manuf. Technol.* 2010, *51*, 639–647. [CrossRef]
33. Nezhadfar, P.D.; Burford, E.; Anderson-Wedge, K.; Zhang, B.; Shao, S.; Daniewicz, S.R.; Shamsaei, N. Fatigue crack growth behavior of additively manufactured 17-4 PH stainless steel: Effects of build orientation and microstructure. *Int. J. Fatigue* 2019, *123*, 168–179. [CrossRef]
34. Kluczyński, J.; Śnieżek, L.; Grzelak, K.; Torzewski, J.; Szachogłuchowicz, I.; Wachowski, M.; Łuszczek, J. Crack Growth Behavior of Additively Manufactured 316L Steel-Influence of Build Orientation and Heat Treatment. *Materials* 2020, *13*, 3259. [CrossRef]
35. Zhang, M.; Sun, C.-N.; Zhang, X.; Wei, J.; Hardacre, D.; Li, H. High cycle fatigue and ratcheting interaction of laser powder bed fusion stainless steel 316L: Fracture behaviour and stress-based modelling. *Int. J. Fatigue* 2019, *121*, 252–264. [CrossRef]
36. Jerrard, P.G.E.; Hao, L.; Evans, K.E. Experimental investigation into selective laser melting of austenitic and martensitic stainless steel powder mixtures. *Proc. Inst. Mech. Eng. B J. Eng. Manuf.* 2009, *223*, 1409–1416. [CrossRef]
37. Baufeld, B.; Brandl, E.; van der Biest, O. Wire based additive layer manufacturing: Comparison of microstructure and mechanical properties of Ti–6Al–4V components fabricated by laser-beam deposition and shaped metal deposition. *J. Mater. Process. Technol.* 2011, *211*, 1146–1158. [CrossRef]
38. Brandl, E. *Microstructural and Mechanical Properties of Additive Manufactured Titanium (Ti-6Al-4V) Using Wire. Evaluation with Respect to Additive Processes Using Powder and Aerospace Material Specifications*; Dissertation, Brandenburg University of Technology Cottbus-Senftenberg: Cottbus, Germany, 2010; ISBN 978-3-8322-9530-1.
39. Khorasani, A.; Gibson, I.; Goldberg, M.; Littlefair, G. On the role of different annealing heat treatments on mechanical properties and microstructure of selective laser melted and conventional wrought Ti-6Al-4V. *Rapid Prototyp. J.* 2017, *23*, 295–304. [CrossRef]
40. Liu, S.; Shin, Y.C. Additive manufacturing of Ti6Al4V alloy: A review. *Mater. Des.* 2019, *164*, 107552. [CrossRef]
41. Riemer, A.; Richard, H.A. Crack Propagation in Additive Manufactured Materials and Structures. *Procedia Struct. Integr.* 2016, *2*, 1229–1236. [CrossRef]
42. Leuders, S.; Lieneke, T.; Lammers, S.; Tröster, T.; Niendorf, T. On the fatigue properties of metals manufactured by selective laser melting—The role of ductility. *J. Mater. Res.* 2014, *29*, 1911–1919. [CrossRef]
43. Tillmann, W.; Hagen, L.; Garthe, K.-U.; Hoyer, K.-P.; Schaper, M. Effect of substrate pre-treatment on the low cycle fatigue performance of tungsten carbide-cobalt coated additive manufactured 316 L substrates. *Mater. Werkst.* 2020, *51*, 1452–1464. [CrossRef]
44. Romano, S.; Patriarca, L.; Foletti, S.; Beretta, S. LCF behaviour and a comprehensive life prediction model for AlSi10Mg obtained by SLM. *Int. J. Fatigue* 2018, *117*, 47–62. [CrossRef]
45. Bressan, S.; Ogawa, F.; Itoh, T.; Berto, F. Low cycle fatigue behavior of additively manufactured Ti-6Al-4V under non-proportional and proportional loading. *Frat. Integrita Strutt.* 2019, *13*, 18–25. [CrossRef]
46. Zhang, S.Q.; Li, S.J.; Jia, M.T.; Prima, F.; Chen, L.J.; Hao, Y.L.; Yang, R. Low-cycle fatigue properties of a titanium alloy exhibiting nonlinear elastic deformation behavior. *Acta Mater.* 2011, *59*, 4690–4699. [CrossRef]
47. Awd, M.; Tenkamp, J.; Hirtler, M.; Siddique, S.; Bambach, M.; Walther, F. Comparison of Microstructure and Mechanical Properties of Scalmalloy®Produced by Selective Laser Melting and Laser Metal Deposition. *Materials* 2017, *11*, 17. [CrossRef]
48. Leuders, S.; Thöne, M.; Riemer, A.; Niendorf, T.; Tröster, T.; Richard, H.A.; Maier, H.J. On the mechanical behaviour of titanium alloy TiAl6V4 manufactured by selective laser melting: Fatigue resistance and crack growth performance. *Int. J. Fatigue* 2013, *48*, 300–307. [CrossRef]
49. Wycisk, E.; Siddique, S.; Herzog, D.; Walther, F.; Emmelmann, C. Fatigue Performance of Laser Additive Manufactured Ti–6Al–4V in Very High Cycle Fatigue Regime up to 109 Cycles. *Front. Mater.* 2015, *2*, 72. [CrossRef]
50. Polozov, I.; Sufiiarov, V.; Popovich, A.; Masaylo, D.; Grigoriev, A. Synthesis of Ti-5Al, Ti-6Al-7Nb, and Ti-22Al-25Nb alloys from elemental powders using powder-bed fusion additive manufacturing. *J. Alloy. Compd.* 2018, *763*, 436–445. [CrossRef]
51. Affolter, C.; Thorwarth, G.; Arabi-Hashemi, A.; Müller, U.; Weisse, B. Ductile Compressive Behavior of Biomedical Alloys. *Metals* 2020, *10*, 60. [CrossRef]

52. Xu, C.; Sikan, F.; Atabay, S.E.; Muñiz-Lerma, J.A.; Sanchez-Mata, O.; Wang, X.; Brochu, M. Microstructure and mechanical behavior of as-built and heat-treated Ti–6Al–7Nb produced by laser powder bed fusion. *Mater. Sci. Eng. A Struct. Mater.* **2020**, *793*, 139978. [CrossRef]
53. Hein, M.; Hoyer, K.-P.; Schaper, M. Additively processed TiAl6Nb7 alloy for biomedical applications. *Mater. Werkst.* **2021**, *52*, 703–716. [CrossRef]
54. Bergmann, G.; Bender, A.; Dymke, J.; Duda, G.; Damm, P. Standardized Loads Acting in Hip Implants. *PLoS ONE* **2016**, *11*, e0155612. [CrossRef]
55. Bergmann, G.; Graichen, F.; Rohlmann, A.; Bender, A.; Heinlein, B.; Duda, G.N.; Heller, M.O.; Morlock, M.M. Realistic loads for testing hip implants. *Biomed. Mater. Eng.* **2010**, *20*, 65–75. [CrossRef] [PubMed]
56. Sallica-Leva, E.; Caram, R.; Jardini, A.L.; Fogagnolo, J.B. Ductility improvement due to martensite α′ decomposition in porous Ti-6Al-4V parts produced by selective laser melting for orthopedic implants. *J. Mech. Behav. Biomed. Mater.* **2016**, *54*, 149–158. [CrossRef] [PubMed]
57. Tao, P.; Li, H.; Huang, B.; Hu, Q.; Gong, S.; Xu, Q. Tensile behavior of Ti-6Al-4V alloy fabricated by selective laser melting: Effects of microstructures and as-built surface quality. *China Foundry* **2018**, *15*, 243–252. [CrossRef]
58. Gil Mur, F.X.; Rodríguez, D.; Planell, J.A. Influence of tempering temperature and time on the α′-Ti-6Al-4V martensite. *J. Alloy. Compd.* **1996**, *234*, 287–289. [CrossRef]
59. Liang, Z.; Sun, Z.; Zhang, W.; Wu, S.; Chang, H. The effect of heat treatment on microstructure evolution and tensile properties of selective laser melted Ti6Al4V alloy. *J. Alloy. Compd.* **2019**, *782*, 1041–1048. [CrossRef]
60. Donachie, M.J. *Titanium. A Technical Guide*, 2nd ed.; ASM International: Materials Park, OH, USA, 2000; ISBN 9780871706867.
61. Bachmann, F.; Hielscher, R.; Schaeben, H. Texture Analysis with MTEX—Free and Open Source Software Toolbox. *Solid State Phenom.* **2010**, *160*, 63–68. [CrossRef]
62. Plumtree, A. Cyclic stress–Strain response and substructure. *Int. J. Fatigue* **2001**, *23*, 799–805. [CrossRef]
63. Rice, R.C. *Fatigue Design Handbook*, 2nd ed.; Society of Automotive Engineers: Warrendale, PA, USA, 1988; ISBN 9780898830118.
64. Skelton, R.P.; Maier, H.J.; Christ, H.-J. The Bauschinger effect, Masing model and the Ramberg–Osgood relation for cyclic deformation in metals. *Mater. Sci. Eng. A Struct. Mater.* **1997**, *238*, 377–390. [CrossRef]
65. Nieslony, A.; Dsoki, C.; Kaufmann, H.; Krug, P. New method for evaluation of the Manson–Coffin–Basquin and Ramberg–Osgood equations with respect to compatibility. *Int. J. Fatigue* **2008**, *30*, 1967–1977. [CrossRef]
66. Basquin, O.H. The exponential law of endurance tests. *Am. Soc. Test. Mater.* **1910**, *10*, 625–630.
67. Coffin, L.F. A Study of the Effects of Cyclic Thermal Stresses on a Ductile Metal. *Trans. Am. Soc. Mech.* **1954**, *76*, 931–950.
68. Manson, S.S. *Behavior of Materials Under Conditions of Thermal Stress*; National Advisory Committee for Aeronautics: Edwards, CA, USA, 1953.
69. Suresh, S. *Fatigue of Materials*, 2nd ed.; Reprint; Cambridge Univ. Press: Cambridge, UK, 2004; ISBN 978-0-521-57847-9.
70. ASM International. *Fatigue and Fracture*, 10th ed.; 3. Print; ASM International: Materials Park, OH, USA, 2002; ISBN 9780871703859.
71. Schupbach, P.; Glauser, R.; Bauer, S. Al$_2$O$_3$ Particles on Titanium Dental Implant Systems following Sandblasting and Acid-Etching Process. *Int. J. Biomater.* **2019**, *2019*, 6318429. [CrossRef]
72. Le Guéhennec, L.; Soueidan, A.; Layrolle, P.; Amouriq, Y. Surface treatments of titanium dental implants for rapid osseointegration. *Dent. Mater.* **2007**, *23*, 844–854. [CrossRef] [PubMed]
73. Wennerberg, A.; Albrektsson, T.; Chrcanovic, B. Long-term clinical outcome of implants with different surface modifications. *Eur. J. Oral Implantol.* **2018**, *11* (Suppl. S1), S123–S136.
74. Yuda, A.W.; Supriadi, S.; Saragih, A.S. Surface modification of Ti-alloy based bone implant by sandblasting. In *The 4th Biomedical Engineering's Recent Progress in Biomaterials, Drugs Development, Health, and Medical Devices, Proceedings of the International Symposium of Biomedical Engineering (ISBE), Padang, Indonesia, 22–24 July 2019*; AIP Publishing: Melville, NY, USA, 2019; p. 20015.
75. Yurttutan, M.E.; Keskin, A. Evaluation of the effects of different sand particles that used in dental implant roughened for osseointegration. *BMC Oral Health* **2018**, *18*, 47. [CrossRef]
76. Gillies, R.J.; Didier, N.; Denton, M. Determination of cell number in monolayer cultures. *Anal. Biochem.* **1986**, *159*, 109–113. [CrossRef]
77. Sercombe, T.; Jones, N.; Day, R.; Kop, A. Heat treatment of Ti-6Al-7Nb components produced by selective laser melting. *Rapid Prototyp. J.* **2008**, *14*, 300–304. [CrossRef]
78. Ajeel, S.A.; Alzubaydi, T.L.; Swadi, A.K. Influence of Heat Treatment Conditions on Microstructure of Ti-6Al-7Nb Alloy as Used Surgical Implant Materials. *Eng. Technol.* **2007**, *25*, 431–442.
79. DebRoy, T.; Wei, H.L.; Zuback, J.S.; Mukherjee, T.; Elmer, J.W.; Milewski, J.O.; Beese, A.M.; Wilson-Heid, A.; De, A.; Zhang, W. Additive manufacturing of metallic components—Process, structure and properties. *Prog. Mater. Sci.* **2018**, *92*, 112–224. [CrossRef]
80. Körner, C. Additive manufacturing of metallic components by selective electron beam melting—A review. *Int. Mater. Rev.* **2016**, *61*, 361–377. [CrossRef]
81. Banerjee, S.; Mukhopadhyay, P. *Phase Transformations. Examples from Titanium and Zirconium Alloys*; Elsevier: Amsterdam, The Netherlands; Oxford, UK, 2007; ISBN 9780080421452.
82. Thijs, L.; Verhaeghe, F.; Craeghs, T.; van Humbeeck, J.; Kruth, J.-P. A study of the microstructural evolution during selective laser melting of Ti–6Al–4V. *Acta Mater.* **2010**, *58*, 3303–3312. [CrossRef]

83. Kobryn, P.; Semiatin, S. Microstructure and texture evolution during solidification processing of Ti–6Al–4V. *J. Mater. Process. Technol.* **2003**, *135*, 330–339. [CrossRef]
84. Lütjering, G. Influence of processing on microstructure and mechanical properties of (α + β) titanium alloys. *Mater. Sci. Eng. A Struct. Mater.* **1998**, *243*, 32–45. [CrossRef]
85. Lütjering, G.; Williams, J.C. *Titanium*, 2nd ed.; Springer: Berlin, Germany, 2007; ISBN 9783540713975.
86. Burgers, W.G. On the process of transition of the cubic-body-centered modification into the hexagonal-close-packed modification of zirconium. *Physica* **1934**, *1*, 561–586. [CrossRef]
87. Peters, M.; Leyens, C. *Titan und Titanlegierungen*, 3rd ed.; Wiley-VCH: Weinheim, Germany, 2002; ISBN 9783527611089.
88. Murr, L.E.; Gaytan, S.M.; Ramirez, D.A.; Martinez, E.; Hernandez, J.; Amato, K.N.; Shindo, P.W.; Medina, F.R.; Wicker, R.B. Metal Fabrication by Additive Manufacturing Using Laser and Electron Beam Melting Technologies. *J. Mater. Sci. Technol.* **2012**, *28*, 1–14. [CrossRef]
89. Rehme, O. *Cellular Design for Laser Freeform Fabrication*, 1st ed.; Cuvillier Verlag: Göttingen, Germany, 2010; ISBN 9783869552736.
90. Sieniawski, J.; Ziaja, W.; Kubiak, K.; Motyk, M. Microstructure and Mechanical Properties of High Strength Two-Phase Titanium Alloys. In *Titanium Alloys—Advances in Properties Control*; Sieniawski, J., Ziaja, W., Eds.; InTech: Rijeka, Croatia, 2014; ISBN 978-953-51-1110-8.
91. Bolzoni, L.; Weissgaerber, T.; Kieback, B.; Ruiz-Navas, E.M.; Gordo, E. Mechanical behaviour of pressed and sintered CP Ti and Ti-6Al-7Nb alloy obtained from master alloy addition powder. *J. Mech. Behav. Biomed. Mater.* **2013**, *20*, 149–161. [CrossRef] [PubMed]
92. ISO 5832-11:2014; Chirurgische Implantate—Metallische Werkstoffe—Teil 11: Titan Aluminium-6 Niob-7 Knetlegierung. DIN Deutsches Institut für Normung e. V.: Berlin, Germany, 2015.
93. Shunmugavel, M.; Polishetty, A.; Littlefair, G. Microstructure and Mechanical Properties of Wrought and Additive Manufactured Ti-6Al-4V Cylindrical Bars. *Procedia Technol.* **2015**, *20*, 231–236. [CrossRef]
94. Attar, H.; Calin, M.; Zhang, L.C.; Scudino, S.; Eckert, J. Manufacture by selective laser melting and mechanical behavior of commercially pure titanium. *Mater. Sci. Eng. A Struct. Mater.* **2014**, *593*, 170–177. [CrossRef]
95. Gorny, B.; Niendorf, T.; Lackmann, J.; Thoene, M.; Troester, T.; Maier, H.J. In situ characterization of the deformation and failure behavior of non-stochastic porous structures processed by selective laser melting. *Mater. Sci. Eng. A Struct. Mater.* **2011**, *528*, 7962–7967. [CrossRef]
96. Blochwitz, C.; Jacob, S.; Tirschler, W. Grain orientation effects on the growth of short fatigue cracks in austenitic stainless steel. *Mater. Sci. Eng. A Struct. Mater.* **2008**, *496*, 59–66. [CrossRef]
97. Rao, K.T.V.; Yu, W.; Ritchie, R.O. Fatigue crack propagation in aluminum-lithium alloy 2090: Part II. small crack behavior. *Metall. Mater. Trans. A Phys. Metall. Mater. Sci.* **1988**, *19*, 563–569. [CrossRef]
98. Riemer, A.; Leuders, S.; Thöne, M.; Richard, H.A.; Tröster, T.; Niendorf, T. On the fatigue crack growth behavior in 316 L stainless steel manufactured by selective laser melting. *Eng. Fract. Mech.* **2014**, *120*, 15–25. [CrossRef]
99. Zerbst, U.; Bruno, G.; Buffiere, J.-Y.; Wegener, T.; Niendorf, T.; Wu, T.; Zhang, X.; Kashaev, N.; Meneghetti, G.; Hrabe, N.; et al. Damage tolerant design of additively manufactured metallic components subjected to cyclic loading: State of the art and challenges. *Prog. Mater. Sci.* **2021**, *121*, 100786. [CrossRef] [PubMed]
100. Afkhami, S.; Dabiri, M.; Alavi, S.H.; Björk, T.; Salminen, A. Fatigue characteristics of steels manufactured by selective laser melting. *Int. J. Fatigue* **2019**, *122*, 72–83. [CrossRef]
101. Schweikl, H.; Müller, R.; Englert, C.; Hiller, K.-A.; Kujat, R.; Nerlich, M.; Schmalz, G. Proliferation of osteoblasts and fibroblasts on model surfaces of varying roughness and surface chemistry. *J. Mater. Sci. Mater. Med.* **2007**, *18*, 1895–1905. [CrossRef] [PubMed]
102. Jeffery, E.H.; Abreo, K.; Burgess, E.; Cannata, J.; Greger, J.L. Systemic aluminum toxicity: Effects on bone, hematopoietic tissue, and kidney. *J. Toxicol. Environ. Health* **1996**, *48*, 649–665. [CrossRef] [PubMed]
103. Daley, B.; Doherty, A.T.; Fairman, B.; Case, C.P. Wear debris from hip or knee replacements causes chromosomal damage in human cells in tissue culture. *J. Bone Jt. Surg. Br.* **2004**, *86-B*, 598–606. [CrossRef]
104. Weng, Y.; Liu, H.; Ji, S.; Huang, Q.; Wu, H.; Li, Z.; Wu, Z.; Wang, H.; Tong, L.; Fu, R.K.; et al. A promising orthopedic implant material with enhanced osteogenic and antibacterial activity: Al2O3-coated aluminum alloy. *Appl. Surf. Sci.* **2018**, *457*, 1025–1034. [CrossRef]
105. Cattoni, D.; Ferrari, C.; Lebedev, L.; Pazos, L.; Svoboda, H. Effect of Blasting on the Fatigue Life of Ti-6Al-7Nb and Stainless Steel AISI 316 LVM. *Proc. Mater. Sci.* **2012**, *1*, 461–468. [CrossRef]
106. Javier Gil, F.; Planell, J.A.; Padrós, A.; Aparicio, C. The effect of shot blasting and heat treatment on the fatigue behavior of titanium for dental implant applications. *Dent. Mater.* **2007**, *23*, 486–491. [CrossRef] [PubMed]
107. Pazos, L.; Corengia, P.; Svoboda, H. Effect of surface treatments on the fatigue life of titanium for biomedical applications. *J. Mech. Behav. Biomed. Mater.* **2010**, *3*, 416–424. [CrossRef] [PubMed]

MDPI
St. Alban-Anlage 66
4052 Basel
Switzerland
www.mdpi.com

Metals Editorial Office
E-mail: metals@mdpi.com
www.mdpi.com/journal/metals

Disclaimer/Publisher's Note: The statements, opinions and data contained in all publications are solely those of the individual author(s) and contributor(s) and not of MDPI and/or the editor(s). MDPI and/or the editor(s) disclaim responsibility for any injury to people or property resulting from any ideas, methods, instructions or products referred to in the content.

www.ingramcontent.com/pod-product-compliance
Lightning Source LLC
LaVergne TN
LVHW070453100526
838202LV00014B/1715